ture Notes in Computer Scienc

d by G. Goos, J. Hartmanis and J. van Lee

Springer
Berlin
Heidelberg
New York
Barcelona
Hong Kong
London
Milan
Paris
Singapore
Tokyo

itrios Soudris   Peter Pirsch
h Barke (Eds.)

tegrated
ircuit Design

wer and Timing Modeling,
timization and Simulation

International Workshop, PATMOS 2000
tingen, Germany, September 13-15, 2000
eedings

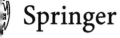 Springer

ries Editors

chard Goos, Karlsruhe University, Germany
is Hartmanis, Cornell University, NY, USA
a van Leeuwen, Utrecht University, The Netherlands

lume Editors

mitrios Soudris
mocritus University of Thrace, Dept. of Electrical & Computer Engineering
100 Xanthi, Greece
mail: dsoudris@demokritos.cc.duth.gr

ter Pirsch
iversity of Hanover, Institute for Communication Theory and Signal Processing
pelstr. 4, 30167 Hanover, Germany
mail: pirsch@mst.uni-hannover.de

ich Barke
iversity of Hanover, Institute for Microelectronic Systems
pelstr. 4, 30167 Hanover, Germany
mail: barke@ims.uni-hannover.de

taloging-in-Publication Data applied for

e Deutsche Bibliothek - CIP-Einheitsaufnahme

egrated circuit design : power and timing modeling, optimization
d simulation ; 10th international workshop ; proceedings / PATMOS
00, Göttingen, Germany, September 13 - 15, 2000. Dimitrios Soudris
(ed.). - Berlin ; Heidelberg ; New York ; Barcelona ; Hong Kong ;
ndon ; Milan ; Paris ; Singapore ; Tokyo : Springer, 2000
(Lecture notes in computer science ; Vol. 1918)
ISBN 3-540-41068-6

R Subject Classification (1998): B.7, B.8, C.1, C.4, B.2, B.6, J.6

SN 0302-9743
BN 3-540-41068-6 Springer-Verlag Berlin Heidelberg New York

ringer-Verlag Berlin Heidelberg New York
member of BertelsmannSpringer Science+Business Media GmbH
Springer-Verlag Berlin Heidelberg 2000
nted in Germany

besetting: Camera-ready by author, data conversion by Steingräber Satztechnik GmbH, Heidelberg
nted on acid-free paper   SPIN: 10722793   06/3142   5 4 3 2 1 0

Preface

This workshop is the tenth in a series of international workshops. This year it takes place in Göttingen, Germany, and is organized by the University of Hannover.

Göttingen has one the most famous German universities, where very well known scientists like Lichtenberg, Hilbert, Gauss and von Neumann studied, worked and taught. It also hosts several research institutes of the Max-Planck-Society. The first electronic tube calculator G1 was built in Göttingen in 1952 by H. Billing. Additionally, Göttingen was selected because it is adjacent to the world exposition EXPO 2000 in Hannover which gives an outlook into the 21st century covering the major topics of humankind, nature and technology.

With respect to these inspiring surroundings the technical program of PAT-MOS 2000 includes 10 sessions dedicated to most important subjects of power and timing modeling, optimization and simulation at the dawn of the 21st century.

The four invited talks address the European research activities in the workshop fields, the evolving needs for minimal power consumption in the area of wireless and chipcard applications and design methodologies of very highly integrated multimedia processors.

The workshop is a result of the joint work of a large number of individuals, who cannot all be mentioned here. In particular, we would like to acknowledge the outstanding work of the reviewers, who did a competent job in a timely manner. We also have to thank the members of the local organizing committee for their effort in enabling the conference to run smoothly. Finally, we gratefully acknowledge the support of all organizations and institutions sponsoring the conference.

September 2000

Peter Pirsch
Erich Barke
Dimitrios Soudris

Organization

Organization Commitee

General Co Chairs:	Peter Pirsch (University of Hannover, Germany)
	Erich Barke (University of Hannover, Germany)
Program Chair:	Dimitrios Soudris
	(Democritus University of Thrace, Greece)
Finance Chair:	Lars Hedrich (University of Hannover, Germany)
Publication Chair:	Achim Freimann
	(University of Hannover, Germany)
Audio-Visual Chair:	Jörg Abke (University of Hannover, Germany)
Local Arrangements Chair:	Carsten Reuter (University of Hannover, Germany)

Program Commitee

D. Auvergne (University of Montpellier, France)
J. Bormans (IMEC, Belgium)
J. Figueras (University of Catalunya, Spain)
C.E. Goutis (University of Patras, Greece)
A. Guyot (INPG Grenoble, France)
R. Hartenstein (University of Kaiserslautern, Germany)
S. Jones (University of Loughborough, United Kingdom)
P. Larsson-Edefors (University of Linköping, Sweden)
E. Macii (Polytechnic of Torino, Italy)
V. Moshnyaga (University of Fukuoka, Japan)
W. Nebel (University of Oldenburg, Germany)
J.A. Nossek (Technical University of München, Germany)
A. Nunez (University of Las Palmas, Spain)
M. Papaefthymiou (University of Michigan, United States)
M. Pedram (University of Southern California, United States)
H. Pfleiderer (University of Ulm, Germany)
C. Piguet (CSEM, Switzerland)
R. Reis (University of Porto Alegre, Brazil)
M. Robert (University of Montpellier, France)
A. Rubio (University of Catalunya, Spain)
J. Sparsø (Technical University of Denmark, Denmark)
A. Stempkowsky (Academy of Sciences, Russia)
T. Stouraitis (University of Patras, Greece)
J.F.M. Theeuwen (Philips, The Netherlands)
A.-M. Trullemans-Anckaert (University of Louvain, Belgium)
R. Zafalon (STMicroelectronics, Italy)

Steering Commitee

D. Auvergne (University of Montpellier, France)
R. Hartenstein (University of Kaiserslautern, Germany)
W. Nebel (University of Oldenburg, Germany)
C. Piguet (CSEM, Switzerland)
A. Rubio (University of Catalunya, Spain)
J. Sparsø (Technical University of Denmark, Denmark)
A.-M. Trullemans-Anckaert (University of Louvain, Belgium)

Sponsoring Institutions

European Commission Directorate – General Information Society
IEEE Circuits and Systems Society

Table of Contents

System-Level Design

Transistor-Level Modeling

Asynchronous Circuit Design

Power Efficient Technologies

Design of Multimedia Processing Applications

Adiabatic Design and Arithmetic Modules

Analog-Digital Circuits Modeling

Constraints, Hurdles and Opportunities for a Successful European Take-Up Action

Rene van Leuken, Reinder Nouta, and Alexander de Graaf

DIMES ESD-LPD, Delft University of Technology
Mekelweg 4, H16 CAS, 2628 CD Delft, The Netherlands
esdlpd@dimes.tudelft.nl, http://www.esdlpd.dimes.tudelft.nl

Abstract. "...Knowledge management is now becoming the foundation of new business theory and corporate growth for the next millennium. The key difference is that it's about networking people not simply processes and PCs..." [1].

1   Introduction

Low power design became crucial with the wide spread of portable information and communication terminals, where a small battery has to last for a long period. High performance electronics, in addition, suffers from a permanent increase of the dissipated power per square millimetre of silicon, due to the increasing clock-rates, which causes cooling and reliability problems or otherwise limits the performance.

The European Union's Information Technologies Programme 'Esprit' did therefore launch a 'Pilot action for Low Power Design', which eventually grew to 19 R&D projects and one coordination project, with an overall budget of 14 million EURO. It is meanwhile known as European Low Power Initiative for Electronic System Design (ESD-LPD) and will be completed by the end of 2001. It involves 30 major European companies and 20 well-known institutes. The R&D projects aims to develop or demonstrate new design methods for power reduction, while the coordination project takes care that the methods, experiences and results are properly documented and publicised.

2   European Low Power Initiative for Electronic System Design

The initiative addresses low power design at various levels. This includes system and algorithmic level, instruction set processor level, custom processor level, RT-level, gate level, circuit level and layout level. It covers data dominated and control dominated as well as asynchronous architectures. 10 projects deal mainly with digital, 7 with analogue and mixed-signal, and 2 with software related aspects. The principal application areas are communication, medical equipment and e-commerce devices.

D. Soudris, P. Pirsch, and E. Barke (Eds.): PATMOS 2000, LNCS 1918, pp. 1–2, 2000.

Instead of running a number of Esprit projects at the same time independently of each other, during this pilot action the projects have collaborated strongly. This is achieved mostly by the novelty of this action, which is the presence and role of the coordinator: DIMES - the Delft Institute of Microelectronics and Submicron-technology, located in Delft, the Netherlands (http://www.dimes.tudelft.nl). The task of the coordinator is to co-ordinate, facilitate, and organize:

- The information exchange between projects.
- The systematic documentation of methods and experiences.
- The publication and the wider dissemination to the public.

3   Constraints, Hurdles and Opportunities

The initiative has been running now for about 3 years. Roughly we can distinguish the next phases:

1. Selection and negotiation phase. Start: 1997. Duration: 6 months.
2. Legal activities, contracts etc. Start: 1997. Duration: 18 months.
3. Start of the initiative and design projects. Start: 1998. Duration: 12 months.
4. Tracking of design project results. Start: 1999. Duration: 18 months and continuing.
5. Start dissemination activities. Start: 1999. Duration: 18 months and continuing
6. Financial administration. Start: 1999. Duration: 18 months and continuing.

Here are some statistics:

1. Number of Associated Contracts: about 60.
2. Number of issued task contracts: about 30.
3. Number of contract amendments: 5 (more planned).
4. Number of contract type changes: 2
5. Number of appendixes of each progress reports: we stopped counting, we ship them in a box.
6. Number of projects on time with deliverables: none.
7. Number of available public deliverables on our web site: about 50.
8. Number planned low power design books: 6. The first has been published.

During the session we will present the audience a number of thesises (5 to 7). Each thesis will a address a topic , for example: "All public deliverables should be written using a defined design document standard", or "There is no knowledge dissemination problem; Only the lack of people is a problem", we will be present to you some historic events, feedback from partners and reviewers. Thereafter we will discuss the thesis with people from the audience and see if we can get some sort of statement which expresses the opinion of the audience.

References

1. C. S. of Management. *The Cranfield and Information Strategy Knowledge Survey.* November 1998.

Architectural Design Space Exploration Achieved through Innovative RTL Power Estimation Techniques

Manuela Anton, Mauro Chinosi, Daniele Sirtori, and Roberto Zafalon

STMicroelectronics, I-20041 Agrate B. (MI), Italy

Abstract. Today's design community need tools that address early power estimation, making it possible to find the optimal design trade-offs without respinning to explore the whole chip.

Several approaches based on a fast (coarse) logic synthesis step, in order to analyze power on the mapped gate-level netlist and then create suitable power models have been published in the last years.

In this paper we present some applications of RTPow, a proprietary tool dealing with the RT-level power estimation. The innovative estimation engine that does not perform any type of on-the-fly logic synthesis, but analyze the HDL description from the functionality point of view, permits a drastic time saving. Besides this top-down estimation, RTPow is able to perform a series of power macromodels and the bottom-up approach that enable an effective power budgeting. The first is an Adaptive Gaussian Noise Filter (28K Eq.Gate), described in VHDL, the second is a Motion Estimation and Compensation Device for Video Field Rate Doubling Application (171K Eq.Gate) also described in VHDL. The third is a micro-processor core (111K Eq.Gate) described using Verilog language.

1 Introduction

The increasing use of portable computing and communication systems makes power dissipation a critical parameter to be minimized during circuit and system design.

Low power design needs efficient and accurate estimation tools at all design abstraction levels. In particular, RT-level power estimation is critical in obtaining short design times and is very important to help the designer in making the right architectural choices.

Nowadays a crucial request is design turnaround time. Allowing the architectural exploration and "what-if" analysis before logic synthesis, the complex design trade-offs can result in a faster time-to-market. Accurate RT-level power estimation allows to reduce the number of design iterations and their cost, making the power budgeting easier.

The approaches proposed in literature (see [1] for a comprehensive survey) can be categorized into two main classes: *top-down* and *bottom-up* methods. While the former class is particularly suited for components with a fixed structure and/or design (e.g., memories, data-path units), the bottom-up methods are based on the idea of building an abstract power model by refining an initial model template through experimental power measurements. In the bottom-up approach, the estimated power is given by a

D. Soudris, P. Pirsch, and E. Barke (Eds.): PATMOS 2000, LNCS 1918, pp. 3-13, 2000.

relation between the significant characteristics of the input stimuli and their weight coefficients determined from power characterization (e.g.: by means of a linear regression or look-up table), performed at a lower level of abstraction and used to match the behavior of the analyzed block.

The power macromodeling techniques [2], [3], [4], [5], [6] can be differentiated by considering the kind of power data they can actually provide: some methods allow a cycle-accurate estimation, while others can manage just the total average power.

RTPow is a dynamic power estimation proprietary tool that operates at RT level and is embedded into the Synopsys Design Environment. It is able to arbitrarily apply both top-down and bottom-up analysis modeling techniques in any combination, to analyze generic sparse logic on one side, and pre-characterized macros and IP's, on the other. In addition, it is able to manage different macromodeling strategies (i.e.: table or regression based) and to take advantages of any macromodel available feature (i.e.: cycle-accurate or cumulative power figures).

The objective of this paper is to validate RTPow capabilities on several industrial applications and to benchmark the results with the corresponding power values obtained at a lower level of abstraction, by means of a gate level reference power simulator (e.g.: Synopsys DesignPower [12]). Another type of monitoring has been done on the actual computer resources requirements, such as the total CPU time and main memory allocated during the data processing (RT-level and gate-level).

2   RTPow Functionality

RTPow is an RT-level power estimation tool that works within the Synopsys' Design-Compiler environment [7], as shown in Figure 1.

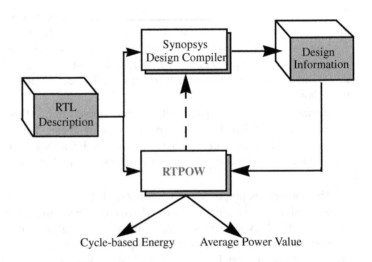

Fig. 1. Conceptual Flow of RTPow

The RTPow software architecture consists of a set of scripts, driven by certain user-specified variables, that outputs information about the current design functionality and structure on a set of files parsed and elaborated by an underlaying C++ main engine. The design functionality and structure are inherited by RTPow from the design database; therefore, the design can be written in any HDL supported by the Synopsys HDL-Compiler (e.g.: VHDL, Verilog) and needs to be previously de-composed into Synopsys generic objects, by running the traditional *analyze* and *elaborate* steps [11]. As a matter of fact, the non-necessity of going through the quite expensive logic synthesis flow need to be emphasized as this feature will enable a true design exploration, by allowing the designer to have a fast estimation and to make an easier finding of the power-optimal architecture.

The power estimator may work in two ways.

1. The first one is the *simulation* mode. In this mode it operates as a co-simulator by using an embedded cycle-accurate internal simulation engine. The user needs to specify the test pattern file which can be obtained either from a formerly written RTL testbench or by providing it on the fly. Provided that all macromodels are cycle-accurate, RTPow can compute a cycle-based energy report, as well as the energy peak and the simulation cycle when that energy peak has been accounted (see Figure 4 as an example). Basically, the cycle-by-cycle plot of energy consumption is useful to identify the operating modes of sequential machines or power peaks related to some specific activity burst who might be more suitable to optimize. In addition, a detailed power log structured on design blocks, down through the hierarchy, is also available. The reported dynamic power is splitted among net and internal power, as those concepts have been widely adopted by the industrial design community. Once the simulation is over, a file which contains switching activity and static probability information about synthesis invariant components (i.e., I/O ports, sub-module boundaries and sequential cell outputs) is written. This file has a ".*saif*" extension (Switching Activity Interchange Format) and can be effectively used either to annotate the switching activity onto the RTL design and then providing an appropriate forward annotation to drive the power-driven synthesis with PowerCompiler or running RTPow in *static* mode (see below).

2. The second mode of operation is *static* (or *probabilistic*). In order to achieve a higher accuracy, this mode formerly requires a RTL design node annotation (i.e.: on the synthesis invariant ports), especially on the sequential cell outputs, since the switching activity propagation engine tends to suddenly loose accuracy when dealing with sequential cells (this is primarily due to the adopted BDD representation of the design functionality, who has an intrinsic limitation in both manageable design size and toggle rate propagation). Node annotation may be either performed by reading an RTL ".*saif*" file (e.g.: from a previous run of RTPow in internal simulation mode or from an external RTL simulation [12]) or by providing a list of *set_switching_activity* commands on the appropriate ports mentioned above.

   However, should the node annotation being unfeasible in the current design under analysis, RTPow is definitely able to operate in a pure probabilistic mode, without any external information. In such a case, the significant switching characteristics at the primary inputs are set to a predefined default value and these values are propagated in circuit, down through the hierarchy, with the aid of the embedded switching activity propagation engine (integrated in RTPow).

Eventually, RTPow will provide the total average power figure for each hierarchical block. Of course, since no input test pattern are involved during the static estimation mode, the cycle-based power log will be not available.

3 More on RTPow Functionality

RTPow is based on circuit functionality and design structure complexity exploration. Some former works ([8], [9], [10]) have proposed to adopt an initial design description from which the circuit topology is directly imported in the estimator by extracting an equivalent functional representations (typically by means of BDDs - Binary Decision Diagrams).

Given the context of RTPow, we wanted an effective method to get the design data base directly from the EDA environment used to analyze the design. After source code analysis and elaboration, a RT-level description is expressed within DesignCompiler as an interconnection of different types of primitives (e.g., Gtech ports, Generic logic blocks, Synthetic operators, DesignWare modules, Generic sequential operators) [11]. The difficulty here is to make this type of description available to the underlying C++ analysis engine. The task is carried out in RTPow by dumping the circuit as a set of equations. Then, connections between the previous components and sequential cells are recognized by parsing the file produced by the *report_cell* command and by including their functionality into the previously created representation.

The top-down approach in RTPow investigates the circuit topology and extracts, from the Synopsys representation before mapping (technology independent components from Synopsys' generic library interconnected with clusters of combinational logic, at their turn expressed by the same Synopsys' generic library components) information readable by the underlaying estimation engine. The generic combinational functionality is represented as a BDDs structure. This is equivalent to a 2-to-1 MUX mapping (see Figure 2), with input signals connected to the selection pin of each MUX. Being a possible library mapping, the area estimation fits the number of BDDs nodes, as a measure for the area occupancy, to the actual area measured on a number of benchmarks mapped onto the target ASIC library. The area allocated by such a BDDs mapping is therefore approximated by the number of fitted BDDs, by means of linear regression, to the actual area obtained on a large number of benchmark circuits implemented with the desired target technology.

Area estimation is then used in power modeling as an approximation of module's capacitance. Therefore, to get the power values, both area and average switching activity estimates are needed. Switching activity is estimated on each virtual net of the equivalent MUX mapping, either in simulation mode or in probabilistic mode, but the estimation is done differently: while simulation mode simply counts the number of transitions provided by a set of input patterns, probabilistic mode uses statistical relation to get the toggle rate on each virtual net of the equivalent MUX mapping.

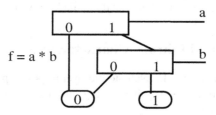

Fig. 2. BDDs as Mux Mapping

The result is tuned to the target technology and to the actual synthesis policy the designer is adopting (e.g.: timing-driven, area-driven or power-driven) by means of area and capacitance scaling factors obtained, once and for all, by characterizing the correlation between the MUX-based design and the actual technology mapping on a large number of pre-defined industrial designs.

Power consumed both on nets and inside combinational cells is processed and reported separately.

A more accurate estimation can be obtained by using dynamically linked power macromodels. In the design flow, intellectual property blocks reuse is a key factor to match the required time-to-market. Macromodeling implies an unique power characterization step, made once and for all and provides power related information to perform a fast and accurate estimation. These IPs can be either soft or hard macros seen as black boxes. In this category fall all DesignWare modules as well as every IP block for whom a macromodel is available.

Each time a block in the design hierarchy is considered, RTPow attempts to run in *bottom-up* approach and first checks if a power macromodel does exist in the macro library and only if it doesn't find one it will use the *top-down* approach.

RTPow is independent of the algorithm used by macromodels to estimate the required parameters, i.e., different implementations can be used for a variety of circuits. For example, if the analyzed macro is a whole processor, an instruction-based estimation could be the most efficient solution, while a look-up table could be good for a bench of control logic.

The macromodel details are hidden to RTPow, in fact the macromodel is built as an external dynamically linked library, implementing a given interface.

The process of macro-block characterization and model building has been fully automated for the table based macromodel (see below), proven to be the most accurate and robust solution. This method provides a look-up table addressed by some concise form of instance's sensitization (e.g., input switching activity) and retains the property of being automatically extracted and general. Moreover, it is robust because look-up table can represent any function with a desired accuracy, provided that the table can be made arbitrarily large.

The automatic table-based macromodel building requires only the mapped implementation on the RTPow reference library of the IP block, no need to re-characterize for different technology target.

Semi-automatic approaches have been developed for other techniques (e.g.: poly-nomial regression), where the dynamic library construction is not yet automatic.

4 RT-Level Library Characterization

RTPow is able to evaluate the overall power performances of the technology cell library and in this way it is taking into account, during estimation, the target library specifications.

During library characterization a set of industrial RTL benchmarks are mapped to the gate level and the corresponding statistical area occupation and power consumption values are inferred. These information are used by RTPow during its top-down estima-tion capabilities and to compute a number of technology scaling factors to be passed to power macromodels.

The process of library characterization is straightforward and fully automatized. A user-friendly interface allows the user to run the task anytime on his site.

In order to address the synthesis based on design constraints, the library can be characterized and its high-level parameters can be computed in relation to different design strategies such as minimum area or maximum speed synthesis, and an average input slope can be specified to improve accuracy.

5 Test Cases and Experimental Results

To evaluate RTPow performances and capabilities, we have chosen three industrial designs.

The first one is an *Adaptive Gaussian Noise Filter* (GNR from now on), written in VHDL, the second one is a *Motion Estimation and Compensation Device for Video Field Rate Doubling Application* (50Hz_to_100Hz) also described in VHDL. The third design is a *Core Processor* (uP) described using Verilog language.

The gate implementations of the designs listed above contain the equivalent num-ber of gates reported in Table1.

Table 1.

Design	Eq. Gate
GNR	28K
50Hz_to_100Hz	171K
uP	111K

5.1 Adaptive Gaussian Filter

GNR is an adaptive intra-field spatial filtering for Gaussian Noise reduction, based on recursive estimation of the noise level. GNR realizes a Pre-Processing filter to improve input video sources quality in order to

- reduce the amount of Gaussian Noise in the image, mainly in its High Spatial Frequency (HSF) components;
  - reduce the HSF components of the video signal in all areas where they are unperceived by the Human Visual System

The applied Pre-Processing is an adaptive low pass filtering. It improves the subjective quality by reducing the Gaussian Noise.

The GNR receives in input three different sets of signals each one related to different information: synchronization, image and filtering.

5.2 Motion Estimation and Compensation Device for Video Field Rate Doubling Application

GNR is part of a larger ST project, named *Motion Estimation and Compensation Device for Video Field Rate Doubling Application* (50Hz_to_100Hz).

50Hz_to_100Hz is a new device for field rate doubling based on a motion-compensation algorithm, where motion information are estimated before their compensation in the final interpolation process. The motion estimation process is based on a recursive block-matching technique. The GNR is introduced as a pre-processing filter to improve the quality of the real motion estimation content.The market introduction of high-end TV sets, based on 100HZ CRTs, required the development of reliable field rate upconversion techniques, to remove artifacts such as large area and line flicker.

5.3 Core Processor (uP)

The core consist of one or more basic execution units (named clusters), an instruction fetch unit and a memory interface (the core memory controller). An uP cluster consist of one more arithmetic units, a register file and an interface to the core memory controller. The design used for experiments contains a single cluster.

5.4 Results

In order to validate the power estimations determined at architectural level by RTPow, each design has been synthesized with a standard cell library realized in 0.25μm CMOS technology and the actual power consumption has been estimated using DesignPower. It is important to highlight how RTPow adopts table-based power macromodels also when dealing with the DesignWare modules of the Synopsys generic objects representation, (i.e., each DesignWare block has been previously mapped and characterized, and a table-based energy model has been generated, thus making the power estimation process faster).

Figure 3 (a) compares pre-synthesis RTL power estimation values resulting from RTPow to the post-synthesis power figures coming from DesignPower, for the first two circuits (*4.28%* difference).

In Figure 3 (b), the processor core (uP) has been tested under two realistic patterns of input stimuli traced from a high level system simulation. The results of RTPow and DesignPower are, then, reported. The first bench of power values address the case when the processor is not executing any operation (i.e.: a sequence of NOP's) while the other refer to the case when the processor is executing two additions simultaneously (2 ADD's). For both these applications, the average power consumption is plotted against the processor stalling percentage (on the X axis), mainly due to cache memory misses.

Regarding the processor core, it may be noticed that, although RTPow is overestimating the absolute power values, the relative power figures predicted at RTL under a wide range of stalling probability show a quite close accordance with the related power performance reported at gate level (always assumed as a reference).

Indeed, the processor core is representing an extreme test case for the architectural estimation since it is strongly based on one only clock domain who is driving all the sequential cells (i.e.: FFs and Latches) of the deeply pipelined internal architecture. As of today, the physical implementation of those kind of heavily loaded networks is usually managed by a set of appropriate clock tree synthesis techniques, whose major goal is the optimal placement and routing of these high fan-out and hierarchical networks in order to meet the severe design constraints on the max delay and max skew between the root and each leaf cells of the interconnection tree, respectively.

While the implication, in terms of power performance, of the clock tree physical implementation is fully tractable at gate level (provided a consistent post-layout back annotation), the prediction of such a structure during the architectural estimation is extremely haphazard and still lacking of a general solution.

In our specific design, while the RTPow's analysis of the clock tree is based on the estimation of the switching energy associated to an equivalent global network with a given fan-out (easily exceeding 10000 leaf cells), the actual implementation of this network is a hierarchical and balanced tree of buffers, necessary to meet the global timing constraints, including the avoidance of any slope degradation. As a matter of fact, the overestimation of the clock's switching energy is due to the large slope degradation induced by assuming an equivalent global network driving an extremely large fan-out.

Our future works we intend to address the development of a robust and more suitable prediction model of those physical structures.

Almost all works addressing RTL power estimation are focused on power models accuracy. Power models on their own are strictly dependent on the evaluation conditions. In a real world, for industrial designs, tuning all characteristics involved in estimation to the actual functional conditions is recognized to be quite hard.

Certainly the goal of an RTL estimator is not to provide sign-off power values but rather to allow designers in exploring, evaluating, comparing and eventually optimizing different architectures using various components and IP blocks, choosing the best candidate for a minimal power consumption. The RTL estimations inherently highlight the "hot" issues, the architectures that should be modified or substituted in order to minimize the overall device power consumption (see [8] and [7] for a survey of RTPow design exploration capabilities).

In order to obtain a substantial increasing in absolute accuracy, the estimator features would need a better matching with real conditions. This issue can't be solved without huge time investment and subsequent severe impact to the demanding time-to-market.

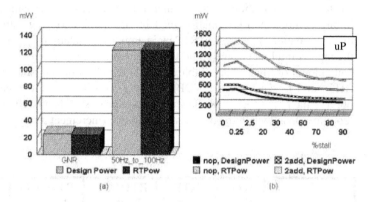

Fig. 3. Power Estimation on three industrial test cases- Absolute Power

Provided that all macromodels are cycle-accurate, RTPow can output, when running in simulation mode, a cycle-based energy report, as well as the energy peak and the simulation time when that energy peak has been registered. Figure 4 illustrates the energy behavior for the GNR, when it has been simulated for a period of 223000ns. The reported energy peak is 1.04048 uJ and the corresponding simulation time when it has been obtained is 882 ns.

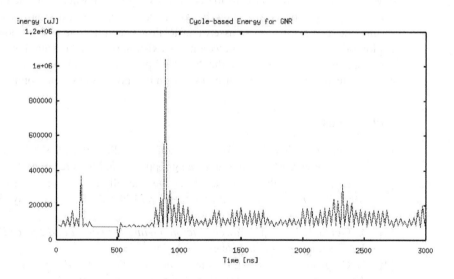

Fig. 4. Cycle-based Energy for GNR

As we mentioned in section 3, the top-down approach in RTPow investigates the circuit topology and extracts, from the Synopsys representation before mapping (technology independent components from Synopsys' generic library interconnected with clusters of combinational logic, at their turn expressed by the same Synopsys' generic library components) information readable by the underlaying estimation engine. All this information are stored by RTPow in an internal database that could be used in case the input stimuli is changed (and the circuit structure is not modified), with an important amount of time saving. Table 2 reports the CPU and memory involved by RTPow (RTP) and by DesignPower (DP) during estimation, for all three circuits. We can observe the high speed of RTPow estimation process, when building the own database (column 1), and when the database is available (incremental mode reported in column 3).

Table 2.

Design	RTP CPU (s)	RTP mem (kB)	RTP CPU database ready(s)	RTP mem database ready(kB)	DP CPU (s)	DP mem (kB)
GNR	170	47984	14	27912	19961	170216
50Hzto100Hz	10611	501280	2487	228640	58009	847400
uP	2115	246304	580	87696	47559	632800

6 Conclusions

We have presented RTPow capabilities on several industrial applications and we have compared the results with the corresponding power values obtained at a lower level of optimization. The RTPow behavior on real industrial designs as well as the results obtained justified us to assert that RTPow is an effective tool for power design exploration, suitable to be integrated into an existing industrial design flow as it allows the designer to quickly evaluate the "what-if" possibilities and to choose the best circuit architecture for a power-conscious design in a pre-synthesis environment.

References

[1]   P. Landman: *High-Level Power Estimation*, ISLPED-96: ACM/IEEE Intl. Symp. on Low-Power Electronics and Design, pp. 29-35, Monterey, CA, Aug. 1996.

[2]   L. Benini, A. Bogliolo, M. Favalli, G. De Micheli: *Regression Models for Behavioral Power Estimation*, PATMOS-96, pp.179-186, Bologna, Italy, September 1996

[3]   S. Gupta, F. N. Najm: *Power Macromodeling for High-Level Power Estimation*, DAC-97, pp. 365-370, Anaheim, CA, June 1997

[4]   R. Corgnati, E. Macii, M. Poncino, *Clustered Table-Based Macromodels for RTL Estimation*, GLS-VLSI-99: IEEE/ACM 9th Great Lake Symposium on VLSI, pp. 354-357, Ann Arbor, Michigan, March 1999

[5]   Z. Chen, K. Roy: *A Power Macromodeling Technique Based on Power Sensitivity*, DAC-98, S. Francisco (CA), June 1998

[6]   S. Gupta, F. N. Najm: *Analytical Model for High-Level Power Modeling of Combinational and Sequential Circuits*, IEEE Alessandro Volta Memorial Workshop on Low Power Design, pp. 164-172, Como, Italy, March 1999

[7]   R. Zafalon, M. Rossello, E. Macii, M. Poncino: *Power Macromodeling for a High Quality RT-level Estimation,* 1st International Symposium on Quality Electronic Design, ISQED 2000, San Jose, CA, March. 2000.

[8]   M. Nemani, F. Najm: *Towards a High-Level Power Estimation Capability*, IEEE Transactions on Computer-Aided Design, Vol. CAD-15, No. 6, pp. 588-598, Jun. 1996.

[9]   D. Marculescu, R. Marculescu, M. Pedram: *Information Theoretic Measures For Power Analysis*, IEEE Transactions on Computer-Aided Design, Vol. CAD-15, No. 6, pp. 599-609, Jun. 1996.

[10] F. Ferrandi, F. Fummi, E. Macii, M. Poncino, D. Sciuto: *Power Estimation of Behavioral VHDL Descriptions*, DATE'98: IEEE Design Automation and Test in Europe, pp. 762-766, Paris, France, Mar. 1998.

[11] Core Synthesis Tools Manual, Synopsys v2000.05

[12] Power Products Reference Manual, Synopsys v2000.05

Power Models
for Semi-autonomous RTL Macros [*]

Alessandro Bogliolo[1], Enrico Macii[2],
Virgil Mihailovici[2], and Massimo Poncino[2]

[1] Università di Ferrara, DI, Ferrara, ITALY 44100
[2] Politecnico di Torino, DAUIN, Torino, ITALY 10129

Abstract. Most power macromodels for RTL datapath modules are both *data-dependent* and *activity-sensitive*, that is, they model power in terms of some *activity* measure of the *data* inputs of the module. These models have proved to be quite accurate for most combinational RTL datapath macros (such as adders and multipliers), as well as for storage units (such as registers). They tend to become inadequate for RTL modules that are *control-dominated*, that is, having a set of control inputs that exercise different operational behaviors. Furthermore, some of these behaviors may be *input-insensitive*, that is, they let the module evolve (and thus consume power) in a *semi-autonomous* way, independently of the input activity. We propose a procedure for the construction of ad-hoc power models for semi-autonomous RTL macros. Our approach is based on the analysis of the functional effect of such control inputs on specific macros. Although the resulting models are tailored to individual macros, the model construction procedure keeps the desirable property of being automatic.

1 Introduction

Most approaches to high-level power estimation specifically target RTL estimation by building abstract power models for the various datapath modules (for a comprehensive survey, see [1,2]). Some of these models [3,4,5,6] may be parameterized with respect to the bit-width of the input data, so that a base model can be scaled according to specific, macro-dependent factors, thus avoiding the characterization of a macro for any possible value of the bit-width size.

Power macromodels are usually built for either combinational RTL modules (such as adders or multipliers), or for storage units (such as registers or register files) with relatively simple I/O behavior. These types of modules share the property of being *data-dominated*, that is, their power is strongly correlated with the activity profile of the input data. The corresponding power models thus relate power to statistical properties of the *data* inputs. For instance, a widely used power model includes an average measure of the input/output switching activity and of the input probability [7,8,9,6]. Average is computed with respect

---

[*] This work was supported, in part, by the EC under grant n.27696 "PEOPLE".

D. Soudris, P. Pirsch, and E. Barke (Eds.): PATMOS 2000, LNCS 1918, pp. 14–23, 2000.
© Springer-Verlag Berlin Heidelberg 2000

to the size of the input/output data. The rationale behind this averaging process is that data inputs have a meaning as whole, and a single quantity is enough to characterize them.

There are other classes of macros, however, for which these types of models may result in significant estimation errors. This is the case of *control-dominated* macros, i.e., macros having a set of control inputs that bring it into totally different operational modes. In addition, some of these modes can be input-insensitive, i.e., the corresponding behavior of the module tends to be totally autonomous. When the macro exhibits such behavior, the traditional "activity-sensitive" model (following the terminology of [4]) becomes inadequate. We call these types of macros *semi-autonomous*, to emphasize the possible insensitivity to the input activity.

A typical example of semi-autonomous macros is a counter with enable or load control signals. If counting is enabled, the counter will actually switch in every clock cycle, in spite of the fact that no switching on the data inputs happens. While it is true that the clock input can be used to track the switching due to counting, it is also true that models that use average switching measures as parameters will hide clock switching inside the average. Furthermore, most models are black-box, so they do not exploit module-specific information such as the semantics of the input signals. Conversely, if the load input is asserted, the counter will switch into a input-sensitive behavior, since the stored value will determine the amount of switching in that clock cycle.

Although the literature on power modeling is vast, the issue of multi-mode, semi-autonomous macros has not been investigated thoroughly. In some applications, however, the power impact of such RTL modules (counters but also shift registers) can be sizable. Designs requiring timeouts, or signal processing applications usually exhibit several instances of such macros. Resorting to traditional black-box models may consequently impair the accuracy of the power estimator.

In this work, we propose a procedure for the construction of ad-hoc power models for semi-autonomous RTL macros. Our approach is based on the analysis of the functional effect of the control inputs on specific macros. This does not simply imply using a straightforward modification of a black-box model, where the control signals (and thus their statistical properties) are explicitly "exposed" in the model as individual paramters. This is, for example, the approach followed in [12], where control signals are used to split the basic model into a set of sub-models (a *regression tree*, in their terminologu), one for each possible assignment of the control signals.

In our case, the model is a single equation, whose form is derived from the inspection of the functional description of the macro. The result is a model which is generally non-linear, because some higher-order terms are used to express the joint effect of some parameters are properly taken into account.

We emphasize that the proposed models are not *black-box*, because they exploit specific functional and behavioral information about the macro. The distinction between data and control inputs is the minimum information required. However, this information can be approximately recovered by simulation, by "measuring" the sensitivity of the outputs to the individual input signals. Techniques like those of [11,6,13] can be used for that purpose. In that case, although with a lower level of confidence, the model can be used as a black-box one.

Regardless of how the functional information is provided, the construction of the proposed models is automatic, and is therefore suitable to be incorporated into a fully automatic estimation tool.

Experimental results on a set of RTL macros with the characteristics described above, taken from the Synopsys *DesignWare* library, demonstrate the increased accuracy of the proposed models with respect to both conventional black-box models and ad-hoc models where control signals are treated separately from the other inputs.

2   Semi-autonomous Sequential Macros

Consider an up-down counter with four modes of operation controlled by three control signals: Ld, Cen and UpDn. When Ld is 1, current input data DataIn are loaded into the internal register. When Ld is 0 and Cen is 0, the counter is idle. When Ld is 0 and Cen is 1, if UpDn is 1 (0) the content of the register is incremented (decremented) by 1 independently of the input data. The internal state is always observable from primary outputs DataOut and an additional terminal-count flag (TerCnt) is raised whenever the all-1 state is reached.

Count-up and count-down modes are *autonomous* operating modes for the up-down counter, because its behavior (and its power consumption) is not affected by the data inputs. We use a Boolean function ($F_{\mathtt{DataIn}}$) to represent the sensitivity of the macro to DataIn, i.e., the set of configurations of the control bits that make its behavior sensitive to DataIn:

$$F_{\mathtt{DataIn}}(\mathtt{Ld}, \mathtt{Cen}, \mathtt{UpDn}) = \mathtt{Ld}$$

The above function expresses the fact that DataIn affects the behavior of the macro if and only if control signal Ld is set to 1.

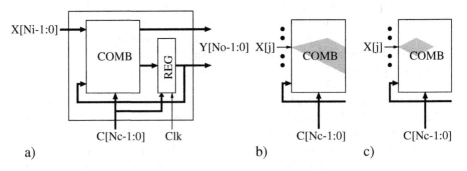

a)   b)   c)

Fig. 1. Schematic Structure of a Generic Semi-Autonomous Macro (a), and Propagation of an Input Signal (b),(c).

The up-down counter example has several interesting properties:

1. It contains a register;
2. The state of the register is directly observable at primary outputs;
3. It has different operating modes controlled by a few control signals;
4. Some configurations of the control signals make it insensitive to (some of) the input data.

We call *semi-autonomous* a macrocell with the four properties listed above. A schematic representation of the structure of a generic semi-autonomous macro is shown in Fig. 1-(a). Data inputs are denoted by X[Ni-1,0], primary outputs by Y[No-1,0], control inputs by C[Nc-1,0], the clock signal by clk.

There are three main structural characteristics of the macro. First, all state bits are also output signals, thus allowing the observability of the internal state. Second, control signals C may feed both the combinational logic and the registers. Finally, some output signals may directly derive from the combinational logic.

Fig. 1-(b) and -(c) show the propagation of a generic signal (namely, X[j]) through the combinational logic, for two different assignments of the control bits. The shaded region within the combinational logic represents the sensitivity to a given input signal X[j]. Depending on the current value of the control inputs C, input signal X[j] may or not propagate to primary outputs and registers.

In Fig. 1(b), the input signal reaches the outputs of the combinational logic, thus affecting the state bits and/or the outputs of the macro. In Fig. 1(c), its propagation is blocked by the control signals, withouth affecting the functionality of the macro. In the latter case, the macro is autonomous w.r.t. X[j].

If we map Fig. 1-(a) onto the up-down counter example, X represents DataIn, Y represents both DataOut and TerCnt, while C is the array of control signals Ld, Cen, UpDn. Fig. 1-(b) and (c) represent the propagation of any bit of DataIn when Ld is set to 1 and 0, respectively.

3   Power Models for Semi-autonomous Macros

In principle, black-box activity-sensitive power models developed for general functional units could be applied to semi-autonomous macros as well. Consider, for example, a simple regression equation relating the power consumption of the macro to the average switching activity at its primary inputs and outputs:

$$P_{BB} = c_0 + c_1 D_{in} + c_2 D_{out} \tag{1}$$

The c's are fitting coefficients, the D's are average transition densities, and P_{BB} is the power estimate given by the black-box model. For a macro with N_{in} inputs simulated for $N_p + 1$ patterns, the input transition density is computed as:

$$D_{in} = \frac{1}{N_{in}N_p} \sum_{i=0}^{N_{in}-1} \sum_{j=1}^{N_p} in[i](j) \oplus in[i](j-1)$$

where $in[]$ represents the generic input signal.

Model (1) can be applied to a semi-autonomous macro by computing the input activity as the weighted average of the activity densities at the data and control inputs (denoted by D_X and D_C, respectively):

$$P_{BB} = c_0 + c_1 \frac{N_X D_X + N_c D_C}{N_X + N_c} + c_2 D_Y \tag{2}$$

where N_c and N_X denote the number of control and data inputs, respectively.

From Equation 2 we observe that the same coefficient (c_1) statically multiplies the activity of data and control inputs. Hence, they are assumed to have the same impact on power consumption; and their contributions are assumed to be independent of the operating mode. Both assumptions are non-realistic and may lead to estimation errors that may be reduced by taking into account the peculiarities of semi-autonomous macros.

To overcome the first limitation of the black-box model, we observe that the activity of data and control signals may have a different effect to the power consumed by the macro. Hence, we split the second term of equation (2) and use two *independent* coefficients for D_X and D_C. We denote by P_{LIN} the power estimates provided by the new linear model:

$$P_{LIN} = c_0 + c_1 D_X + c_2 D_C + c_3 D_Y \tag{3}$$

Second, we observe that the power contribution of each input signal may depend on the operating mode of the macro. In particular, we expect the activity of input signal X[j] to have a sizeable impact on power consumption when it affects the functionality of the macro, and little or no impact when its propagation through the combinational logic is blocked by the control signals.

In the up-down counter example of Section 2, propagation of input data is conditioned to Ld $= 1$. Hence, we could characterize two coefficients for D_X (i.e., the activity density of DataIn) to be used alternatively depending on the value of control signal Ld:

$$P_{CND} = c_0 + c_{1,\text{Ld}} D_{X,\text{Ld}} P_{\text{Ld}} + c_{1,\text{Ld}'} D_{X,\text{Ld}'} P_{\text{Ld}'} + c_2 D_C + c_3 D_Y \tag{4}$$

where subscripts Ld and Ld$'$ denote quantities referring to operating modes with Ld$=1$ and Ld$=0$, respectively. In particular, $D_{X,\text{Ld}}$ denotes the transition density computed on the subset of input patterns with Ld$=1$, while P_{Ld} denotes the signal probability of Ld. Notice that $D_{X,\text{Ld}}$ is the conditional probability of having a transition on a data input when control signal Ld$=1$.

If we assume data and control inputs to be independent of each other, we may replace conditional probabilities with total probabilities ($D_{X,\text{Ld}} = D_{X,\text{Ld}'} = D_X$), thus obtaining:

$$P_{CND} = c_0 + c_{1,\text{Ld}} D_X P_{\text{Ld}} + c_{1,\text{Ld}'} D_X (1 - P_{\text{Ld}}) + c_2 D_C + c_3 D_Y \tag{5}$$

where $(1 - P_{\text{Ld}})$ has been used in place of $P_{\text{Ld}'}$. The power estimates provided by Equation (5) are denoted by P_{CND}, since they are conditioned to the sensitivity function of the data inputs: $F_{\text{DataIn}} = \text{Ld}$.

We can then start from Equation (5) to develop a general model for semi-autonomous macros. We need two forms of generalization: First, we need to extend the model to the case of general sensitivity functions; second, we need to extend the model to handle cases where different sensitivity functions are associated to disjoint subsets of data inputs.

We partition the set of data inputs X[Ni-1,0] into K disjoint subsets X_1, ..., X_K. All data inputs in the same subset (say, X_j) have the same sensitivity function F_{Xj}, i.e., they affect the behavior of the macro in the same operating modes. In the most general case, each data input has a different sensitivity function ($K = N_i$), in most cases of practical interest, however, two subsets are sufficient. The generalized power model we propose has the following form:

$$P_{CND} = c_0 + \sum_{j=1}^{K} \left(c_{(j,F_{Xj})} D_{Xj} P_{F_{Xj}} + c_{(j,F'_{Xj})} D_{Xj} (1 - P_{F_{Xj}}) \right) + \qquad (6)$$
$$c_{K+1} D_C + c_{K+2} D_Y$$

where $P_{F_{Xj}}$ is the probability of the j-th sensitivity function. Notice that equation (6) is a family of ad-hoc models, that may have different number of terms, different input subsets and different sensitivity functions depending on the macro. Nevertheless, the power models can be automatically constructed and characterized starting from the functional specification of the macro.

4   Experimental Results

We applied the proposed power model to instances of all the sequential soft macros in the *Synopsys' DesignWare* library that meet the definition of semi-autonomous macros. Each macro was mapped onto a reference technology library characterized for power and simulated by means of *Synopsys VSS* with *DesignPower* to obtain reference power values to be used for characterization and validation. Estimation results are collected in Table 1.

For each macro, sensitivity functions (and subsets) were directly obtained from the functional specification. Each model was characterized using the results of a large set of simulation experiments, sampling different input statistics and different operating modes. In particular, for a macro with K sensitivity functions, $2K + 1$ sets of experiments were used. Each set of experiments consists of 10 simulations of 50 patterns each, and was conceived to exercise a given operating mode (characterized by a fixed value of a given sensitivity function) under different data statistics. To this purpose, input streams were generated by assigning fixed values to the control inputs appearing in a given sensitivity function, and changing the remaining (data and control) inputs according to the given input statistics.

For each experiment, input/output transition densities, control signal probabilities, and the probability of all sensitivity functions were computed and stored in a row of a characterization matrix. Finally, the power model was automatically built and characterized to fit the data in the matrix. Black-box and linear models were also characterized for comparison.

Table 1. Experimental Results for DesignWare Semi-Autonomous Macros.

Macro	Stream Type	P_{BB}		P_{LIN}		P_{CND}	
		AvgErr	StdDev	AvgErr	StdDev	AvgErr	StdDev
DW03_shift_reg	Char	8.68	7.36	7.72	6.78	6.16	4.80
	Rand	9.14	4.79	4.59	3.03	3.07	2.67
	Ld=0	20.81	2.79	19.38	2.83	10.99	5.73
	Ld=1	3.73	2.97	3.43	2.52	5.29	3.37
	Shift=0	4.77	3.79	4.86	3.15	4.83	3.55
	Shift=1	5.04	3.29	5.81	3.80	5.46	3.81
Average		8.70	4.17	7.63	3.69	5.97	3.99
DW03_lfsr_dcnto	Char	4.37	3.12	3.75	3.15	3.59	3.01
	Rand	4.46	2.76	3.68	3.13	3.86	2.96
	Ld=0	3.74	3.44	3.87	3.30	3.45	2.89
	Ld=1	4.91	3.34	3.69	3.05	3.46	3.22
	Cen=0	33.75	12.41	31.40	11.74	28.57	13.28
	Cen=1	11.17	2.60	10.00	2.19	11.39	1.94
Average		10.40	4.61	9.40	4.43	9.05	4.55
DW03_bictr_dcnto	Char	6.18	5.48	5.07	4.69	3.56	3.37
	Rand	5.72	3.80	4.23	2.63	3.05	2.11
	Ld=0	6.02	4.01	4.36	3.62	3.34	3.04
	Ld=1	6.79	7.75	6.62	6.58	4.29	4.47
	Cen=0	5.19	5.32	5.25	4.80	4.60	3.72
	Cen=1	8.53	4.40	10.05	3.83	7.94	3.46
Average		6.41	5.13	5.93	4.36	4.46	3.36
DW03_bictr_decode	Char	8.64	7.51	8.23	7.57	6.87	6.99
	Rand	8.91	5.58	7.27	6.09	6.46	5.71
	Ld=0	11.41	10.27	10.76	10.61	9.87	9.18
	Ld=1	5.58	4.14	6.66	3.86	4.28	3.91
	Cen=0	8.95	5.84	10.25	5.51	9.43	5.36
	Cen=1	10.03	4.30	12.12	4.45	10.33	4.29
Average		8.92	6.27	9.22	6.35	7.87	5.91
DW03_lfsr_scnto	Char	5.50	3.86	4.61	3.58	4.53	3.56
	Rand	5.56	3.58	4.52	3.12	4.66	3.19
	Ld=0	4.81	3.77	4.68	3.59	4.12	3.21
	Ld=1	6.13	4.16	4.64	4.05	4.80	4.22
	Cen=0	60.51	28.66	53.67	26.51	50.17	27.22
	Cen=1	14.73	4.52	13.55	3.82	14.79	3.90
Average		16.21	8.09	14.28	7.45	13.85	7.55

Similar experiments have been performed for evaluation, yet with different input statistics from the characterization phase, so that also the out-of-sample accuracy has been evaluated. Evaluation streams have been synthesized so as to represent realistic situations as much as possible, that is, meaningful alternation of different operational modes. Accuracy has been measured in terms of average error and standard deviation, defined as:

$$AvgErr(\%) = \frac{1}{N} \sum_i \frac{|P_{est}(i) - P_{real}(i)|}{P_{real}(i)} \cdot 100 \tag{7}$$

$$StdDev(\%) = \sqrt{\sum_i \left[\frac{|P_{est}(i) - P_{real}(i)|}{P_{real}(i)} - StdErr \right]^2 \cdot 100} \tag{8}$$

For each macro, we have reported the estimation error and standard deviation for various streams, each one corresponding to a different statistical profile. Stream *Char* refers to a stream with similar properties to the one used for the characterization. This row reports then the *in-sample* error.

Stream *Rand* is constructed without separately exercising control and data inputs, and by applying uniform white noise to all the input bits. This stream represents the worst case for our model, since the advanatage of exposing control variables is lost.

The other streams clearly depend on the specific macro. Most have the form *input_name = value*, denoting the fact that the stream has been built with that input signal stuck to that particular value.

Results show that the *CND* model consistently yields higher accuracy than the two other models, in terms of both error and standard deviation. It is important to emphasize that, although the improvements appear to be limited, data and control inputs are *not correlated* in the streams we have considered for testing the model. In fact, even test streams with a fixed input, do not actually have an effect on the switching of the data inputs. We can thus claim that the evaluation conditions used for Table 1 represent the worst case improvement in accuracy.

To further observe where the proposed model improves over the others, we analyze the results for a specific macro, namely a Linear Feedback Shift Register with parallel load (corresponding to the DesignWare macro *DW03_lfsr_load*).

We compare a conventional linear regression model as the one of Equation 3 with a model based on the derivation of Equation 6:

$$P_{LIN} = C_0 D_{clk} + C_1 D_{reset} + C_2 D_{load} + C_3 D_{cen} + C_4 D_{count} + C_5 D_{data};$$

$$P_{CND} = C_0 D_{clk} + C_1 D_{reset} + C_2 D_{load} + C_3 D_{cen} + C_4 D_{count} + \\ C_5 D_{data} P_{load} + C_6 D_{data} (1 - P_{load}).$$

After characterization, the two models are extracted, yielding the following regression coefficients:

LIN	CND
$C_0 = 1685$	$C_0 = 1651$
$C_1 = 2448$	$C_1 = 2404$
$C_2 = 3302$	$C_2 = 3369$
$C_3 = 3068$	$C_3 = 2902$
$C_4 = 5733$	$C_4 = 6207$
$C_5 = 1999$	$C_5 = 1286$
	$C_6 = 2936$

From inspection of the model, we observe that the main difference between P_{LIN} and P_{CND} lies in the way the dependency between power and the input data switching D_{data} is modeled. In P_{LIN}, D_{data} is considered as an independent variable (and thus depending on a single coefficient), whereas in P_{CND}, the joint effect of D_{data} and the control signal load is considered. This amounts to splitting the contribution $C_5 \cdot D_{data}$ of P_{LIN} in two parts, depending on the value of load.

This is reflected in the values of the coefficients. While $C_0, \ldots, C_4$, that refer to the non-controlled inputs and outputs, have similar values in both models, coefficient C_5 of the LIN model is actually the average of C_5 and C_6 in the CND model.

5   Conclusions

We have proposed a new power macromodel for *control-dominated* RTL macros. The control inputs may activate input-insensitive behaviors that let the macro evolve in a *semi-autonomous* way.

The proposed model overcomes the limitations of conventional black-box, activity-sensitive power models, because it explicitly represents the *correlation* between some of the control and data inputs by adopting a higher-order model.

The model, although macro-specific, can be automatically generated because it only requires the specification of what control signals affect a set of data inputs.

Results are promising, and have better accuracy over conventional models, even for stream that do not enforce the existing correlation between control and data signals.

References

1. P. Landman, "High-Level Power Estimation," *ISLPED-96*, pp. 29-35, Monterey, CA, August 1996.
2. E. Macii, M. Pedram, F. Somenzi, "High-Level Power Modeling, Estimation, and Optimization," *IEEE Transactions on CAD*, pp. 1061-1079, Nov. 1998.
3. S. Powell, P. Chau, "Estimating Power Dissipation in VLSI Signal Processing Chips: The PFA Technique," *VLSI Signal Processing IV*, pp. 250-259, 1990.
4. P. E. Landman, J. Rabaey, "Activity-Sensitive Architectural Power Analysis", *IEEE Transactions on VLSI Systems*, Vol. 15, no. 6, pp. 571-587, 1995.

5. G. Jochens, L. Kruse, E. Schmidt, W. Nebel, "A New Parameterizable Power Macro-Model for Datapath Components," *DATE'99*, Munchen, Germany, pp. 29-36, Mar. 1999.

6. A. Bogliolo, R. Corgnati, E. Macii, M. Poncino, "Parameterized RTL Power Models for Combinational Soft Macros," *ICCAD'99*, S. Jose, CA, pp. 284-287, Nov. 1999.

7. S. Gupta and F. Najm, "Power Macromodeling for High Level Power Estimation", *DAC-37*, Anaheim, CA, pp. 365-370, Jun. 1997.

8. S. Gupta, F. Najm, "Analytical Model for High Level Power Modeling of Combinational and Sequential Circuits", *IEEE Alessandro Volta Memorial Workshop on Low Power Design*, pp. 164-172, Como, Italy, Mar. 1999.

9. M. Barocci, A. Bogliolo, L. Benini, B. Riccò and G. De Micheli, "Lookup Table Power Macro-Models for Behavioral Library Components", *IEEE Alessandro Volta Memorial Workshop on Low Power Design*, pp. 173-181, Como, Italy, Mar 1999.

10. L. Benini, A. Bogliolo, M. Favalli, G. De Micheli, "Regression Models for Behavioral Power Estimation", *PATMOS'96*, pp. 179-187, Bologna, Italy, Sep. 1996

11. R. Corgnati, E. Macii, M. Poncino, "Clustered Table-Based Macromodels for RTL Power Estimation," *GLS-VLSI'99*: Lafayette, LA, Mar. 1998.

12. L. Benini, A. Bogliolo, G. De Micheli, "Adaptive Least Mean Square Behavioral Power Modeling," *EDTC-97*, pp. 404-410, Paris, France, Mar 1997.

13. Z. Chen, K. Roy, T.-L. Chou, "Power Sensitivities – A New Method to Estimate Power Dissipation Considering Uncertain Specification of Primary Inputs," *ICCAD-97*, pp. 40-44, San Jose, CA, Nov. 1997.

14. F. Brglez, D. Bryan, K. Kozminski, "Combinational Profiles of Sequential Benchmark Circuits", *ISCAS'89*, pp. 1929-1934, May 1989.

15. A. Salz, M. Horowitz, "IRSIM: An Incremental MOS Switch-Level Simulator" *DAC-26*, pp. 173-178, Las Vegas, NV, Jun. 1989.

Power Macro-Modelling for Firm-Macro[1]

Gerd Jochens, Lars Kruse, Eike Schmidt, Ansgar Stammermann, and Wolfgang Nebel

OFFIS Research Institute, Oldenburg
Jochens@OFFIS.DE

Abstract. An approach for power modelling of parameterized, technology independent design components (firm-macros) is presented. Executable simulation models in form of C++ classes are generated by a systematic procedure that is based on statistical modelling and table look-up techniques. In contrast to other table look-up based approaches the proposed model separately handles the inputs of a component, and with this it allows to model the effects of corresponding joint-dependencies. In addition, a technique for the generation of executable models is presented. The generated models are optimized with respect to simulation performance and can be applied for power analysis and optimization tasks on the behavioral and architectural level. Results are presented for a number of test cases which show the good quality of the model.

1 Introduction

Recent years have brought an enormous increase in integration of circuit elements on a single chip. This trend of higher performance and smaller device sizes comes with enormous physical challenges. One of these challenges is the power dissipation. High power consumption means high power costs and short battery life-time of mobile applications. Consequently power dissipation is an important part of the cost function of modern designs and tools that allow to analyze the power consumption already on high levels of abstraction are in high demand.

Meanwhile, techniques for power analysis and low power synthesis on the behavioral level have come up [1,2,3,4]. Given a behavioral description of an algorithm, the techniques allow an efficient estimation of upper and lower bounds of the power consumptions and even suggest power optimal allocation and binding of datapath components. Usually, these components are combinational arithmetic and logic units which are provided as so called 'firm-marcos' (VSI Alliance recommendation) in a component library (e.g. DesignWare®-library from Synopsys®). These firm-macros have a defined module architecture and are parametric in terms of the word- length. They are provided as technology-independent descriptions which can be mapped onto a specific technology by logic synthesis. To guide analyses and optimizations, these estimation and optimization techniques require power models for the datapath components which describe the dependency of the power consumption on significant macro parameters and support typical optimization steps.

---

[1] This work is founded by the BMBF project EURIPIDES under grant number 01M3036G. and by the Commission of the European Community as part of the ESPRIT IV programme under contract no. 26796

D. Soudris, P. Pirsch, and E. Barke (Eds.): PATMOS 2000, LNCS 1918, pp. 24–35, 2000.

Already, a number of techniques for power modelling of combinational and sequential design-components have been proposed. A good overview of existing approaches of power modelling techniques is given in 5.. Unfortunately, most of these techniques focus on power modelling of so called 'hard-macros'. The structure of these components is fixed and a low-level implementation on gate- or transistor-level is available. Only a few of these techniques can principally be extended to the handling of datapath components, as for this word-length independent model parameters and variables are necessary. For one of these techniques such an extension has been presented in [5].

A power model dedicated to parameterized datapath components was presented by Landman [11]. The model is derived under the assumption of certain signal statistics. Unfortunately, statistics of real application data may significantly differ from these assumptions, especially in the case of resource sharing. In Section 3 we will further comment on this approach and on the limitations.

In the following, we propose a new approach for power modelling of firm macros and suggest a corresponding technique for automatic model generation. The generated models describe the dependency of the power consumption on significant input characteristics and macro parameters. Different module inputs are regarded separately. With this the influence of input-data joint-dependencies and of the mapping of the data streams onto module inputs is considered. The dependency on the module word-lengths is handled by a regression technique that considers architecture informations to minimize the number of prototypes which are necessary to fit the model to a specific technology. Furthermore, a technique for a systematic generation and integration of executable simulation models is suggested. Evaluation results for a number of test cases are presented which demonstrate the good quality of the model.

The rest of this article is structured as follows. In Section 2 we start with a definition and separation of the modelling problem. Section 3 describes our concept for statistical modelling and presents the approach for modelling the data-dependency. In Section 4 we focus on modelling the word-length dependency and explain the handling of control-inputs and logic-optimizations. Section 5 explains our technique for generating and integrating the models into a behavioral power analysis and optimization tool. This is followed by an explanation of our evaluation process and the presentation of results. The paper concludes with a brief summary in Section 6.

2   Problem Definition and Separation

The problem of modelling the power consumption of combinational firm macros is the problem of identifying a functional relationship between the power consumption P and 1) significant characteristics of two consecutive input vectors $G(X[n-1], X[n])$, 2) the vector of input word-lengths BW of a component instance, 3) the architecture of a component instance A and 4) the mapping technology T.

$$P[n] = f(G(X[n-1], X[n]), BW, A, T). \qquad (1)$$

This problem can be separated into smaller sub-problems. Without loss of generality, separate models can be used for different architectures and technologies while the technique used to generate the model is the same. With this, A and T disappears from the equation. By heedfully choosing a set of model variables the dependency of P from $G(X[n-1], X[n])$ and BW is approximately statistical independent (we will show this in

Section 4), so that the problem reduces into the separate issues of: 1) modelling the dependency of the power consumption in terms of the input vector characteristics $G(X[n-1], X[n])$ and 2) modelling the dependency of the component word-lengths:

$$P[n] = f(G(X[n-1], X[n])) \cdot h(BW). \qquad (2)$$

A wide variety of statistical modelling techniques exist that help to systematically ascertain such relationships (for an overview see [15,16]). The quality of this functional relationship can be measured by proper statistical measures, e.g. the mean square error, etc. which describe the discrepancy between the true power values $P[n]$ and the estimated values $\hat{P}[n]$.

3 Modelling Data Dependencies

In this section we describe our approach for systematically deriving a statistical model that describes the data dependency of the power consumption. The process of model derivation that has been used can be separated into the steps of *model identification* and *model fitting*. Model identification is the process of finding or choosing an appropriate functional relationship for the given situation. Model fitting is the stage of moving from a general form to a numerical form. Because of the limited space of the paper we can not go into the details of all modelling steps, but instead present some basic methodologies and underlying statistical techniques that are used.

3.1 Model Identification

The process of model identification contains the steps of data identification, model parameter selection and the identification of a functional relationship. Data identification is the process of analyzing the input data with the aim of deriving hints for the selection of model variables, parameters and forms of the functional relationship.

Data Identification and Model Parameter Selection
On high levels of abstraction stationary signals are represented as abstract values. Characteristics of the signals are mean μ, variance σ^2 and temporal correlations ρ [8]. On lower levels, statistics of bits and bit-vectors are of interest. Bit-characteristics are signal probability p, switching activity t and temporal correlations. Characteristics of bit-vector streams are the Hamming-distance Hd, average values of switching activity and signal probabilities as well as measures of spatio-temporal correlations [9,10]. In addition to data models, techniques for empirical and analytical estimation of bit-level statistics from word-level statistics exist [11,12]. These models are usually restricted to simple Gauß-distributions or simple AR-models. Assuming such input streams leads to bit-level statistics with some typical characteristics, which can be used to derive a power model. Landman and his dual-bit-type model was the first who consequently used this technique to develop a power model for datapath components [11].

The disadvantage of this methodology is that it restricts the application of the power model to applications where the assumption on the distribution can be assured. Unfortunately, this assumption does not hold for a number of real applications (e.g. [17]) especially in the case of resource sharing, as different input streams are mixed here.

So, for not restricting the application, we have considered model variables and parameters during the selection process that do not require any high-level data models. Instead we consider bit-level statistics which have a high significance with respect to the modules power consumption and which can efficiently be captured during a functional simulation of the design. Nevertheless, our model variables values can be estimated *analytically* from high-level statistics for a number of typical cases, e.g. Gauß- or Laplace distributions. Hence, our modelling approach can also be used for fast probabilistic simulation techniques on RT-level, where high level statistics are propagated through the design. The methodology for calculating our model parameters from high level statistics has been presented in [13].

For selecting our model variables and parameters we used a mixture of empirical and analytical techniques. Different sets of model variables have systematically been chosen and have been evaluated for typical datapath components using statistical model selection techniques and significance analyses [15]. Because of the limited space we will not go into further detail, but focus on the result of this step.

To characterize a sequence of two consecutive bit-vectors $X[n-1], X[n]$ at *one* module input, we use the Hamming-distance Hd and the number of digits with a fixed value equal to zero #0 or one #1 . Consequently, a transition T of two consecutive bit-vectors is characterized by:

$$T[n] = (Hd[n], \#0[n], \#1[n]) \tag{3}$$

with

$$Hd[n] = Hd(X[n-1], X[n]) = \left| \{i | (X[n-1]_i \neq X[n]_i)\} \right|$$

$$\#0[n] = \left| \{i | (X[n-1]_i) = X[n]_i = 0\} \right|$$

$$\#1[n] = \left| \{i | (X[n-1]_i) = X[n]_i = 1\} \right|$$

for $1 \leq i \leq m$ and m as the vector word-length.

As for an instantiated component the word-length of an input is fixed ($m = Hd + \#0 + \#1 = const$), only two of the three variables are independent. Furthermore, we normalize these values to the word-length m to get a word-length independent value-range, so that we choose the normalized values $\underline{Hd} = Hd/m$ and $\underline{\#0} = \#0/m$ as model variables. We desist from the usage of module output statistics as model parameters in general (except for multiplexer), as it is difficult to generate output values with a specific combination of statistics, which is necessary during the model fitting.

The significance of the chosen parameters is exemplified in Figure 1. The figure shows the average charge consumption per transition over the Hamming-distances at the multiplier-inputs A and B. Two settings of the non-switching bits are distinguished. Points on a line have a constant Hd-sum ($Hd_A + Hd_B = Hd_{Sum} = const$), so that only the distribution of the total value (Hd_{Sum}) onto the inputs varies.

From the figure it is clear that the charge consumption strongly depends on the Hamming-distance *and* on the distribution onto the module inputs (values differ by a factor of up to 3 for constant values of Hd_{Sum}). Furthermore, it can be seen that the power consumption for a certain input stream at one input, strongly depends on the input data of the other input. It is important, that these influences can only be handled by separately regarding the data at the module inputs. Furthermore, the significance of the signal-values of the non-switching bits is of interest. The figure contains the average charge consumption per transition for different combinations of Hd's at the inputs for the case that all non switching bits are '0' and '1'. It can be seen that the charge consumption differs by a factor of 4 to 10 for corresponding points.

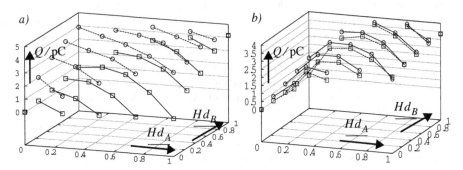

Fig. 1: Average power consumption of a 24x24 bit CSA-multiplier (a) and Booth-Coded Wallace-Tree multiplier (b) over the normalized Hamming-distances at inputs A and B for all non-switching bits '0' and '1', respectively

Summing up, the variables chosen allow a good distinction of transitions in terms of power for a wide variety of different design components and architectures. The capturing of the variable values from the signal values can efficiently be processed by a table look-up technique. To calculate a variable value for two consecutive input vectors, only an exor-command and an array access is required, i.e. no time-consuming loops and if-commands are necessary.

Identification of Functional Relationships

As it is difficult to infer a functional relationship that holds for the complete value-range, we decided to use an interpolation technique for localized approximations. Because of performance reasons, we apply a multi-dimensional linear interpolation technique. Values between neighboring grid-points are approximated by first-order Taylor-rows f. The differential-coefficients of the Taylor-row are approximated by difference-coefficients, which are calculated from the function values p of the nearest grid-points [18].

This technique allows a fast calculation for a given set of variable values, as it only requires to select and calculate a corresponding function f. The selection of an adequate grid-size and the calculation of the approximation-functions is explained in Section 3.2. As an alternative to the presented interpolation technique, we suggest the usage of a technique presented in [20]. This technique has a higher flexibility and smaller memory demands, but leads to higher computational cost.

3.2   Model Fitting

Model fitting is the process of moving to a numerical form. For our approach this includes the definition of the grid-size and the determination of the interpolation functions f. For the interpolation functions it is necessary to estimate the functional values at the grid points and the difference-coefficients.

Grid-Size Identification

The grid-size can iteratively be determined by an analysis of appropriate error measures. We use an empirical technique, where the n samples are regarded that are located in the geometric centre of neighboring grid-points (intermediate points). The 'true' values at these points (taken from a simulation of a component prototype) and values calculated by interpolation functions are evaluated by the mean square error :

$$MSE = \frac{1}{n}\sum_{k=1}^{n} (P[n] - \hat{P}[n])^2 \qquad (4)$$

with:

$P[n]$: values from simulation,
$\hat{P}[n]$: values from interpolation.
The procedure for grid-size estimation is then as follows:
1) define an error-limit MSE_{limit}
2) set grid-size Δ to initial value Δ_{init}
3) estimate 'true' function values for grid-points and calculate interpolation-functions $f_{i,j}$
4) estimate 'true' function values for intermediate points,
5) calculate MSE,
6) if $MSE > MSE_{limit}$ set $\Delta = \Delta/2$ and repeat 3) to 6), otherwise stop procedure.

Instead of globally reducing the grid-size it is also possible to locally reduce the grid-size, if a more detailed data analysis is processed. Furthermore, it is important that the grid-size must comply to the word-length's of the component prototype, e.g. for a 16 x 16 bit component only multiples of 1/16 are possible.

For all components we have analyzed until now (see Section 5) the interpolation technique works very well and a step-size $\Delta = 0.25$ was sufficient to achieve average interpolation errors which are less than 5-10%. As an example, Figure 2 illustrates the quality of the interpolation technique for an 16x16bit carry-save-array-multiplier.

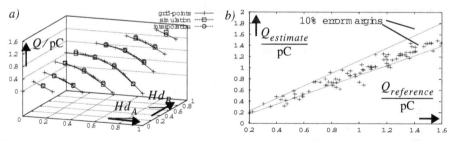

Fig. 2: a) Comparison of values at intermediate grid-points; b) Comparison of estimated and 'true' intermediate grid-point values

Model-Parameter Estimation

The process for parameter estimation is as follows:
1) generate a stream of input-patterns for all grid-points, i.e. a stream where consecutive input vectors have a defined characteristic,
2) perform a power simulation of a component prototype for each pattern stream,
3) use the average charge consumption per transition as estimation for the 'true' model parameter (corresponding to the method of smallest squares, the average is the best estimate),

4) calculate error margins based on the central limit theorem,
5) if error margins are not met, extend the pattern stream by a number of new vectors which can be estimated from the central limit theorem [19].

It has to be mentioned that steps 4) and 5) can only be executed if a cycle accurate power simulator is used.

For the components we have characterized, we have found that in 95% of all cases a vector-stream of 100 patterns leads to error margins of less than 5%. So, using a fixed set of 200 vectors is sufficient in practice. It is important that the patterns for fitting our model can be generated as a continuous stream, which can be simulated in one run (for each grid point). I.e., it is not necessary to run the simulator n times to simulate n vector-pairs, which is usually very time consuming. Because of the limited space, we omit the presentation of the algorithm for the generation of characterization pattern streams.

4   Modelling Word-Length Dependencies

The process of modelling the word-length dependency also consists of the sub-problems of model-identification and fitting. In contrast to the procedure of model-identification which has been used for modelling the data dependency and that mainly relies on empirical techniques for data analysis, the procedure for word-length dependency modelling is based stronger on conceptual techniques. If available, we extract the form of the functional dependency from the architecture of the component, i.e. we use the knowledge about the component structure.

4.1   Model Identification

The problem of describing the dependency on the word-length can be mapped to the problem of describing the influence of the word-length on the interpolation function or function parameters p (values at grid-points), respectively:

$$p_{i,j} = k_{i,j}(BW),\qquad(5)$$

where i, j denotes a certain grid point.
As:

$$p_{i,j} = \frac{1}{2} \cdot V_{dd}^2 \cdot C \cdot \alpha_{i,j}$$

with:
V_{dd}: supply voltage,
C: module capacitance and
α: an average activity factor, which is constant for a certain grid-point,
it follows that the model parameters $p_{i,j}$ are proportional to the module capacity, which is a function of the input word-length's for fixed architecture:

$$p_{i,j} \propto C \propto k_{i,j}(BW)\qquad(6)$$

As suggested originally in [11] the form of this dependency can be derived from the architecture of a component. For example, the dependency for an carry-save-array multiplier with the input word-lengths $BW = \{bw_A, bw_B\}$ is of the form:

$$p_{i,j} = k_{i,j}(BW) = r_2 \cdot (bw_A \cdot bw_B) + r_1 \cdot (bw_A + bw_B) + r_0$$

with: $R = [r_2 \; r_1 \; r_0]^T$ the vector of function parameters. It is import that these dependency functions also allow the handling of components with multiple inputs, that have different word-lengths.

Since for small grid-sizes the number of parameters and with this the number of corresponding functions k might be large, it is necessary to use an additional approximation. So, instead of regarding the dependency of each parameter we use an average dependency. This is done as follows:

1) Normalize all parameters:

$$\underline{p_{i,j}^{BW}} = \frac{p_{i,j}^{BW}}{p_{norm}^{BW}}$$

with:

$p_{i,j}^{BW}$:a parameter $p_{i,j}$ for a component prototype with fixed word-length BW,

p_{norm}^{BW} :the average of all parameters $p_{i,j}^{BW}$ as norming value.

2) Average the normalized parameters over the word-length:

$$\underline{p_{i,j}^{avg}} = \frac{1}{size(C)} \cdot \sum_{BW \in C} \underline{p_{i,j}^{BW}}$$

with:

C a set of component prototypes with different word-lengths.

With this the problem reduces to the problem of fitting the function that describes the dependency of the norming value on the word-length BW:

$$p_{norm} = k(BW).$$

If a numerical form is determined, the parameters $p_{i,j}(BW)$ can be estimated by

$$p_{i,j}^{approx}(BW) = \underline{p_{i,j}^{avg}} \cdot p_{norm} = \underline{p_{i,j}^{avg}} \cdot k(BW). \tag{7}$$

The effects of this approximation step can be evaluated by analyzing the differences of the 'true' and the approximated values, in terms of average deviation or mean square errors. As an example, Figure 3a shows some 'true' normalized parameters and the approximated values, again for the CSA-multiplier. For a grid-size of $\Delta = 0.25$ the average difference of true and approximated parameter

$$d_{i,j}^{avg} = \frac{1}{size(C)} \sum_{BW \in C = \{8x8, 16x16, 24x24, 32x32\}} (\underline{p_{i,j}^{BW}} - p_{i,j}^{approx}(BW))$$

is illustrated in Figure 3b. It can be seen, that the (relative) deviation are less than 10% in 90% of all parameters. The large error peeks seen in Figure 3b are for parameter values which have very small absolute values, so that the impact on the estimation accuracy is usually small.

Nevertheless, if this can not be accepted, it is possible to use local approximations instead of a global one to reduce the deviations.

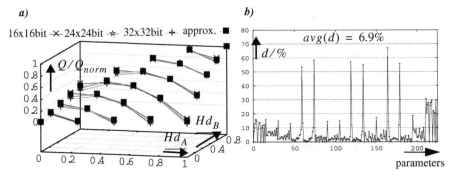

Fig. 3: a) Comparison of true and approximated values for transition power (CSA-mult.); b) Deviations of approximated an true parameters, for prototypes with wordlength {8x8,16x16,24x24,32x32}

4.2   Model Fitting

For model fitting we use a regression approach, which calculates the model parameters based on the method of least squares. This is done as follows:
1) select a set of component prototypes C with different word-lengths,
2) extract the parameters $p_{i,j}$ as described in Section 3,
3) use approximation technique described in Section 4.1
   to calculate normalized parameters $p_{i,j}^{BW}$,
4) define the form of function $k(BW)$ based on architecture information,
5) start a regression process to determine the parameters R,
6) evaluate the quality of the regression, and adapt the number of prototypes if necessary.
The quality of the regression can be evaluated in terms of correlation coefficients and risk functions. With this it can be evaluated how well the values used as input to the regression (estimation values) are approximated by the regression function. This measure can also be used to measure the quality for values not used within the regression process (test values).

From our experiences we have found, that a small number of components is sufficient to fit the functions $k(BW)$. To fit the function for word-lengths in the range from 8 to 32 bits using four component prototypes (8,16,24 and 32 bits) is sufficient to achieve deviations of less than 5%. Nevertheless, for components with unknown architectures it is possible to use eclectic techniques to select an adequate functional relationship [15,14]. The price for this is an increase of the number of prototypes to achieve a defined accuracy and confidence.

The handling of control inputs and the influence of logic optimizations is straight forward. As most of the datapath components only have one control input, for each setting of this input ('0', '1' or toggling) a separate model is generated. Logic-optimizations are considered by generating minimum area and delay variants of a component and adapting the norming value and word-length dependency function.

5   Evaluation

The proposed technique for power modelling has been realized as an interactive modelling tool, which is a part of the OFFIS behavioral-level power estimation tool-suite ORI-

NOCO®. The tool generates C++ classes which encapsulate the complete parameterizable model. The realization as C++ allows a simple and flexible integration into power analysis and optimization tools. An open interface methodology simplifies the integration of third party models for IP components.

Until now we have generated power models for a number of components that are relevant for behavioral VHDL power analysis. For each component a set of prototypes used for characterization and a set used for validation was generated. For each prototype the estimation accuracy has been analyzed for different sets of evaluation data (test cases to stress the model) and some sets of real application data. Figure 4 gives an overview of the components and data-sets involved in the evaluation procedure. The quality of the model has been evaluated in terms of absolute and relative accuracy of the average- and cycle-accurate power estimates. Furthermore, we have compared our approach to the DBT-model and a technique presented in [7] which allows cycle accurate estimates for components with a fixed word-length, and achieved better results for a number of practical designs, especially where input-streams are mixed due to resource sharing.

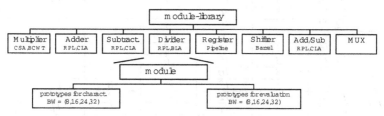

Fig. 4: Overview of components used within the evaluation procedure

Because of the limited space we only present a sub-set of the evaluation results. Table 1 presents the estimation errors for a number of different components and test-data sets. Estimation errors are delivered for the case that a model is built for a single instance (instance model) and for the case that a word-length parameterizable model (BW-param. model) is used. This allows to evaluate the effects of the word-length parameterization. Estimation errors are presented for randomly generated input streams (rand.) and for streams with data characteristics, so that the corresponding model variable values lie on intermediate grid points (i.m). These intermediate points are equally distributed over the complete value range, and with this they stress the model over a wide variety of possible input streams. Furthermore, these intermediate points are a worst case to the modelling approach, as the total difference between intermediate and (characterized) grid points is maximum, which is crucial for the interpolation. The deviations for this case are given as the average of the absolute values of relative errors in the table:

$$\varepsilon_{abs} = \frac{1}{size(IM)} \sum_{i \in IM} \left| \frac{q_i^{model} - q_i^{ref}}{q_i^{ref}} \right| \tag{8}$$

with:

IM : set of all intermediate grid points,
q_i^{model} : cycle charge consumption estimated by the model,
q_i^{ref} : cycle charge consumption from logic level simulation

and as average relative estimation error (i.m. avg.). From the table, it can be seen that, except for the case of small word-length of the csa-multiplier and rpl-divider, the impact of the word-length dependency modelling is very small. Even for the intermediate grid points, the estimation accuracy is acceptable and within error-bounds of 10 to 15% over the complete value range (cf. Figure 2b).

Table 1. Estimation errors in % for different components and input patterns compared to logic level simulation

compo-nent	BW	instance model			BW-param.model		
		rand.	i.m.	i.m. avg	rand.	i.m.	i.m. avg
mult-Csa	8x8	-1	8	-5	-18	12	9
	16x16	-2	9	-5	-12	16	12
	24x24	-3	9	-6	-1	10	-4
	32x32	-2	11	-5	2	8	2
mult-Bcwt	8x8	-3	12	-2	-6	15	8
	16x16	3	12	-7	-1	14	-11
	24x24	-1	13	-9	-3	15	-9
	32x32	4	13	-9	-5	15	7
add-Rpl	8x8	-3	12	-4	-7	14	-11
	16x16	-5	12	-1	-5	13	1
	24x24	-3	9	-5	-1	11	-3
	32x32	-3	10	-4	-2	11	-1
addCla	8x8	-1	8	-6	-5	11	-8
	16x16	-1	9	-3	-1	9	-2
	24x24	-3	10	-6	-2	9	-5
	32x32	-3	10	-5	3	8	-3
subRpl	8x8	-3	12	-5	-8	14	-9
	16x16	-2	12	-6	-2	12	-6
	24x24	-3	11	-8	1	11	-5
	32x32	-3	12	-6	2	11	-5
subCla	8x8	-1	13	-8	6	12	-6
	16x16	-2	11	-8	-4	11	-8
	24x24	-1	12	-7	9	12	-7
	32x32	-4	11	-6	12	11	-5
divRpl	8x8	6	15	1	19	19	3
	16x16	4	17	-1	8	16	-1
	24x24	-1	15	2	-8	16	2
	32x32	-4	17	3	-7	19	-3
divBla	8x8	1	12	1	1	16	1
	16x16	5	13	1	7	14	1
	24x24	1	11	4	2	18	4
	32x32	4	13	3	3	17	4

6   Summary

In this paper we have presented a concept for power modelling of parameterized datapath components. The approach is based on statistical modelling techniques and allows a separate handling of module inputs. With this, the model allows to consider the influence of input-data joint-dependencies. The separate handling of inputs furthermore allows to model the influence of the mapping of data streams onto inputs, which is especially of interest for non-symmetric module structures. This information can be used for commutative operations to optimize the binding in terms of power. The significance of the proposed model variables and the adequateness of the modelling form has been shown for several examples.

Furthermore, we proposed a technique for model generation, which produces executable simulation models as C++ classes that are optimized with respect to simulation efficiency and flexibility. In combination with table look-up techniques for capturing the model variables from simulation data, a very efficient and simple integration into high-level power analysis tools is possible.

The high simulation performance, the parameterizability, the separate consideration of module inputs and the automatic model generation in form of executable C++ model make the model attractive for high level power estimation and optimization tasks. The results that have been presented for a number of different test cases show the good quality of the approach.

References

1. Chang, J.-M.; Pedram, M.: Module Assignment for Low Power, DAC, 1995
2. Ohm, S. Y.; Whitehouse, H. J.; Dutt, N.D.: A Unified Lower Bound Estimation Technique for High-Level Synthesis, IEEE Trans. on CAD, 1997
3. Kruse, L.; Schmidt, E.; Jochens, G.; Nebel, W.: Lower and Upper Bounds on the Switching Activity in Scheduled Data Flow Graphs, ISLPED, 1999
4. Kruse, L.; Schmidt, E.; Jochens, G.; Nebel, W.: Low Power Binding Heuristics, PATMOS, 1999
5. Macii, E.; Pedram, M.; Somenzi, F.: High level power modelling, estimation and optimization. Trans. on Design Automation of Electronic Systems, 1998
6. Bogliolo, A.; Corgnati, R.; Macii, E.; Poncino, M.: Parameterized RTL Power Models For Combinational Soft Macros, ICCAD, 1999
7. Benini, L.; Bogliolo, A.; Favalli, M.; De Micheli, G.: Regression Models for Behavioral Power Estimation, PATMOS, 1996
8. Rabiner, L. R.; Schafer, R. W.: Digital Processing of speech Signals, Prentice-Hall, 1978
9. Marculescu, R.; Marculescu, D.; Pedram, M.: Vector Compaction Using Dynamic Markov Models, IEICE Transactions on Fundamentals of Electronics, Communications, and Computer Sciences, Vol. E80-A, No. 10, Oct. 1997
10. Marculescu, R.; Marculescu, D.; Pedram, M.: "Efficient Power Estimation for Highly Correlated Input Streams", DAC, 1995
11. Landman, P. E.; Rabaey, J. M.: Architectural power analysis: The dual bit type method. IEEE Trans. VLSI Syst., Vol. 3, 1995
12. Ramprasad, R.; Shanbhag, N. R.; Hajj, N.: Analytical Estimation of Signal Transition Activity from Word-Level Statistics. IEEE Trans. on CAD, Vol. 16, No.7, 1997
13. Jochens, G.; Kruse, L.; Nebel, W.: A New Parameterizable Power Macro-Model for Datapath Components, DATE, 1999
14. Linhart, H.; Zucchini, W.: Model Selection, J. Wiley & Sons, 1986
15. Gilchrist, W.: Statistical Modelling, John Wiley & Sons, 1984
16. Hjorth, J.S.U.: Computer Intensive Statistical Methods, Chapman & Hall, 1993
17. Tsui, C.-Y. ;Chan, K.-K. ;Wu, Q.; Ding, C.-S.; Pedram, M:: A Power Estimation Framework for Designing Low Power Portable Video Applications, DAC, 1997
18. Davis, Ph. J.: Interpolation & Approximation, Dover Publications, 1975
19. Yeap, G. K.: Practical Low Power Digital VLSI Design, Kluwer Academic Press, 1998
20. Rovatti, R.; Borgatti, M.; Guerrieri, R.: A Geometric Approach to Maximum-Speed n-Dimensional Continuous Linear Interpolation in Rectangular Grids, IEEE Trans. On Comp., Vol. 47, No. 6, 1998

RTL Estimation of Steering Logic Power [*]

Crina Anton[1,4], Alessandro Bogliolo[2], Pierluigi Civera[1],
Ionel Colonescu[3], Enrico Macii[3], and Massimo Poncino[3]

[1] Politecnico di Torino, DELEN, Torino, ITALY 10129
[2] Università di Ferrara, DI, Ferrara, ITALY 44100
[3] Politecnico di Torino, DAUIN, Torino, ITALY 10129
[4] ST Microelectronics, Central R&D, Agrate, ITALY 20041

Abstract. Power dissipation due to the steering logic, that is, the multiplexer network and the interconnect, can usually account for a significant fraction of the total power budget. In this work, we present RTL power models for these two types of architectural elements. The multiplexer model leverages existing scalable models, and can be used for special complex types with re-configurable numbers of data bits and ways. The interconnect model is obtained by empirically relating capacitance to circuit area, that is either estimated by means of statistical models or extracted from back-annotation information available at the gate level.

1 Introduction

Although several works have addressed the problem of RTL power estimation (see [1] for a survey), most have proposed power models for either the datapath modules (instantiated in the RTL description as a result of behavioral synthesis) or for the control logic driving those modules.

Besides the contribution of such elements, that are explicitly exposed in the RTL description as either HDL statements (the controller) or synthetic operators (the datapath modules), also the *steering logic*, that is, the multiplexer network and the interconnect, can usually account for a significant fraction of the total power budget.

In spite of their potential impact, especially for design with a large amount of shared resources, only a few works have addressed the problem of estimating the power due to the steering logic. Different motivations are at the basis of this limited analysis.

Multiplexers are not usually considered during RTL estimation essentially because, unlike datapath operators, they are not explicitly instantiated in the specification; rather, they are generated during the high-level synthesis as a result of resource sharing.

Similarly, the impact of the interconnect is usually neglected because it requires information on the physical implementation of a design. This implies that, unlike datapath modules, wire capacitances (and thus power) cannot be pre-characterized. Some approaches have dealt with this problem by leveraging existing statistical models [2,3,4,5] that relate the length of the interconnect to macroscopic parameters that can be more easily inferred from a high-level specification [6].

Regardless of its complexity, the impact of steering logic cannot be simply ignored; it has to be carefully accounted for to achieve absolute power estimates during design

---

[*] This work was supported, in part, by the EC under grant n. 27696 "PEOPLE".

D. Soudris, P. Pirsch, and E. Barke (Eds.): PATMOS 2000, LNCS 1918, pp. 36–46, 2000.

validation, and to make significant comparisons during design exploration. The following example shows the impact of MUXes and wires on power dissipation of a design with different amounts of sharing.

Example 1 *The behavioral specification of an elliptic filter (*ellipf*), taken from [7] contains 26 additions. We synthesized three alternative RTL implementations of the filter:* ellipf26, *with 26 adders and latency 1;* ellipf10, *with 10 adders and latency 3;* ellipf1 *with 1 adder and latency 26. Using Synopsys' DesignCompiler, we mapped the three implementations onto a gate-level library characterized for power and we performed gate-level simulation and power estimation by means of VSS and DesignPower. The same clock period (of 50ns) and data stream (of 100 patterns) were used for all implementations. The energy budgets are reported in the following table:*

	Energy			Percentage		
	ellipf26	ellipf10	ellipf1	ellipf26	ellipf10	ellipf1
ADDER	28670	31950	26910	68.61	32.85	11.92
RANDOM LOGIC	505	4185	52390	1.21	4.30	23.20
MUX	0	23670	110630	0.00	24.33	48.99
WIRES	12610	37470	35880	30.18	38.52	15.89
TOTAL	41785	97275	225810	100.00	100.00	100.00

We notice that: i) The total energy consumption is significantly different; ii) The energy spent in performing the sums is almost the same (the only difference being due to the different signal statistics at the inputs of shared resources); iii) Wiring and MUXes may be responsible of more than 50% of total energy.

In this work, we address the problem of estimating the power contribution of the steering logic. We refer to a high-level-synthesis flow that takes a behavioral specification and builds an RTL implementation, based on a given library of functional macros (hereafter called *RTL library*). We consider RTL descriptions that consist of a set of (both hard and soft) macros belonging to the RTL library, some sparse logic implementing the controller, and the steering logic that is used to properly connect and drive the datapath modules and the controller itself.

In the flow, RTL simulation is used to evaluate the power consumption of the proposed implementation. This is realized in two steps: First, a specific *power macromodel* for each macro belonging to the RTL library is built. These models are meant to express a relation between actual power and some higher-level quantities such as input statistics. Second, power is obtained by summing the result of a context-dependent evaluation of the power models for each macro, plus the contribution of the control logic, which is modeled separately, using a different approach. The power macromodels for the RTL macros are either pre-characterized, or constructed and characterized online, during high-level synthesis, whenever a power estimate is required for a not-yet characterized macro. In both cases, power characterization requires fast synthesis of the macro and mapping on a reference technology library. Both the pre-characterization paradigm and the direct link to the synthesis flow are essential for the discussion of the models for the steering logic we propose in this paper.

It is important noticing the different nature of the models proposed in this work. Multiplexers are soft macros that can be specialized by specifying the number of ways, the bit-width and the encoding used for selection. Any instance can be synthesized and pre-characterized for power. The model we present is general (it is applicable to any MUX with any number of ways) and scalable with respect to the bit-width of the data path (the same model can be scaled to be used for MUXes with different bit-widths without re-characterization). Wiring power cannot be pre-characterized. Moreover, wiring is usually unknown at the RTL. We do not propose a new power model for wiring; rather, we show how existing gate-level wire models can be exploited to obtain accurate estimates of wiring power while working at the RTL.

2   Wiring

The power consumed in charging and discharging wiring capacitance is expressed as:

$$P_{wiring} = \sum_i \frac{\alpha_i}{2T} V_{dd}^2 C_i \tag{1}$$

where V_{dd} is the supply voltage, T is the clock cycle, α_i and C_i are the switching activity and the total capacitance of the i-th net, and the sum is extended to all RTL nets (i.e., to all nets connecting RTL modules). We assume T and V_{dd} are specified by the designer, while α_i is computed, for each net, during RTL simulation. Hence, the task of modeling wiring power reduces to that of modeling wiring capacitance.

The parasitic capacitance associated with the interconnection between two or more modules is the sum of many contributions: the output capacitance of the driving component C_{out}, the input capacitance of all driven components C_{in} and the actual wiring capacitance C_{wire}. In general, however, the power contribution of C_{out} is usually implicitly modeled by the power model of the driving macro. If this is the case, it doesn't need to be also ascribed to the net, or otherwise the total power would be overestimated. On the contrary, the input capacitance of a macro does not contribute to its power consumption when it is simulated in isolation for characterization. In fact, the power estimates provided by gate-level or circuit-level simulation represent the power drawn from the supply net, while input capacitors are directly charged by primary-input lines.

Figure 1 schematically shows the parasitic capacitors connected to a net. All capacitors represented within the boundary of the driving macro contribute to its output capacitance C_{out}, all capacitors represented within a driven macro contribute to its input capacitance C_{in}^j, while the sum of all external capacitors is C_{wire}. We denote by C_{in} the sum of the input capacitance of all driven modules. According to the above observations, we neglect C_{out} and we compute the actual capacitance to be associated

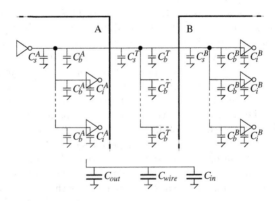

Fig. 1. Hierarchical Topology of a Generic Wire Connecting RTL Modules.

with a net as the sum of C_{wire} and C_{in}:

$$C = C_{wire} + \sum_{j=1}^{FO} C_{in}^j = C_{wire} + C_{in} \tag{2}$$

In the following we discuss the estimation of C_{wire} and C_{in}^j within the high-level-synthesis flow described in the introduction.

2.1   Wiring Topology

The generic wire shown in Figure 1 has a hierarchical structure with three fan-out points: the first fan-out point, internal to the driving macro, distributes the output signal to some internal gates that take it as an input to realize more complex output functions; the second fan-out point, at the RTL, distributes the output signal to all driven macros; the third fan-out point, internal to each driven macro, distributes the primary input to several internal gates.

The actual structure of a wire, however, is unknown before placement and routing and is not necessarily hierarchical. In many cases, placement and routing tools take a flattened gate-level netlist and break the RTL structure to find optimal solutions. Nevertheless, in the following treatment we assume the hierarchical structure depicted in Figure 1 to provide an early estimate of wiring capacitance at the RTL. This assumption, though arbitrary, makes the estimation of wiring power consistent with the estimation of the pre-characterized power models used for functional units, and enables consistent comparisons between alternative design solutions.

2.2   Wiring Model

The topology shown in Figure 1 can be viewed as the hierarchical composition of fan-out points. A wire with a single fan-out (or, equivalently, with a flattened fan-out topology) can be viewed as a basic block for building any hierarchical structure. Without loss of generality, in this section we focus on modeling the capacitance of a wire with a single fan-out point. With respect to a fan-out point, we call *stem* the incoming segment, and *branch* each out-coming edge. The wiring capacitance associated with a wire is the sum of its stem and branch capacitances:

$$C_{wire} = C_s + \sum_{j=1}^{N_{FO}} C_b^j = C_s + N_{FO} C_b \tag{3}$$

where C_b denotes the average branch capacitance, C_s the stem capacitance and N_{FO} the number of fan-out branches. While N_{FO} is available at the RTL, C_s and C_b are not. From a practical point of view, C_s and C_b are the coefficients of a high-level linear model for C_{wire}. The values of C_s and C_b depend both on the technology and on the area of the circuit: the wiring capacitance per unit length is a technology parameter, while the average length of a stem/branch segment depends on the total area. In our tool flow, the values of C_s and C_b can be read from the Synopsys technology file used for mapping. Each wiring model includes the unit capacitance and the average lengths of a stem and branch segment. In addition, a look-up table is provided to associate each model with a range of area values. In summary, the capacitance of a wire with N_{FO} fan-out branches is estimated as:

$$C_{wire} = C_s(tech, area) + C_b(tech, area) N_{FO} \tag{4}$$

2.3   Estimating RTL Wiring Capacitance

Equation (4) can be directly applied to estimate RTL wiring capacitance C_{wire}. The wire model (including the values of C_s and C_b tabulated as functions of technology and area) is taken from the Synopsys technology library specified by the user. The number of fan-out branches N_{FO} is directly obtained from the RTL netlist. The total area is computed as the sum of the back-annotated area estimates for all RTL components.

The only point that needs to be further discussed is the estimation of the area associated with each RTL component. In our tool-flow, fast synthesis is automatically performed whenever a new power model needs to be constructed and characterized for a functional macro. Characterization is based on the results of the power simulation of the gate-level implementation of the RTL macro. In principle, the same paradigm could be used for area estimates: whenever a new macro is instantiated, fast synthesis can be performed to characterize (and back-annotate) its area. On the other hand, if the area has already been characterized, the back-annotated value is used directly without repeating synthesis and characterization.

The problem with this process is efficiency. Suppose the designer is using a new library (without pre-characterized power/area models) and he/she wants to estimate only wiring power to evaluate the impact of sharing. According to the above approach, all macros instantiated within the design should be synthesized with the only purpose of evaluating area. In many cases, this process is very expensive in terms of CPU-time and tool licenses. On the other hand, the dependence of wiring power on the total area is a step function (the same wiring model is associated with a range of area values) whose accurate evaluation does not require accurate area estimates.

According to the above observations, we developed a hybrid approach for area estimation that realizes a better trade-off between accuracy and performance:

1. Fast synthesis of a macro is performed only if a power model for the macro needs to be characterized;
2. Whenever a new macro is synthesized the area of its gate-level implementation is annotated;
3. When computing total area, gate-level area estimates are used only if already available;
4. For macros whose area has not been pre-characterized at the gate-level, a high-level area estimator is used based only of RTL information.

The high-level area estimator we use is derived from Rent's rule. The area A of a macro is expressed as a power function of its pin count N_{IO}:

$$A = cN_{IO}^r \tag{5}$$

Coefficient c and exponent r are the parameters of the model that need to be characterized. Characterization is based on the knowledge of the actual area of all macros that have been synthesized and mapped onto the current library. In the four-step process outlined above, the general estimator is refined (re-characterized) whenever a new macro is synthesized (step 2) in order to take advantage of the new area information.

2.4   Estimating Input Capacitance

The input capacitance C_{in} to be associated with a wire is the sum of the input capacitance of all macros fed by the wire. We assume that the input capacitances viewed at the inputs of a macro are computed at the gate-level during characterization and stored in an array to be used at the RTL. If back-annotated input capacitances are not available (because the macro has never been synthesized) a high-level estimator is used similar to the area estimator introduced in the previous subsection. The average input capacitance C_{in_avg} of a macro is assumed to be related to its area and to its pin count. Since the area is, in its turn, related to the number of pins, we use the model:

$$C_{in\_avg} = dN_{IO}^s \tag{6}$$

where parameters d and s need to be characterized in order to fit available data.

3 Multiplexers

Multiplexers have two peculiarities that make them different from most other macros: First, they have a regular structure; second, they are bit-sliced elements (i.e., a n-bit macro can be viewed as an array of n 1-bit macros, independently processing individual bits). In this section, we exploit the first property to build ad-hoc models that improve upon the accuracy of the general-purpose power models developed for functional macros, and the second property to make the models scalable with the bit-width. Model scaling reduces significantly the characterization effort, allowing us to characterize (i.e., synthesize and simulate at the gate-level) only 1-bit macros, while using the models for arbitrary bit-widths.

Multiplexers are usually specified as soft macros that can be specialized by the designer by setting not only the bit-width (W) of the data inputs, but also the number of input ports, the number of control inputs and the encoding used for input selection. In principle, the concept of model scaling could be applied to scale the power model of a soft macro with respect to all its generics. Generalized model scaling trades off some accuracy to save characterization time. In this context, however, we are not investigating general accuracy-efficiency tradeoffs, rather, we are interested in exploiting the bit-sliced structure of MUXes to scale their models with negligible (if any) accuracy loss. Hence, bit-width W is the only parameter we consider for scaling. From a practical point of view, instances of the same soft macro that differ only for the value of W will share the same (scaled) power model, while instances that differ (also) for some other generics will be treated as different macros with different power models. The model we will derive has the form:

$$Power = S(W)P(stats) \tag{7}$$

where W is the bit-width, $stats$ represents generic boundary signal statistics, $S(W)$ is a scaling function and $P(stats)$ is the power model for the 1-bit instance of the macro. The modeling task is thus partitioned into two sub-tasks: First, deriving a power model $P(stats)$ for a 1-bit MUX; second, determining $S(W)$ for scaling the model.

3.1 Preliminary Analysis

We performed preliminary experiments to verify the disjoint dependence of power on I/O statistics and bit-width. We used as benchmark the universal multiplexer taken from the *Synopsys' DesignWare* library. A larger set of benchmarks was obtained by generating different instances of the MUX by specifying different generics. Each benchmark (i.e., each macro with assigned generics) was then synthesized for different bit-widths, mapped onto a library characterized for power and simulated for different input statistics using *Synopsys' VSS* with *DesignPower*.

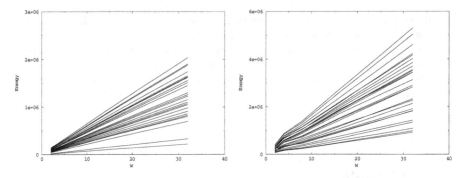

Fig. 2. Energy as a Function of Bit-Width, for a 2-Port and a 4-Port Multiplexer.

From the first set of experiments we observed that the signal probability (i.e., the probability that an I/O signal takes value 1) has a negligible impact on the power consumption of all benchmarks. Based on this observation, we used only the transition probability (i.e., the probability that an I/O signal has a transition) to represent input/output statistics. From a practical point of view, we performed power simulations with input transition probabilities (denoted by D_{in}) ranging from 0.01 to 0.99, while keeping the average signal probability at 0.5.

Figure 2 plots on the left the power consumption of a 2-port MUX as a function of W. Piece-wise linear curves have been obtained by connecting points corresponding to the same input statistics. Hence, curves are parameterized with respect to the input statistics, namely D_{in}.

From the plots we notice that: *i)* The relative distance between the curves is almost independent of the bit-width; *ii)* The relation between bit-width and energy is almost linear. Observation *i)* suggests that the dependence on the bit-width can be de-coupled from the dependence on input statistics, thus motivating the development of power models of the form of Equation 7. Observation *ii)* shows the beneficial effect of the bit-sliced nature of the macros: The power consumption of an n-bit macro is approximately n times the power consumption of the 1-bit macro evaluated for the same input statistics. A similar behavior has been observed for general datapath macros [8].

Unfortunately, there are exceptions to this linear dependence, as shown in the right-hand diagram of Figure 2, that plots the same curves for a 4-port MUX. The reason for this non-linearity, that violates the bit-slice composition rule, can be found by looking at the gate-level implementation of the macros. Though it is always possible to build a n-bit MUX from n 1-bit components, width-dependent design choices can be taken by the synthesis tool to optimize the implementation, resulting in different netlists. If this is the case, we cannot rely on linearity and we need a refined scaling criterion.

3.2   Scaling

To further verify the disjoint dependence between input statistics and bit-width, we analyzed the behavior of:

- $R_1(W_1, W_2, D_{in}) = P(W_1, D_{in})/P(W_2, D_{in})$, representing the ratio between the energy consumption of two macros with different bit-widths, for the same input statistics;
- $R_2(W, D_{in1}, D_{in2}) = P(W, D_{in1})/P(W, D_{in2})$, representing the ratio between the energy consumption of the same macro for different input statistics.

Figure 3 shows, on the left, the behavior of R_1 as a function of D_{in} (curves are parameterized on the values of W_1 and W_2) and on the right the behavior of ratio R_2 as a function of W (curves are parameterized on the values of D_{in1} and D_{in2}). It is apparent that both R_1 and R_2 are almost independent of D_{in} (their standard deviation being 0.029% and 0.0053%, respectively). The scaling factor that has to be applied to a power model characterized for a reference macro with bit-width W_{ref} in order to estimate the power consumption of a different instance of the same macro with bit-width W, is nothing but ratio R_1 computed for $W_1 = W$ and $W_2 = W_{ref}$, that is, $S(W) = R_1(W, W_{ref})$. We refer to this scaling as *analytical*, since it does not require the synthesis of the macro to be scaled.

Under the ideal bit-slice composition assumption, the value of R_1 can be directly obtained at no cost as the ratio between the two bit-widths: $S(W) = W/W_{ref}$. Though in principle we could use $W_{ref} = 1$ for characterization, we achieved better accuracy by using $W_{ref} = 8$. In general, the larger the value of W_{ref}, the lower the scaling factor that multiplies the inherent characterization noise.

If the bit-slice assumption doesn't hold (as shown in Figure 3), using the analytical scaling factor W/W_{ref} may lead to unacceptable errors. In this case, we resort to fast synthesis and simulation (for a fixed value of D_{in}) of the scaled macro in order to obtain the term $P(W, D_{in})$ used to compute $S(W)$ as $R_1(W, W_{ref}, D_{in}) = P(W, D_{in})/(W_{ref}, D_{in})$. We refer to this scaling as *synthesis based*.

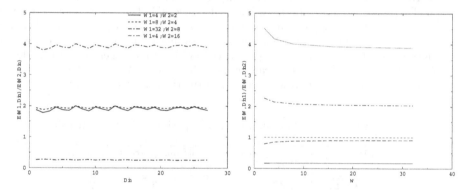

Fig. 3. Plots of R_1 and R_2 for a 2-port Multiplexer.

3.3 Power Model

The model we use to represent the dependence of the power consumption of a multiplexer on boundary statistics is based on the following observations: *i)* Signal probabilities have a negligible impact (and can be neglected); *ii)* The activity at each I/O port is positively correlated with power consumption (and should appear in the model); *iii)* All data inputs have similar fanout cones (and similar effect on power consumption); *iv)* All data outputs have similar fanin cones (and similar correlation with internal power). These observations lead to the following power model for $P(stats)$:

$$P(stats) = C_{in}D_{in} + C_sD_s + C_{out}D_{out} \qquad (8)$$

where D_{in} is the average transition probability of data inputs, D_s is the average transition probability of the selection signals, D_{out} is the average output activity, and C_{in}, C_s and C_{out} are fitting coefficients to be determined by regression analysis.

4 Experimental Results

4.1 Wiring

Our approach allows the evaluation at the RTL of gate-level wire models. Hence, when all RTL components have been pre-characterized for area and input capacitance, the accuracy provided by our model is, by construction, the target gate-level accuracy. What needs to be tested is the approximation introduced by the lack of gate-level information about the area and input load of (some of) the design components.

We present three sets of results that assess the accuracy of the area estimates provided by Equation 5, the input capacitance estimates provided by Equation 6, and the overall wiring power estimates provided by our approach.

For our experiments we considered all the DesignWare macros, with multiple instances of each soft macro. A first area model is obtained by individually characterizing each type of macro. Another model is obtained according to the value of the exponent r used in the generic model; in particular, we built three clusters of macros with similar A-N_{IO} relations, and we associated a unique area model with each cluster. Finally, we characterized a general, unified area model for all macros. We use the terms `specific`, `clustered` and `general` to denote the three types of area models above. The accuracy provided by these models is reported in the left hand-side of Table 1, expressed in terms of average relative error and standard deviation. As expected, the more general is the model, the less accurate are the estimates it provides. In fact, the relation between pin count and area strongly depends on the type of macro. For instance, multipliers have a quadratic relation ($r = 2$) while adders have a linear relation ($r = 1$). Trying to use the same model for all macros impairs accuracy. The same experiment was performed for the estimation of the input capacitances. Results as similar as those obtained for area estimates, and they are summarized on the right hand-side of Table 1.

Table 1. Area and Input Wiring Capacitance Estimate Results.

Model	Area		Input Wiring Capacitance	
	Avg.Error	*Err.St.Dev.*	*Avg.Error*	*Err.St.Dev.*
`specific`	1.37	1.18	1.16	1.23
`clustered`	23.51	97.37	7.46	44.93
`general`	66.26	77.03	31.53	31.44

Finally, we tested the entire approach on the case study of Example 1. When area and input capacitance have been pre-characterized at the gate-level for all the adders and the MUXes instantiated within the elliptic filter, our approach provides the same estimates for wiring power reported in Example 1. In other terms, the RTL accuracy is the same as the gate-level one.

If no pre-characterization is performed and the general models are used to estimate area and input capacitance, the average error is around 31%. If specific area and capacitance models are used for adders and MUXes, the average error on wiring power estimates reduces to 6%.

4.2   Multiplexers

We tested our power model on different MUXes obtained by specifying different port numbers and encoding styles for the universal multiplexer taken for the *Synopsys' DesignWare* library. Each MUX was characterized using a reference bit-width of 8 ($W_{ref} = 8$). For characterization, the 8-bit instance of the MUX was first synthesized and mapped on a gate-level library characterized for power. Gate-level power simulation was then repeatedly performed by *DesignPower* for different input statistics to obtain data for linear regression. Least square fitting was finally performed to fix the three coefficients of the linear equation.

The accuracy was evaluated through concurrent RTL and gate-level simulation for 25 input streams. The average error obtained for the reference bit-width (i.e., without scaling) was below 10% for all benchmarks, with a standard deviation around 5%.

The accuracy loss caused by scaling is reported in Figure 4 for 2-port and 4-port MUXes. Two series of results are reported on each graph, that refer to the analytical and synthesis-based scaling. As expected, synthesis-based scaling improves upon the accuracy of analytical scaling. The advantage is negligible for 2-port MUXes (because of the good linear relation between power and bit-width), while it is remarkable for 4-port MUXes with bit-width of size 1 and 2, i.e., when the power consumption does not scale linearly with the bit-width.

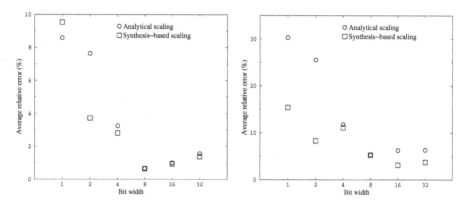

Fig. 4. Experimental Results on a 2-Port and 4-Port Multiplexer.

5   Conclusions

We have presented RTL power models for the steering logic (multiplexers and wiring) that are usually not accounted for during RTL power estimation because they are not explicitly instantiated into the RTL description, in spite of their potentially high impact on the total power budget. The power model for MUXes can be used for special complex types with re-configurable numbers of data bits and number of ways. The interconnect model is obtained by empirically relating capacitance to area, that is either estimated by means of statistical models or extracted from back-annotation information available at the gate level. Experimental results have demonstrated the good accuracy of the models, yielding estimation errors with respect to gate-level estimates below 10%.

References

1. E. Macii, M. Pedram, F. Somenzi, "High-Level Power Modeling, Estimation and Optimization," *IEEE TCAD*, Vol. 17, pp. 1061-1079, 1998.
2. B. S. Landman, R. L. Russo, "On a Pin vs. Block Relationship for partitions of Logic Graphs," *IEEE TCOMP*, Vol. 20, pp. 1469-1479, 1971.
3. W. E. Donath, "Wire Length Distribution for Placements of Computer Logic," *IBM J. of Res. and Dev.*, Vol. 25, pp.152-155, 1981.
4. S. Sastry, A. C. Parker, "Stochastic Models for Wireability of Analysis of Gate Arrays," *IEEE TCAD*, Vol. 5, pp. 52-65, 1986.
5. F. J. Kurdahi, A. C. Parker, "Techniques for Area Estimation of VLSI Layouts," *IEEE TCAD*, Vol. 8, pp. 81-92, 1989.
6. P. E. Landman, J. Rabaey, "Activity-Sensitive Architectural Power Analysis," *IEEE TVLSI*, Vol. 15, pp. 571-587, 1995.
7. *High-Level Synthesis Benchmarks*, CAD Benchmarking Laboratory (CBL), North Carolina State University, 1992.
8. A. Bogliolo, R. Corgnati, E. Macii, M. Poncino, "Parameterized RTL Power Models for Combinational Soft Macros," *ICCAD-99*, pp. 284-287, 1999.

Reducing Power Consumption through Dynamic Frequency Scaling for a Class of Digital Receivers

N.D. Zervas, S. Theoharis, A.P. Kakaroudas, D. Soudris[†],
G. Theodoridis, and C.E. Goutis

Dep. of Electrical & Computer Engineering, University of Patras, Rio 26500, Greece.
[†] Dep. of Electrical and Computer Engineering, Democritus University of Thrace,
Greece

Abstract. In this paper, a power management technique based on dynamic frequency scaling is proposed. The proposed technique targets digital receivers employing adaptive sampling. Such circuits over-sample the analogue input signal, in order to succeed timing synchronization. The proposed technique introduces power savings by forcing the receiver to operate only on the "correct" data for the time intervals during which synchronization is achieved. The simple architectural modifications, needed for the application of the proposed strategy, are described. As test-vehicle a number of FIR filters, which are the basic components of almost every digital receiver, are used. The experimental results prove that the application of the proposed technique introduces significant power savings, while negligibly increasing area and critical path.

1 Introduction

Nowadays, sophisticated handsets, with wireless communication capabilities, have invaded the world market. In such applications low power consumption is of great importance to allow for extended battery life but also to reduce the packaging and cooling related cost.

One of the most efficient low-power techniques, applicable at all levels of abstraction of the design flow, is the dynamic power management [1,2]. The most common approach to dynamic power management is to selectively shutdown a resource, when it performs useless operations. Techniques based on the previous concept have been proposed in [2,3,4,5,6].

In this paper the power management concept is applied in a class of digital receivers, namely digital receivers employing adaptive sampling. Adaptive sampling is very commonly met in a variety of receiving applications but especially into wireless telecommunication terminals. An analysis of the behavior of such receivers indicates that the percentage of data, contained in the incoming stream, that is necessary to be processed for the correct operation of the receiver depends on whether or not synchronization is achieved. A novel technique based on dynamic frequency scaling, that introduces power savings by forcing the receiver to operate only on the "correct" data for the time intervals during which

D. Soudris, P. Pirsch, and E. Barke (Eds.): PATMOS 2000, LNCS 1918, pp. 47–55, 2000.

synchronization is achieved, is presented here. The application of the proposed technique is rather simple, since it requires minor modifications with respect to conventional architectures, while it introduces significant power savings.

The rest of this paper is organized as follows: Section 2 is dedicated to basic background. In section 3 the target architecture models are described. In section 4 the proposed power management technique is presented. In section 5 the proposed technique is applied in demonstrator applications and its effect on the various design parameters is analyzed. Finally, in section 6 some conclusions are offered.

2   Basic Background

In a typical digital telecommunication system, the transmitter modulates the binary data into phase, and/or frequency, and/or amplitude, differences of an analogue signal (carrier) [10]. *Symbol frequency (f_S)* is the ratio of the number of transmitted symbols per second. For the receiver the symbol frequency is known. However, the exact instance, that the modulated input signal must be sampled, is not known. Furthermore, in the general case the receiver knows the width of the data bursts, but the starting position of them is not known. *Symbol timing synchronization* is the process of deriving at the receiver timing signals indicating where in time the transmitted signals are located [9,10]. If symbol timing synchronization is not achieved, then even a small shifting in time of the sampling instances can result to receive erroneous data. *Frame synchronization* is the process of locating at the receiver the position of a synchronization pattern (marker), periodically inserted in the data stream by the transmitter [9,10].

Adaptive sampling through oversampling is a very commonly met synchronization method in mobile telecommunication systems [e.g.7, 8], since it eliminates the need for a power hungry, analog VCO (Voltage-Controlled Oscillator) and provides both symbol timing and frame synchronization in one step [9]. According to adaptive sampling, the sampling instance during the symbol period is selected separately for each data burst. *Oversampling* is a mechanism employed to choose the correct sampling instance during the symbol period for each data burst. Specifically, instead of sampling the analogue input signal once, the input signal is sampled N (*oversampling ratio*) times during each symbol period. This way, instead of one input data stream, we have N input data streams: one input data stream per sampling instance during the symbol period. A block responsible for synchronization decides which of the N input data streams corresponds to the correct sampling instance. The input stream that corresponds to the correct sampling instance is the one that includes the synchronization pattern. In this way symbol timing and frame synchronization are jointly performed.

3   Target Architecture Model

Adaptive sampling through oversampling imposes one of the following receiver architectures styles:

1. The receiver is designed in the same way as it would be designed in the case that it process only one input stream and after that the data registers are replaced with N-position shift-registers. In this way the data-path must operate at N-times the symbol frequency, in order to produce output at symbol frequency.

2. N-parallel identical data paths operating at the symbol frequency are implemented, one for each input stream.

3. P-parallel identical data-paths are implemented each one of them processing N/P input streams and operating at N/P times the symbol frequency.

Area constraints usually prohibit the use of the second architecture style. So, in order to be realistic only the first and third design styles are studied here.

The following analysis assumes that two signals are generated with in the receiver: The first is the signal TS indicating that timing synchronization is achieved. This signal is usually an interface signal to the system of which the receiver is a component. The second signal is the *correct_sample* that denotes which of the N input streams is the one that corresponds to the correct sampling instance. This signal is present in receivers incorporating adaptive sampling though oversampling, since it is needed in order to select the input stream that will be fed in the output.

An abstract model of the $1^{s}t$ architecture style is illustrated in fig. 1. The functional units (FUs) implement the digital filtering and the demodulation algorithm. In order to reuse the resources for all input streams and perform the operations between samples belonging to the same input streams, N-stage shift-registers are used to store the data at the inputs and output of each resource. The shifters are clocked with N times the symbol frequency in order to produce output with symbol frequency. The output of the receiver can be in high impendence when T_S indicates that there is no timing synchronization. When timing synchronization is established the *correct_sample* signal selects the stream that must be fed in the output and the output register is clocked with symbol frequency. In cases of pipelined FUs, the data of the same stream are not stored in the same stage at the input and output shift registers of the FU. Specifically, for the stream J, if the data are positioned in the Kth stage of the output shift register of an L-stage pipelined FU, then the data that correspond to the same data stream are mapped at the input shift register of the same FU according to the equation: $(K + L) mod N$. Thus, in order to perform operations between samples of the same stream, special care must be taken. For example in fig.1, lets name the output of FU1 $f(X_n)$ and assume that FU2 is a multiplier with pipeline depth=2, which performs the operation $f(X_n) \times X_{n-1}$. In any case the $f(X_n)$ is fed to FU2 by the Nth stage of the shift register SR-2. In order for the FU2 to perform the multiplication between samples of the same stream, the sample X_{n-1} must be fed to FU2 by the $(N + 2) mod N = 2^{nd}$ stage of the shift register SR-1. The effect of the pipelined FUs is ignored in the figures, for clarity reasons, and it won't be referenced again during the rest of this paper, due to space limitations and since it is almost straightforward.

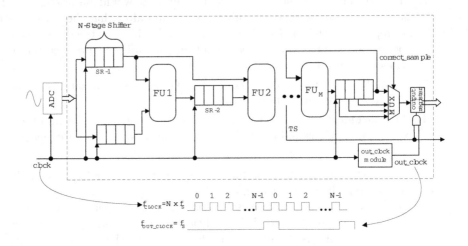

Fig. 1. Architecture Model (1).

An abstract view of the third architecture style is given in fig. 2. Lets assume that P parallel data paths are used and that the sample ratio is N. According to the third architecture style, the input stream (containing N different sample streams) is demultiplexed in time and each of the derived (P) streams (containing N/P different sample streams) is fed to the inputs of one of the P parallel data-paths. Alternatively, P parallel ADCs can be used. Each of the parallel data-paths operates at N/P times the symbol frequency and is implemented based on the principles of the first architecture style. Again, the output of the receiver can be in high impendence when TS indicates that there is no timing synchronization. During the interval that the receiver is synchronized the *correct_sample* signal selects the stream that must be fed in the output. Also here, the output register is clocked with symbol frequency. The latter architecture model actually consists of P parallel data-paths compatible to the first architecture model (each one operating at N/P times the symbol frequency). For this reason, the rest of this paper focuses on the first architecture model. The proposed technique is applied in an analogous way on the third architecture model as well.

4   Dynamic Frequency Scaling

Adaptive sampling through oversampling imposes that the whole receiving algorithm is computed on N input data streams, each one of which corresponds to a different sampling instance during the symbol period, while only one stream corresponds to the correct sampling instance. This introduces a significant power overhead. The power overhead cannot be avoided for the time intervals during which synchronization is not achieved. However, after the detection of the syn-

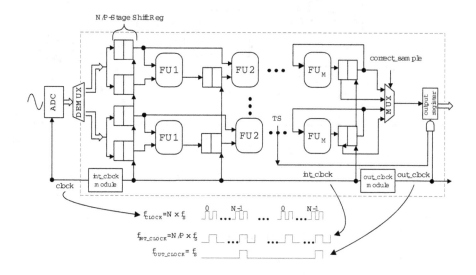

Fig. 2. Architecture Model (3).

chronization pattern and up to the end of the frame the power overhead can be removed by operating only to the data that correspond to the correct sampling instance.

For this purpose a frequency scaling technique is proposed in this paper. Specifically, the operation frequency can be reduced to symbol frequency after the synchronization pattern detection and up to the end of the frame. During this time interval, the receiver is forced to operate only on the input stream that corresponds to the correct sampling instance. After the end of each frame the receiver operates again at oversampling frequency.

From the architecture point of view, some modifications of the original architectures model are needed in order to preserve the correct functionality of the receiver. In fig. 3, the proposed first architecture model is illustrated. The main differences between the original (fig. 1) and proposed architecture model (fig. 3) are the following:

1. The shift-registers of the original architecture model are replaced, in the proposed architecture model, with shift-registers, whose output is either their N^{th} or their 1st stage. The signal T_S (indicator of synchronization) selects which of the two stages is fed as output.
2. In the proposed architecture model, the *clock_select* module is added while the *out_clock* module is removed. The *clock_select* module produces the *select_clock*, which is the clock with symbol frequency. When the *select_clock* is used to trigger the ADC, then the analogue input is sampled only at the instances indicated by the *correct_sample* signal. In fig. 3 the waveform of

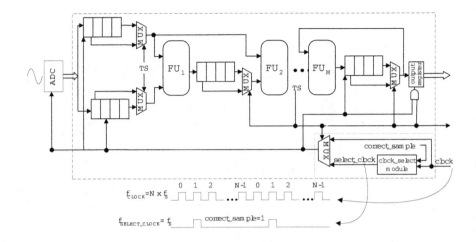

Fig. 3. Proposed Architecture Model.

the *select_clock*, when the *correct_sample* signal indicates that the correct input stream is the stream₁, is given.

For the proposed architecture model, when the signal T_S indicates that synchronization is not achieved, the clock with N times the symbol frequency is used. Additionally, the Nth stage of shift-registers is fed to their output. For this case, the proposed architecture model operates exactly as the original one. However, when T_S indicates that synchronization is achieved, the *select_clock* (symbol frequency clock) is used throughout the receiver. The *select_clock* waveform is such that forces the ADC to sample only at the correct instance during the modulation period. Since the data path should manipulate only one stream, there is no need to shift the data on the inputs and outputs of the functional units. Thus, the 1st stage of the shift registers is used as a single register in order to feed and store the inputs and outputs of the functional units. The rest N-1 stages of the shift-registers are bypassed and for this reason their clock can be disabled. The same are valid for the receiver's output shift-register.

5   Experimental Results

As stated earlier in this paper, the proposed frequency scaling technique is applicable in receivers employing adaptive sampling through oversampling. One of the main parts of such receivers is the digital filtering stage. For this reason, as demonstrator application Finite Impulse Response (FIR) filters are chosen. Four- and eight-taps FIR filters are implemented. A low-power implementation of the FIR filters is considered, where multiplications are replaced with shift-add operations. For this reason no resource reuse is considered. The filter coefficient bit-width is 10 bits, while the symbol bit-width is 9 bits. For each FIR filter three

#taps	N	Power (mW)		Diff. (%)
		Conv.	Prop.	
4	4	10.405	3.320	68.09
	8	21.549	6.502	69.83
	16	47.460	12.937	72.74
8	4	27.032	8.478	68.64
	8	54.945	16.630	69.73
	16	116.328	33.166	71.49

Table 1: Power for $T_S/T_F = 1/128$

#taps	N	Power (mW)		Diff. (%)
		Conv.	Prop.	
4	4	10.403	3.354	67.76
	8	21.322	6.539	69.33
	16	47.240	12.929	72.63
8	4	26.997	8.569	68.26
	8	54.468	16.747	69.25
	16	115.754	33.115	71.39

Table 2: Power for $T_S/T_F = 2/128$

#taps	N	Power (mW)		Diff. (%)
		Conv.	Prop.	
4	4	10.287	3.381	67.13
	8	21.101	6.506	69.17
	16	46.647	12.919	72.30
8	4	26.735	8.722	67.38
	8	53.869	16.709	68.98
	16	114.308	33.019	71.11

Table 3: Power for $T_S/T_F = 4/128$

#taps	N	Power (mW)		Diff. (%)
		Conv.	Prop.	
4	4	10.043	3.391	66.24
	8	20.572	6.532	68.25
	16	45.432	12.882	71.65
8	4	26.064	8.764	66.38
	8	52.472	16.739	68.10
	16	111.204	32.793	70.51

Table 4: Power for $T_S/T_F = 8/128$

#taps	N	Area (mils^2)		Diff. (%)
		Conv.	Prop.	
4	4	5344.91	5456.64	2.09
	8	5957.43	6058.57	1.70
	16	7181.44	7270.18	1.24
8	4	10802.74	11011.79	1.94
	8	11832.08	12022.89	1.61
	16	13895.81	14602.52	3.19

Table 5: Area measures

#taps	N	Critical Path (ns)		Diff. (%)
		Conv.	Prop.	
4	4			
	8	33.60	34.04	1.31
	16			
8	4			
	8	38.76	39.20	1.14
	16			

Table 6: Critical path measures

different oversampling ratios are considered, namely 4, 8 and 16. The frame is considered to consist of 128 symbols. Finally, four different values for the width of the synchronization pattern are considered, namely 1, 2, 4 and 8 symbols.

For each different FIR filter configuration, two implementations are considered: one according to the first architecture model of section 3 and one according to the proposed architecture model. For the FIR implementation, the SYNOPSYS and Mentor-Graphics CAD tools were employed and the 0.6 micron AMS cell-library was used for mapping. For each different FIR filter the power consumption, the area and the critical path were measured. Power measurements were acquired with toggle-count during logic-level simulation under real delay model and capacitance estimates provided by the CAD tools, assuming 5V power supply and 4K highly-correlated, 9-bit, valid input vectors. The later means that if the oversampling ratio is N then the number of input vectors used is $4K \times N$. The $4K \times (N-1)$ vectors that correspond to the wrong sampling instances were generated with a random variation from the corresponding valid vector in the range of $\pm 10\%$. For area and critical path measurements the reports provided by the CAD tools were used.

Tables 1 up to 4 illustrate the power consumed for both the original and proposed implementations for four different cases. Each case corresponds to a different ratio T_S/T_F of synchronization pattern width to frame width. The ratio T_S/T_F determines the relation between the sizes of the time intervals operating at oversampling and symbol frequency. So it was expected that as the later ratio increases, power savings decrease. Experimental results indicate that the effect of the ratio T_S/T_F on power savings is weak. For example, the power savings for the case that $T_S/T_F = 1$ are on average 1.57% greater than the power savings for the case that $T_S/T_F = 8$.

Additionally, from tables 1-4, it can be observed that the amount of power saved by the proposed technique, mainly depends on the oversampling ratio. This is rational, since the oversampling ratio determines the difference between the power dissipated for the time intervals operating at oversampling and symbol frequency. Furthermore the percentage of power savings does not seem to depend on the design size since the results are very close for both 4-taps and 8-taps FIRs. In any case, the proposed architecture model consumes significantly less power (on average 69.43%) than the original architecture model.

Tables 5 and 6 illustrate the area and critical path measures respectively, for both the original and proposed implementations. As it can be observed, the area overhead, introduced by the multiplexers, the additional control logic and interconnections of the proposed architecture model, is on average 1.96%. Furthermore, the proposed architecture increases the critical path of the original design by the delay of the multiplexer that is added at the output of the shift registers. This increase is less than $0.5ns$ in any case. The overheads in area and critical path introduced by the proposed technique are considered to be negligible compared to the corresponding power savings. It must be stressed here, that the latency of the design is not affected by the proposed technique.

Finally an interesting observation is the following: the power consumption of the original case when $N = 4$ and 8 is greater than the power consumption of the proposed case when $N = 8$ and 16 respectively. This means that the application of the proposed technique enables the use of higher oversampling ratios (which means higher reception quality), while not exceeding the initial power budget.

6   Conclusions

This paper focuses on the strategy of power management through dynamic frequency scaling, which is based on the adaptation of operation frequency to the computational load. A technique for the application of the above strategy on receiver applications, employing adaptive sampling through oversampling, is proposed. With the proposed technique, the operation frequency of the receiver can be reduced for the time intervals during which timing synchronization is achieved. The architectural modifications needed for the application of the proposed technique are described. The proposed technique is applied in a number of

FIR filters, which are part of every digital receiver, and the experimental results prove that significant power savings are introduced, with a very small area and critical path overhead.

References

1. J. M. Rabaey and M. Pedram, Low Power Design Methodologies, *Kluwer Academic Publishers*, 1995.
2. L. Benini, G. De Micheli, DYNAMIC POWER MANAGEMENT: Design Techniques and CAD tools, *Kluwer Academic Publishers*, 1998.
3. L. Benini, P. Siegel, G. De Micheli, *"Automatic Synthesis of Gated Clocks for Power Reduction in Sequential Circuits"*, IEEE Design & Test of Computers, vol. 11, no. 4, pp. 32-40, 1994.
4. L. Benini, G. De Micheli, *"Transformation and Synthesis of FSMs for Low Power Gated Clock Implementation"*, IEEE Transaction on CAD, vol. 15, no. 6, pp. 630-643, 1996.
5. M. Aldina, J. Monteiro, S. Devadas, A. Ghosh, M. Papaefthymiou, *"Precomputation-Based Sequential Logic Optimization for Low Power"*, IEEE Tran. on VLSI Systems, vol. 2, no. 4, pp. 426-436, 1994.
6. V. Tiwari, S. Malik, P. Ashar, *"Guarded Evaluation: Pushing Power Management in Logic Synthesis/Design"*, Int'l Symposium on Low Power Design, pp. 221-226, Dana-Point, CA, April 1995.
7. K. Murota, K. Hirade, *"GMSK Modulation for Digital Mobile Radio Telephony"*, IEEE Transactions on Communications, Vol. Com 29, No 7, pp. 1044-1050, July, 1981.
8. E. Metaxakis, A. Tzimas and G. Kalivas, *"A low complexity baseband receiver for direct conversion DECT-based portable communications"*, in Proc of IEEE Int'l Conf. on Universal Personal Communications, pp. 45-49, Florence, Italy, 1998.
9. J. D. Gibson, The communications handbook, *CRC Press and IEEE Press*, 1997.
10. J. G. Proakis, Digital Communications, 3rd edition, *McGraw-Hill*, New York, NY 1995

Framework for High-Level Power Estimation of Signal Processing Architectures *

Achim Freimann

Universität Hannover
Institut für Theoretische Nachrichtentechnik und Informationsverarbeitung
freimann@mst.uni-hannover.de
http://www.mst.uni-hannover.de/~freimann

Abstract. A framework for high-level power estimation dedicated to the design of signal processing architectures is presented in this work. A strong emphasis lies on the integration of the power estimation into the regular design-flow and on keeping the modeling overhead low. This was achieved through an object-oriented design of the estimation tool. Main features are: an easy macromodule extension, the implementation of a Verilog HDL subset, and a moderate model complexity. Estimation results obtained using the framework for development of a discrete cosine transform compare to the deviation of power consumption imposed by their data dependency.

1   Introduction

The emerging market for wireless communication and mobile computing forces manufacturer and designers to pay special attention to low-power aspects of their products. Additionally, due to the short life cycle of these mobile devices, time-to-market has become a crucial factor. This leads to the requirement of power estimation on a high level of abstraction in the design process.

In this paper, an object-oriented framework is presented which addresses the problem of finding a low-power realization of signal processing architecture alternatives in an early design phase with considering the time-to-market issue. A high-level power estimation tool build around a centrally macromodule database was developed which integrates into the normal design-flow. This was achieved by the implementation of a subset of the Verilog hardware description language in the power estimation tool. The design-time factor was addressed through an easy extensibility of the framework and a moderate model complexity.

Approaches for high-level power estimation often use a two stage design-flow [1] – [4]: In a characterization phase accurate simulations on gate- or transistor-level netlists for the implementation of a macromodel are performed. From these simulations, power consumption coefficients for a later power estimation phase

---

* The work presented is supported by the German Research Foundation, Deutsche Forschungsgemeinschaft (DFG), within the research initiative "VIVA" under contract number PI 169/14.

D. Soudris, P. Pirsch, and E. Barke (Eds.): PATMOS 2000, LNCS 1918, pp. 56–65, 2000.
© Springer-Verlag Berlin Heidelberg 2000

are extracted. For an estimation on a higher level (RT-, architecture-level), signal characteristics on the module interfaces are taken and used to calculate the power consumption from the coefficients previously extracted during the characterization phase. All these methods have in common that a module or a class of similar modules has to be characterized once and can afterwards be used in the high-level design process as often as necessary. Adequate precision in the estimation process is ensured through the accurate characterization on a lower level of abstraction.

During the design process, the distinction between characterization and evaluation phase has one drawback: Each module has to be characterized once before its use. Therefore, the model creation and characterization of a new module should be easy and reasonable fast. Reuse of already created models is of great interest for a rapid development cycle. Furthermore, the power estimation should integrate easily in the design process.

The rest of the paper gives a short review of two high-level power estimation models and a description of the implemented power estimation framework with macromodels exploiting the Dual Bit Type (DBT) method. As an example, a signal processing application is given and the results are discussed.

2   Power Macromodeling

For high-level power estimation with the two stages *characterization* and *estimation*, different modeling methods can be applied. One method treats macromodules as black boxes with no knowledge of the functionality and realization of the module [1]. Another approach is the white box modeling, where an indepth knowledge of the structure and functionality is required [2]. Other methods lie between both approaches and require a moderate knowledge of the module's functionality [3].

As mentioned earlier, creation and characterization of new modules during the design process should require only moderate effort in model creation and simulation time. Therefore, simulation intensive models and models where an indepth knowledge of structure and functionality is required may not be accepted by circuit designers in time-to-market critical projects.

In [1], a macromodel approach relying on probabilistics for combinational CMOS circuits is proposed. Their model captures four input/output signal switching statistics leading to a four dimensional look-up table for the estimation of power consumption. As the four indices result from continuous values, a discretization has to be performed prior table look-up. With their choosen discretization, the table for one module comprises 10^4 entries. Filling this table during the characterization process is a time consuming task and takes hours to days to generate the coefficients for just one module. On the other hand, the independence of the characterization phase from the type of module is advantageous.

The authors of [3] use an approach where number formats and word lengths are considered. They reported first the so called Dual Bit Type (DBT) method, which especially aims at signal processing applications. The main idea behind the model is to identify regions of similar behaviour within the data words of module in- and outputs. One region behaves like the sign in a number representation. Another region of the word has random data characteristics. For modules modeled using the DBT method, up to 73 coefficients and an additional architecture parameter matrix have to be determined during the characterization. In [4], an approach to refine the estimation accuracy for macromodels with input signals differing from the DBT data model is given. It uses a training phase for model improvement.

If a fast architecture exploration without the neccessity to get accurate figures for the absolute power consumption is required, all overhead should be reduced to get a first idea which architecture implementation is a candidate for the final design. Additionally, the influence of the processed data on the power consumption can be remarkable high. This in turn shows, that there is often no need to achieve the highest possible estimation accuracy, as the power consumption of the application is not a single value but a fuzzy one.

With the given constraints for a fast creation of new macromodels during the design phase, the DBT method is better suited than the pure probabilistic model in [1]. In the following, the model derived from the original DBT method and its application for rapid prototyping of video signal and image processing architectures is described.

2.1   DBT Macromodel

The choosen DBT macromodel is especially beneficial for signal processing architectures, which are built on numerical operations performed on data words. It is quite evident that the least significant bits (LSBs) in a data word tend to behave randomly, and can be describe by a uniform white noise (UWN) process. In contrast, the most significant bits (MSBs), in two's-complement number representation, correspond to sign bits. The signal and transition probabilities in the MSB region do not behave like a random process due to the temporal correlation of the data words.

In Fig. 1, the decomposition of a data word in the dual bit type model into sign- and random-like data regions is shown. The high-order bits from the MSB down to breakpoint $BP1$ is the sign region. Starting at the LSB up to $BP0$ is the UWN region. The intermediate region between $BP1$ and $BP0$ can be modeled quite well by linear interpolation of the sign and UWN activity.

The definitions of the breakpoints $BP1$ and $BP0$ is given in (1) and (2), which can be computed from the word-level statistics variance(σ^2), and temporal

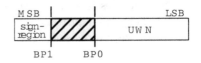

Fig. 1. Decomposition of data word in sign and UWN region

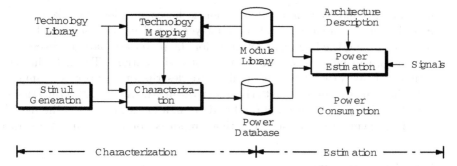

Fig. 2. Overview of power estimation framework

correlation (ρ):

$$BP1 = log_2 6\sigma \tag{1}$$

$$BP0 = log_2 \sigma + log_2 \left[\sqrt{1 - \rho^2} + |\rho|/8 \right] \tag{2}$$

The definition of $BP1$ found in [3] additionally uses the mean value (μ), whereas the choosen definition (1) reflects the one reported in [5]. For modules with two inputs and more, the original method introduced *misaligned* breakpoints for the case where the breakpoints of the inputs differ. E.g., this distinction led not to a relevant model improvement. Omiting the misaligned breakpoints from the model results in a much faster characterization process. In practice, the characterization of a module with two inputs speeds up by a factor of approximately five.

Here, in this case of a module with two inputs a and b, the calculation of the resulting breakpoints is performed by the equations (3) and (4). Where $BP0_{a,b}$ beeing the breakpoints $BP0$ of input a and b, respectively, and $BP1_{a,b}$ both breakpoints $BP1$:

$$BP0 = \max(BP0_a, BP0_b) \tag{3}$$

$$BP1 = \max(BP1_a, BP1_b) \tag{4}$$

3   Power Estimation Framework

An overview of the high-level power estimation framework is shown in Fig. 2. Central elements of the framework are the *module libray* and the *power database*. The module library consists of parameterizable high-level module descriptions required for the design of the architecture. In the power database, coefficients for the computation of the power consumption of each module are stored. While the module descriptions in the module library are technology independent, the power coefficients in the power database are especially computed for the target technology.

The design of a signal processing architecture follows the usual design flow for high-level designs: In a first step the functional view of the architecture is created, where the circuit description is based on modules taken from the module library. In case a module is not yet stored in the library, it has to be functionally modeled and characterized for the power estimation process. Then, the design can be verified through functional simulations. The following power estimation takes the high-level architecture description, performes a simulation with the same signals applied to the design during the functional verification, and calculates the power consumption. In addition, area estimates and timing delays based on the module's worst case path are reported for architecture comparisons.

3.1 Characterization

Before using a module in the design process for the first time, it has to be characterized for determining the power coefficients for a given technology. In a first step, the module is implemented as a Verilog high-level description with parameterizable input and output port sizes and, if applicable, for parameterizable implementation architectures. This high-level description is mapped for the required port sizes and implementation onto the target technology using a synthesis tool, e.g., the Synopsys Design Compiler. Both module views, the high-level and the technology-mapped, are stored in the module library for module characterization and for the verification of the circuit during the architecture design phase.

The module characterization itself uses a gate-level power simulator. Simulation of the module is performed with specifically generated stimuli pattern. Characteristic power coefficients obtained from the simulation are stored in the power database. Area and delay timing estimates are directly taken form the technology mapping and also stored in the database.

3.2 Estimation

The estimation part of the framework reads the high-level architecture description previously created for the verification and simulates the architecture together with the input signals. The simulation is cycle accurate and tracks the signal statistics σ^2, ρ, and sign activity on the module interfaces.

The power consumption of the architecture can be computed at any simulation time using the DBT model described in section 2.1. The power consumption P_{Module} is computed from (5), where P_U and $P_S(SS)$ are the power coefficients for the UWN region and the power coefficients for sign transitions, all taken from the power library. N is the number of bits of the input, N_U and N_S are the number of bits for the UWN and sign region.

$$P_{Module} = \frac{N_U}{N} P_U + \frac{N_s}{N} \left(\sum_{SS} P_S(SS) \right) \tag{5}$$

The number of bits in the sign region N_S computes from (7) and comprises one half of the bits found in the intermediate region N_I (6). The second part of the linear interpolated region accounts to the number of UWN bits N_U from (8):

$$N_I = BP1 - BP0 - 1 \qquad (6)$$
$$N_S = (N - BP1) + N_I/2 \qquad (7)$$
$$N_U = (BP0 + 1) + N_I/2 \qquad (8)$$

The total estimated power consumption P_{Total} for the architecture is the sum over all P_{Module}, where M is the number of modules:

$$P_{Total} = \sum_{m=0}^{M-1} P_{Module}(m) \qquad (9)$$

3.3 Implementation

The implementation of the framework consists of the two parts characterization and estimation.

Controlled by a set of scripts, the characterization process is semi-automatic, where one ore more modules are mapped onto the target technology and characterized in one run. The scripts control the generation of stimuli pattern and generate the testbench for the module which is afterwards simulated using a Verilog based simulator. Power coefficients resulting from the simulation are extracted by the scripts and stored into the power database.

This power database is implemented on a SQL (structured query language) database and allows for concurrent read and write accesses from several host computers.

The power estimation part of the framework is implemented in Java which gives several advantages compared to other programming languages: Simple access to SQL databases, strong object-oriented programming and run-time linkage. To ease the creation of new modules, the estimation process is implemented using an object-oriented design. In Fig. 3, the object hierarchy of the macromodules is given. Through reuse of already existing components, the framework can be easily extended for additional modules. Because Java uses run-time linkage, the estimation framework need not to be recompiled when adding new macromodules. Thus, a designer needs not to have access to the sourcecode of the framework, but is still able to extend it.

With the subset of Verilog HDL implemented in the estimation program, the same high-level description of the architecture can be used for the verification of the design using a Verilog simulator and its power consumption, area, and timing estimation.

4 Application

As an example for a signal processing application, a one-dimensional discrete cosine transformation (DCT) was examined. The 8×1-DCT is the basis for a

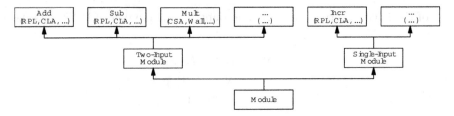

Fig. 3. Object hierarchy of macromodules

two-dimensional 8×8 cosine transform which is a major part of modern image and video compression schemes.

As the DCT requires several numerical operations to calculate the transform, many different implementation alternatives were proposed in the past. The DCT architecture under investigation was reported by Zhang and Bergmann in [6]. Their variant requires 11 multiplications, 15 additions, and 14 subtractions to perform the necessary operations for an one-dimensional transform.

5   Results

This section discusses the results achieved with the implemented DBT macro-model. First, the model accuracy for stand-alone modules is given. Next, the influence of data on the power consumption of an application is investigated. Finally, results for the DCT signal processing applications consisting of serveral macromodules are discussed.

All simulations and computations were performed on a Sun Ultra 10 work-station with a 440MHz clocked processor and 512MB main memory.

5.1   Characterization Times

The time to perform the characterization of the macromodules is given in Table 1. For the modules "Add", "Sub", and "Mult" the range of the word size for which the module was characterized and simulation time is shown. It can be seen that even to characterize the whole set of modules, the time to perform the characterization is low. Characterization was performed for a carry-look-ahead achitecture of the adders and subtractors, whereas a carry-save-adder implementation was choosen for the multiplication.

Table 1. Characterization times for macromodels

Add		**Sub**		**Mult**	
word size	time	word size	time	word size	time
4...32	161 sec	4...32	173 sec	4...16	214 sec

Table 2. Model accuracy for addition, subtraction, and multiplication macromodules depending on the input word size

Add			**Sub**			**Mult**		
size	rme	rmse	size	rme	rmse	size	rme	rmse
8	6.4 %	7.3 %	8	5.7 %	6.9 %	8	30.9 %	30.7 %
12	5.1 %	5.6 %	12	2.6 %	3.6 %	10	19.9 %	20.8 %
16	3.2 %	3.6 %	16	1.9 %	2.7 %	12	13.8 %	14.8 %
20	2.5 %	2.8 %	20	1.7 %	2.2 %	14	10.9 %	11.9 %
24	1.9 %	2.1 %	24	0.9 %	1.5 %	16	8.8 %	9.7 %

Table 3. Simulation times for model accuracy verification

Module	gate-level		high-level	
	total	mean	total	mean
Add	11.5 hrs	3.6 sec	16.0 min	83.3 msec
Sub	13.1 hrs	4.1 sec	16.0 min	83.3 msec
Mult	46.2 hrs	14.4 sec	16.0 min	83.3 msec

5.2   Model Accuracy

To verify the model accuracy, single modules were simulated using gate-level netlists and the estimation framework. For each module, different experiments with different pseudo-random normal-distributed sequences and varying statistical characteristics and input word sizes were performed.

The experiments used 5 different input word sizes, 4 variations of the mean, 3 different standard deviations, and a set of 4 temporal correlation values. In total, each of the macromodels for addition, subtraction, and multiplication was simulated 11.520 times using 10.000 input patterns on each input port.

In Tab. 2, results for the experiments concerning these three macromodules are given. For each module, the word size, relative mean error (rme), and relative mean square error (rmse) is shown. For the modules, the same implementation architecture was taken as during the characterization.

The simulation times for the model accuracy experiments are shown in Table 3. It took several hours up to days to simulate the experiment on gate-level netlists. The average for one simulation is in the order of seconds for the three types of modules. But it should be considered that the simulation time on gate-level netlists depends on the circuit complexity, thus modules with smaller word sizes simulate faster than the same module with a larger word size. In contrast, the run-time of the high-level estimation framework is module and word size independent and results in 16 min simulation time for all runs. Thus, one simulation is completed in 83.3 msec.

5.3   Data Influence

Simulations on the gate-level netlist of the DCT with different input sequences were performed to analyze the effect of data dependency on the power con-

Table 4. Deviation of power consumption for gate-level simulation of DCT with different video sequences

Add			Sub			Mult		
min	max	mean	min	max	mean	min	max	mean
9.8 %	35.7 %	23.9 %	9.1 %	35.5 %	23.0 %	10.7 %	29.2 %	20.7 %

sumption. These simulation results were obtained using the video test sequences "coastguard", "news", and "weather".

In Tab. 4, the deviation of power consumption is given. For addition, subtraction, and multiplication the minimum, maximum, and mean deviations calculated by (10) are given.

$$
mean = \left(\frac{1}{M} \sum_{m=0}^{M-1} \frac{\max(\forall sequence \ P_{Module}(m, sequence))}{\min(\forall sequence \ P_{Module}(m, sequence))} \right) - 1 \qquad (10)
$$

From these results, it can be seen that the influence of the processed data on the power consumption is remarkable high. Thus, a deviation of the estimated power for a certain data stream is tolerable.

5.4   DCT

Estimation accuracy results for the 1D-DCT are given in Tab. 5. For each of the modules, the name of the instance, the input word size, and the relative mean error (rme) is given. The module instance name is composed of the type of operation (e.g. "add") followed by the stage number (1 . . . 4) in the DCT and an additional identifier. Thus, the module "sub2_65" is a subtraction operation located in the second stage of the circuit.

The mean error for all adders is 9.3%, for the subtracters 0.8%, and 19.5% for the multiplieres. Compared to the deviations in power consumption for different video sequences shown in Tab. 4, the achieved accuracy for the DCT application is sufficient.

Running a gate-level netlist simulation using the Verilog based power simulator takes 138 sec, whereas the high-level estimation tool takes 8 sec. A high-level simulation with the Verilog simulator needs 12 sec to complete. Thus, the power estimation framework compares well to functional simulation.

6   Conclusion

This paper presented a high-level power estimation framework with emphasis on an easy integration into the design-flow for signal processing architectures. Through an object-oriented design of the framework and the reduction of unnecessary overhead in the model, extending the module library requires moderate additional extra work and a short module characterization time during the design process. Simulations determining the model accuracy by comparing results

Table 5. Estimation results for one-dimensional DCT

Module instance	word size	rme	Module instance	word size	rme	Module instance	word size	rme
add1_07	9	-4.8 %	sub1_07	9	-13.5 %	mults2_4	10	-6.2 %
add1_16	9	-5.5 %	sub1_16	9	-13.2 %	mults2_7	10	0.4 %
add1_25	9	-5.7 %	sub1_25	9	-13.4 %	mults4_2	11	15.3 %
add1_34	9	-5.9 %	sub1_34	9	-13.4 %	mults4_23	12	-12.2 %
add2_03	10	-7.6 %	sub2_03	10	-13.2 %	mults4_3	11	-4.2 %
add2_12	10	-7.3 %	sub2_12	10	-13.5 %	mults4_4	16	0.5 %
add2_65	10	-15.0 %	sub2_65	10	-15.8 %	mults4_47	16	-5.8 %
add3_01	11	-8.8 %	sub3_01	11	-13.6 %	mults4_5	16	38.7 %
add3_45	18	12.3 %	sub3_45	18	11.8 %	mults4_56	16	-5.6 %
add3_76	18	13.7 %	sub3_76	18	10.6 %	mults4_6	16	12.1 %
add4_23	11	-13.3 %	sub4_2	21	-8.4 %	mults4_7	16	22.3 %
add4_47	19	-4.6 %	sub4_3	21	-4.2 %			
add4_56	19	-7.4 %	sub4_4	26	-9.6 %			
add4_6	26	-8.1 %	sub4_5	26	-8.3 %			
add4_7	26	-7.5 %						

obtained by time consuming gate-level simulations to the power consumption estimated by the framework show reasonable good quality. The application of the power estimation to the design of a DCT architecture give deviations to the gate-level simulation that are of the same order like the deviations imposed by applying different data sequences to the architecture. Hence, using the implemented modeling framework, it could be shown that the investigated approach is of sufficient accuracy.

References

1. Gupta, S., Najm, F. N.: Power Modeling for High-Level Power Estimation, IEEE Transactions on Very Large Scale Integration (VLSI) Systems, Vol. 8, No. 1, (2000) 18–29
2. Theoharis, S., Theodoridis, G., Soudris, D., Goutis, C.: Accurate Data Path Models for RT-Level Power Estimation, International Workshop on Power and Timing Modeling, Optimization and Simulation (PATMOS), (1998) 213–222
3. Landman, P. E., Rabaey, J. M.: Architectual Power Analysis: The Dual Bit Type Method, IEEE Transactions on Very Large Scale Integration (VLSI) Systems, Vol. 3, No. 2, (1995) 173–187
4. Tsui, C.-Y., Chan, K.-K., Wu, Q. Ding, C.-S., Massoud, P.: A Powerestimation Framework for Designing Low Power Portable Video Applications, Design Automation Conference (DAC), (1997) 421–424
5. Ramprasad, S., Shanbhag, N. R., Hajj, I. N.: Analytical Estimation of Signal Transition Activity from Word-Level Statistics, IEEE Transactions on Computer-Aided Design of Integrated Circuits and Systems, Vol. 16, No. 7, (1997) 718–733
6. Zhang, J., Bergmann, N. W.: A New 8x8 Fast DCT Algorithm for Image Compression, IEEE Workshop on Visual Signal Processing and Communications, (1993) 57–60

Adaptive Bus Encoding Technique for Switching Activity Reduced Data Transfer over Wide System Buses *

Claudia Kretzschmar, Robert Siegmund, and Dietmar Müller

Dpt. of Systems and Circuit Design
Chemnitz University of Technology
09126 Chemnitz, Germany
{clkre,rsie}@infotech.tu-chemnitz.de

Abstract. In this paper, we describe a new encoding technique which reduces bus line transition activity for power-efficient data transfer over wide system buses. The focus is on data streams whose statistical parameters such as transition activity are either non-stationary or a priori unknown. The proposed encoding technique extends the Partial Businvert encoding method [1] with a dynamic selection of the bus lines to be encoded. In this work, we present the encoding algorithm and a low power implementation of a corresponding coder-decoder system. Experiments with real-life data streams yielded a reduction in transition activity of up to 42 % compared to the uncoded data stream.

1 Introduction

The minimization of on-chip power dissipation is nowadays a key issue in the design of highly integrated electronic systems. There are two main reasons for this: First, the prolongation of operating time of battery powered mobile applications and second, the reduction of on-chip heat generation.

The power dissipated on a clocked system bus of a CMOS circuit is approximated by the following equation: $P_V = \frac{1}{2}V_{dd}^2 f \sum_{i=0}^{n-1} C_{L_i} \alpha_i$, where n is the bus width, f the bus clock frequency, V_{dd} the operating voltage, C_{L_i} the parasitic capacitance and α_i the transition activity of bus line i, respectively. Usually parasitic capacitances of bus lines exceed module-internal capacitances by some orders of magnitude, therefore up to 80 % of the total power dissipated on a chip are dissipated on system buses. At higher levels of design abstraction the designer has usually no influence on the choice of parameters such as operating voltage and bus clock frequency and cannot affect intrinsic parasitic capacitances. In most cases the only parameter in the equation given above that can be optimized at higher levels of design abstraction is the transition activity.

In this work we present a new technique for system bus encoding in order to minimize bus line transition activity. We refer to it as *Adaptive Partial Businvert*

---

* This work is sponsored by the DFG within the VIVA research initiative.

D. Soudris, P. Pirsch, and E. Barke (Eds.): PATMOS 2000, LNCS 1918, pp. 66–75, 2000.

Encoding (APBI). Our technique is based on the Businvert encoding scheme [2]. We extend the method of Partial Businvert encoding published in [1] with an adaptive component. Based on the statistics of the data stream observed during system operation APBI dynamically selects a subset of bus lines to be encoded using the Businvert encoding scheme. Our encoding technique requires one additional bus line and data are transmitted over the bus each cycle (e.g. we do not exploit spatial redundancy) with a delay of one clock cycle.

In contrast to all static encoding schemes that have been published so far, the ability of our encoding technique to adapt to a changed characteristics of the transmitted data stream eliminates the necessity of a priori knowledge of its statistical parameters for selecting an appropriate encoding scheme. Therefore our method is especially suited for system buses that transport data streams with unknown or strongly time-varying distribution of transition activity. For such data streams our method yielded a reduction in transition activity of up to 42 %.

The paper is structured as follows: Section 2 gives an overview of related work and the motivation of this work. Some preliminaries are given in Sect. 3. In Sect. 4 we describe the algorithm of APBI encoding. An efficient, power-optimized implementation of a corresponding coder-decoder system is given in Sect. 4.3. Experimental results are shown in Sect. 5. Section 6 summarizes the paper.

2   Related Work and Motivation

Different application-specific methods for system bus encoding have been published, that exploit the characteristics of the transmitted data stream. In microprocessor systems, typical streams can be grouped into data and instruction streams and address bus streams. The Businvert encoding scheme [2] is applicable for both kinds of streams and minimizes the Hamming distance between the current state of the bus and the following data word to be transmitted. If more than half of the bits would change, the data word is inverted. An additional bus line is used to signal the data sink if the word has been inverted. The T0 encoding scheme [3] exploits the high in-sequence portion of address bus data streams generated by microprocessors. Consecutive addresses are transmitted by setting an increment signal on an extra bus line, while at the same time freezing the bus state. The data sink calculates the new address by adding a constant increment value to the last address. The "Beach Solution" [4] uses a statistical analysis of an application specific address data stream, followed by the generation of a transition minimizing bus code. Combined encoding schemes published in [5] optimize the encoding for different data and address streams multiplexed over a single bus.

In the case of uncorrelated bus lines which have uniformly distributed switching activity the Businvert method is optimal [2]. However, on *real* buses, switching activity is often distributed in a non-uniform fashion or bus lines are spatially correlated. In [1] it was shown, that for these cases the performance of Businvert encoding can be improved if bus lines that have a lower switching activity and spatial correlation than other lines are excluded from encoding. Encoding these lines would rather increase total transition activity than reducing it. The corresponding published technique is called *Partial Businvert Encoding (PBI)*. Selecting k lines to be encoded out of a n-bit wide bus has a complexity of $\mathcal{O}(2^n)$. Therefore, for wide buses in [1] a heuristic approach of complexity $\mathcal{O}(n)$ is described that selects a sub bus which includes lines of high transition activity and high spatial correlation.

All encoding schemes mentioned so far presume the knowledge of the statistics or the characteristical nature of the data streams to be transmitted. For many applications, benchmark data streams are not available for all possible operating conditions or the streams have non-stationary statistical parameters such as a time-varying switching activity on the bus lines. For these cases static encoding schemes are inefficient. Rather, encoding techniques are needed that have the ability to *automatically adapt* to a priori unknown or time-varying statistical parameters of data streams. An approach to bit-wise adaptive bus encoding is published in [6]. We refer to it as IAEB (Implementation of Adaptive Encoding presented by Benini et al). Based on the analysis of the number of state changes on a bus line in a sampling window of fixed size, from four possible, simple encoding schemes the one that minimizes average line activity is chosen. Unfortunately, the power dissipation of the corresponding coder-decoder system over-compensates the achieved reduction in transition activity. So, in this work our focus is on a less power consuming approach for adaptive encoding, that yields a higher effective reduction in power consumption.

3   Preliminaries

For characterization of the efficiency of bus encoding schemes we define the following equations: The efficiency E_α of an encoding scheme describes the reduction of switching activity α on the bus. It is defined as follows:

$$E_\alpha = 1 - \frac{\alpha_{coded}}{\alpha_{uncoded}}; \qquad -\infty \le E_\alpha \le 1. \tag{1}$$

E_α describes the performance of the encoding algorithm and is independent of implementational aspects such as the target technology. Because implementations of coder-decoder systems dissipate power themselves, an effective power reduction is only achieved after compensation of that portion of dissipated power. This is illustrated with the following power-balance equation:

$$P_{V,uncoded} = P_{V,coded} + P_{V,Codec} + P_{V,saved} \tag{2}$$

where $P_{V,uncoded}$, $P_{V,coded}$ is the power dissipated on the uncoded and the coded bus, respectively, and $P_{V,Codec}$ represents the power consumption of the coder-decoder system. We now define the efficiency E_p of an encoding scheme regarding the reduction of the power dissipated on a bus by:

$$E_P = \frac{P_{V,saved}}{P_{V,uncoded}} = 1 - \frac{P_{V,coded} + P_{V,Codec}}{P_{V,uncoded}}; \qquad -\infty \le E_P \le 1. \tag{3}$$

E_P depends on the target technology the coder-decoder system is implemented with. In order to effectively reduce the power dissipated on the bus and the coder-decoder system, E_P must have a value greater than 0. From (2), the *effective capacitance* $C_{i,eff}$ results which is the average minimum capacitance of a bus line for $E_P > 0$:

$$C_{L,i} \geq C_{i,eff} = \frac{P_{Codec}}{\frac{1}{2}fV_{dd}^2(\alpha_{uncoded} - \alpha_{coded})}. \tag{4}$$

4   Adaptive Partial Businvert Encoding

4.1   Overview

For a n-bit system bus whose lines have uniformly distributed switching activity, a one-probability of $p = 0.5$ and whose lines are not correlated, e.g. it resembles an identical, independently distributed source (i.i.d. source), the Businvert encoding method is optimal [2], and, as we have shown in [7], it has a coding efficiency of $E_\alpha = 1 - \frac{1}{n2^{n-1}}\sum_{k=1}^{\frac{n}{2}} k\binom{n+1}{k}$. E.g., for a 32 bit bus a reduction in switching activity of 12 % is achieved. The Businvert encoding scheme is defined as follows

$$X_{Bus}^t = \{X_d^t, INV\} = \begin{cases} \{\overline{Q}^t, 1\} : \mathcal{W}(X_{Bus}^{t-1} \oplus \{Q^t, 0\}) > \frac{n+1}{2} \\ \{Q^t, 0\} : \text{else.} \end{cases} \tag{5}$$

Q^t and X_{Bus}^t represent the uncoded and encoded data words, respectively, and $\mathcal{W}(x)$ is the weight (number of ones) of a binary vector. In *real* applications, as for example image processing systems, switching activity is usually distributed over bus lines in a non-uniform fashion and bus lines are more or less correlated. If the activity distribution and line correlation are known and stationary, PBI combined with the heuristics for selection of the sub bus to be encoded as described in [1] represents an efficient encoding technique. However, if activity distribution and line correlation are unknown or non-stationary, the algorithm for static selection of the sub bus for PBI can not be applied or yields an inefficient solution with respect to the resulting coding efficiency. For these cases E_α can be improved if the choice of the sub bus to be encoded is adapted to the statistical parameters in certain time intervals. This can be interpreted as the dynamic adaptation of a coding mask $mask(t) = \{m_0^t, m_1^t, ..., m_{n-1}^t \mid m_i^t \in \{0,1\}\}$, where $m_i^t = 1$ at time t means that the i-th bus line is included into the encoded subset of bus lines. Considering this, we can derive the APBI encoding algorithm through extension of the Businvert encoding algorithm:

$$X_{Bus}^t = \{X_d^t, INV\}$$
$$= \begin{cases} \{Q^t, 0\} & : \mathcal{W}(X_{Diff}^t) \leq \frac{\mathcal{W}(mask(t))+1}{2} \\ \{Q^t \oplus mask(t), 1\} : \text{else.} \end{cases} \tag{6}$$
$$X_{Diff}^t = (X_{Bus}^{t-1} \oplus \{Q^t, 0\}) \cdot mask(t).$$

The bus data stream is then decoded in the following way:

$$Q^t = \begin{cases} \{X_d^t\} & : INV = 0 \\ \{X_d^t \oplus mask(t)\} & : INV = 1. \end{cases} \tag{7}$$

The Businvert encoding method is a special case of APBI with all mask bits constantly set to $m_i^t = 1$. Figure 1 visualizes the concept of an Adaptive Partial Businvert coder-decoder system. The encoder consists of a Businvert coder, a mask computation logic and bus line selection logic. The mask computation logic calculates from the input data stream the encoding mask $mask(t)$, which is then used to select the bus lines to be encoded. The encoding mask is also fed into the Businvert coder block because, according to (6), the encoder output is a function of both the input data and the weight of the encoding mask bit vector. The APBI decoder consists of a Businvert decoder, a bus line selection logic and a mask computation logic which is the same as in the encoder. In order to be

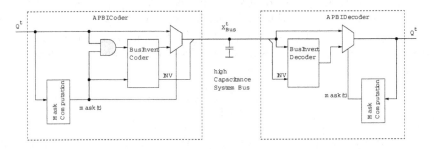

Fig. 1. Block Diagram of an APBI Coder-Decoder System

able to decode the encoded bus data stream correctly, the APBI decoder has to have knowledge about the encoding mask that was used to encode the data. For that purpose, the mask computation logic in the APBI decoder extracts the encoding mask from the decoded data. The mask value is then used to select from the encoded bus all lines that were left uncoded and combines them with the output of the Businvert decoder for all decoded lines in order to obtain the decoded data. For correct extraction of the value of the encoding mask in the decoder, both encoder and decoder have to use the same initial mask (e.g after system reset) and the same size of the sampling window which is used for mask computation.

4.2   Encoding Mask Computation

The coding efficiency E_α of APBI directly depends on the proper selection of the coding mask $mask(t)$. In general $mask(t)$ is a function of switching activities α_i, spatial and temporal correlation $\rho_{i,j}$ of the bus lines and the total switching activity α_{tot} of the uncoded bus. A hardware implementation for estimation of correlation coefficients would be extremely costly, because for every pair of bus lines one counter counting joint one-states would be required, which are $\binom{32}{2} = 496$ counters for a 32 bit bus. The implementation would be that power consuming that no effective reduction in power dissipation could be achieved. For that reason and because experiments showed that, compared to activities α_i, the correlation coefficients $\rho_{i,j}$ have a negligible influence on the choice of the coding mask, line correlation is not considered in the mask computation algorithm. We rather restrict ourselves to determine the i-th mask bit from the switching activity of the corresponding bus line and the total switching activity α_{tot} of the uncoded bus:

$$
\begin{aligned}
mask(t) &= \{m_0^t, m_1^t, ..., m_{n-1}^t\} \\
&= \{F_0(\alpha_0, \alpha_{tot}), F_1(\alpha_1, \alpha_{tot}), ..., F_{n-1}(\alpha_{n-1}, \alpha_{tot})\}.
\end{aligned}
\tag{8}
$$

where n is the bus width. The choice of the functions F_i is based on the following considerations: The average switching activity per line can be calculated by $\overline{\alpha}_i = \alpha_{tot}/n$. Having an i.i.d. source, switching activity is nearly identically distributed over all bus lines, so $\alpha_i \approx \overline{\alpha}_i$. For an i.i.d. source the Businvert encoding method is optimal, that means, all lines should be included into encoding to achieve the best reduction in transitions. If switching activity is non-uniformly distributed over bus lines $\overline{\alpha}_i$ will serve as boundary to decide whether a bus line is included into encoding or not. So we determined the following functions F_i

$$
m_i^t = F(\alpha_i, \alpha_{ges}) = \begin{cases} 1 : \alpha_i \geq \alpha_{tot}/n \\ 0 : \alpha_i < \alpha_{tot}/n. \end{cases}
\tag{9}
$$

The values for α_i and α_{tot} are determined by windowing the data stream using N samples of input data per window and counting the transitions within each window. At the end of every window the new coding mask is calculated from α_i and α_{tot}. There is a tradeoff between the window size N and accuracy of α_i and α_{tot}. These values become more accurate with larger windows. On the other hand a larger window size results in higher implementation costs for the coder-decoder system and increases its response time to a change in statistical parameters of the input data stream. Experiments showed that a window size of $N=32$ is a good compromise.

4.3   Implementation

Due to limited space, we will not show the implementation of the BI coder and decoder blocks in the APBI coder-decoder system (these can be found e.g. in [2]) but restrict ourselves to present a power-efficient implementation for the mask computation logic. An efficient implementation according to (9) is shown in the block diagram in Fig. 2. The selection algorithm requires the determination of the switching activities of each uncoded bus line as well as the total transition activity of the uncoded bus. Transitions are detected by xoring the current and last data word which is stored in a n-bit register. The resulting signal

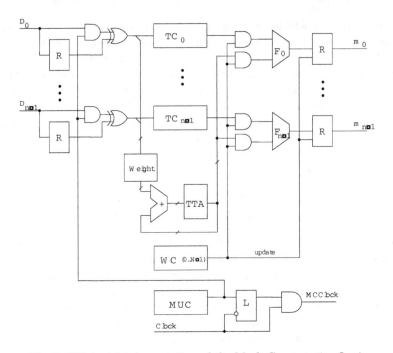

Fig. 2. Efficient Implementation of the Mask Computation Logic

serves as count enable for the bit line transition counters $TC_0..TC_{n-1}$. At the same time the weight function *Weight* computes the number of total transitions between two consecutive data words that is added up in the *Total Transition Accumulator (TTA)*. In order to reduce glitches, the weight function block is implemented with a balanced Carry-Save Adder tree. At the end of each window the *windows counter (WC)* produces an update signal for the registers that store the current coding mask. The new coding mask is calculated by the functions $F_0..F_{n-1}$, which compare according to (9) the counter results $TC_0..TC_{n-1}$ with the contents of TTA, divided by the bus width. In order to simplify the division operation, we restrict ourselves to bus widths that are a power of 2, so the division can be replaced by a much simpler shift by $log_2 n$ operation. The resulting encoding mask will then be stored in the mask registers. In order to minimize the power dissipation of the APBI coder-decoder system we integrated a feature which allows to increase the mask update interval from every window up to every k-th window. The mask computation logic is not required to be active for the windows $0...k-1$. In this time its clock (MC clock) is turned off, which is realized by the *Mask Update Counter (MUC)* and the clock gate consisting of a latch and an AND-gate. The power dissipation of the mask computation logic could be further minimized by isolating all major asynchronous logic blocks such as the selection functions F_i and the weight computation logic during cycles of inactivity.

5   Experimental Results

The proposed APBI coder-decoder system has been implemented as a synthesizable VHDL model for a bus width of 32 bits. With this model, switching activities of coded and uncoded buses for the following set of test data streams have been measured by simulation:

- **gen**: A random, segmented data stream, generated with Mathematica®, with varying distribution of switching activity over the bus lines in every segment
- **ascii**: An ASCII file in EPS format
- **binary**: Example for an executable file (gzip binary)
- **image**: 4 different concatenated images with varying characteristics in PPM format
- **noise**: White Gaussian noise

For all APBI simulations a window size of 32 samples was used. The encoding mask was updated in intervals of 1, 2, 4, 8 or 16 windows. APBI has been compared with BI, PBI and IAEB encoding, because BI and IAEB are the only encoding schemes which do not require any a priori knowledge of the statistics of the unencoded data stream. PBI was chosen since our method is derived from this encoding scheme. The mask for PBI was separately optimized for every test case using the proposed bus line selection heuristics[1].

Table 1. Relative Reduction in Switching Activity Regarding $T_{uncoded}$

Sequence	T uncod.	T coded							
		$APBI_{32,1}$	$APBI_{32,2}$	$APBI_{32,4}$	$APBI_{32,8}$	$APBI_{32,16}$	BI	PBI	IAEB
gen	2130360	37.05 %	41.96 %	41.95 %	41.99 %	42.00 %	12.70 %	12.78 %	44.08 %
ascii	221309	10.57 %	10.55 %	10.51 %	10.43 %	10.18 %	4.49 %	11.43 %	11.73 %
binary	154620	8.78 %	8.58 %	8.00 %	6.98 %	6.60 %	5.53 %	7.75 %	21.17 %
image	2878651	12.26 %	11.72 %	11.39 %	11.15 %	11.60 %	4.42 %	10.08 %	-9.37 %
noise	4086760	7.08 %	7.08 %	7.11 %	7.14 %	7.13 %	11.15 %	11.15 %	0.45 %
Average Reduction		15.15 %	15.98 %	15.79 %	15.54 %	15.50 %	7.66 %	10.64 %	13.61 %

Table 1 presents the reduction of transition activity at the coded bus in percental figures, compared to the unencoded bus. As expected APBI gave the best reduction in transitions for the **gen** data stream. For **binary** and **image** APBI outperforms BI and PBI while it is slightly less effective for **ascii**. Compared to IAEB, APBI yielded a higher reduction for **image** and **noise**. For the other test streams it achieved less reduction in transition activity. The **noise** example shows that BI is optimal for an i.i.d. source which can not be outperformed by any other encoding scheme. But on average APBI has a higher reduction in transition activity than every other investigated scheme.

In a second experiment we determined the power dissipation for implementations of BI, PBI, IAEB and APBI coder-decoder systems using our test suite of data streams. For that purpose the VHDL models have been synthesized with Synopsys Design Compiler for Fujitsu CE71 technology because it was the only available library with cells characterized for internal power. Other libraries which were available for our experiments, such as LSI10k or XILINX XC4000, did not have that feature, so the measures of the power dissipation of the coder-decoder systems would in general be too low. The switching activities of all internal nodes in the resulting netlists were determined by simulation with the test data streams using timing annotated VITAL models. Table 2 lists the resulting power dissipation at $f=50$MHz und $V_{dd}=2.5$V determined with Synopsys Design Power, and Table 3 lists the area and critical paths of the implementations of the corresponding coder-decoder systems. It has to be pointed out, that

Table 2. Power Dissipation of Coder-Decoder Systems in Fujitsu CE71 Technology

Sequence	$P_{V,Codec}$							
	APBI$_{32,1}$	APBI$_{32,2}$	APBI$_{32,4}$	APBI$_{32,8}$	APBI$_{32,16}$	BI	PBI	IAEB
gen	31.70 mW	19.64 mW	13.33 mW	10.22 mW	8.66 mW	5.52 mW	4.72 mW	47.31 mW
ascii	28.96 mW	18.64 mW	12.80 mW	9.96 mW	8.58 mW	4.76 mW	4.05 mW	48.58 mW
binary	24.86 mW	16.48 mW	11.28 mW	8.65 mW	7.40 mW	3.83 mW	2.08 mW	48.17 mW
image	18.75 mW	13.39 mW	9.15 mW	7.06 mW	6.04 mW	2.62 mW	1.58 mW	48.76 mW
noise	33.03 mW	20.19 mW	13.69 mW	10.47 mW	8.88 mW	5.25 mW	5.25 mW	49.25 mW

these figures completely depend on the target technology the coder-decoder systems are implemented with. Using other technologies may possibly result in a lower power dissipation. Finally, Table 4 shows the effective capacitances C_{eff} calculated according to (4) for the investigated coder-decoder implementations.

Table 3. Area and Delay for Implementations of Coder-Decoder Systems

Measure	APBI$_{32,1}$	APBI$_{32,2}$	APBI$_{32,4}$	APBI$_{32,8}$	APBI$_{32,16}$	BI	PBI	IAEB
Critical Path (ns)	10.4	10.7	10.8	10.8	10.8	7.2	5.7	3.1
Area (BC)	16125	17783	17829	17869	17939	1343	1177	29173

Table 4. Effective Average Capacitances C_{eff}

Seq	T uncod.	C_{eff}							
		APBI$_{32,1}$	APBI$_{32,2}$	APBI$_{32,4}$	APBI$_{32,8}$	APBI$_{32,16}$	BI	PBI	IAEB
gen	2130360	33.69 pF	18.43 pF	12.51 pF	9.58 pF	8.12 pF	17.11 pF	14.54 pF	42.27 pF
ascii	221309	134.53 pF	86.73 pF	59.79 pF	46.89 pF	41.39 pF	52.07 pF	17.40 pF	203.44 pF
binary	154620	177.54 pF	120.48 pF	88.39 pF	77.76 pF	70.31 pF	43.47 pF	16.85 pF	142.75 pF
image	2878651	163.33 pF	122.01 pF	85.76 pF	67.62 pF	55.58 pF	63.29 pF	16.74 pF	-
noise	4086760	186.97 pF	114.39 pF	77.17 pF	58.81 pF	49.93 pF	18.87 pF	18.87 pF	4370 pF

6   Conclusions

The high efficiency of the APBI encoding technique for system buses with strongly time-varying activity profile could be demonstrated through the experimental results. In contrast to most static encoding schemes such as PBI that only have a good encoding performance E_α for streams they are explicitly optimized for, APBI has the ability to adapt to a changing activity profile of the data stream to be transfered. While IAEB achieves a higher reduction in switching activity for particular data streams, on average APBI outperformed all other investigated encoding schemes regarding reduction of transition activity or coding efficiency E_α. In all test cases APBI coder-decoder implementations had a lower power dissipation than their IAEB counterparts. The resulting effective capacitances C_{eff} show, that partly higher reductions in switching activity achieved by IAEB cannot compensate the higher power dissipation of the coder-decoder system. Tolerating a slight deterioration in coding efficiency E_α, the power dissipation of the APBI coder-decoder system can be further reduced by enlarging the mask update interval. Update intervals of 2, 4 and 8 gave an acceptable reduction in switching activity with essentially reduced power dissipation of the coder-decoder system. Our coding scheme can be applied for highly capacitive system buses, e.g. bus lines which cross chip boundaries, whose activity profile is heavily changing over time or is a priori unknown.

References

1. Youngsoo Shin, Soo-Ik Chae, and Kiyoung Choi. Partial Bus-Invert Coding for Power Optimization of System Level Bus. In *ISLPED*, pages 127–129, 1998.
2. Mircea R. Stan and Wayne P. Burleson. Bus-Invert Coding for Low-Power I/O. In *Transactions on VLSI Systems*, volume 3, pages 49–58, March 1995.
3. L. Benini, G. De Micheli, E. Macii, D. Sciuto, and C. Silvano. Asymptotic Zero-Transition Activity Encoding for Address Busses in Low-Power Microprocessor-Based Systems. In *Great Lakes VLSI Symposium*, pages 77–82, March 13-15 1997.
4. L. Benini, G. De Micheli, E. Macii, M. Poncino, and S. Quer. System-Level Power Optimization of Special Purpose Applications: The Beach Solution. In *ISLPED*, pages 24–29, 1997.
5. L. Benini, G. De Micheli, E. Macii, D. Sciuto, and C. Silvano. Address Bus Encoding Techniques for System-Level Power Optimization. In *DATE*, 1998.
6. L. Benini, A. Macii, E. Macii, M. Poncino, and R. Scarsi. Synthesis of Low-Overhead Interfaces for Power-Efficient Communication over Wide Buses. In *DAC*, 1999.
7. C. Kretzschmar, R. Siegmund, and D. Mueller. Theoretische Untersuchungen zur verlustleistungsminimierten Informationsuebertragung auf Bussen integrierter digitaler Systeme. Technical report, TU Chemnitz, Professur Schaltungs- und Systementwurf, 1999.

Accurate Power Estimation of Logic Structures Based on Timed Boolean Functions[1]

G. Theodoridis, S. Theoharis, N.D. Zervas, and C.E. Goutis

VLSI Design Lab., Dept. of Elect. and Comp. Eng.,
University of Patras, Rio 26110, Greece
{theodor, theohari, zervas, goutis}@vlsi.ee.upatras.gr

Abstract. A new probabilistic method to estimate the switching activity of a logic circuit under a real delay gate model, is introduced. Based on Markov stochastic processes and generalizing the basic concepts of zero delay-based methods, a new probabilistic model to estimate accurately the power consumption, is developed. More specifically, a set of new formulas, which describe the temporal and spatial correlation in terms of the associated zero delay-based parameters, under real delay model, are derived. The chosen gate model allows accurate estimation of the functional and spurious (glitches) transitions, leading to accurate power estimation. Comparative study and analysis of benchmark circuits demonstrates the accuracy of the proposed method.

1 Introduction

Power dissipation is recognized as a critical parameter in modern VLSI design. Thus, efficient low power design techniques have been developed to solve certain issues at all design levels [1]. Also, a number of power estimation methods for combinational logic circuits have been developed [2]. Recently, a number of probabilistic estimation methods, considering zero gate delay model [3,4,5] and real gate delay model [6,7], were proposed. The method presented in [5] is the most accurate assuming zero-delay gate model since all types of correlations among the circuit signals are considered. The temporal correlation was captured by modelling the behaviour of a signal as a two state Markovian stochastic process, while the spatial correlation by the introduction of the concepts of the spatiotemporal transition correlation coefficient and the signal isotropy.

Assuming arbitrary gate delay model, a few probabilistic power estimation methods have been published [6,7]. In [6], a symbolic simulation algorithm has been proposed. Given the switching activities of the primary inputs and using OBDDs, the transition probability of a node at time t resulted by XORing the Boolean functions that correspond to two successive switching time instances, i.e. t and $t+1$. The structural and the first-order temporal correlations are handled, but the input pattern dependency is not captured, since the primary inputs are assumed uncorrelated. Based

---

[1] This work was partially supported by European Union in context of ESPRIT IV project 25256 "LPGD".

D. Soudris, P. Pirsch, and E. Barke (Eds.): PATMOS 2000, LNCS 1918, pp. 76-87, 2000.
© Springer-Verlag Berlin Heidelberg 2000

on the signal probability calculation method of [8], a new method for calculating the transition probabilities was proposed in [7]. To manipulate large circuits, an efficient methodology, which reduces the support set of an internal circuit node, has been developed. This method is parameterised in terms of the depth of the circuit levels. The structural and the first order temporal correlations were captured, but the primary inputs were considered uncorrelated.

Considering simultaneous input transitions, structural, temporal and input pattern dependencies we propose a probabilistic method for accurate power estimation of combinatorial circuits assuming real-delay date model. It is proved that the switching activity estimation under real-delay gate model is transformed to switching activity estimation at specific time instances assuming zero-delay gate model. Based on the concepts of the transition probability (i.e. temporal correlation) and transition correlation coefficient [5] (i.e. spatial correlation), new formulas for calculating the transition probabilities for different time sub-intervals and the transition correlation coefficients of any pair of signals in terms of two time instances are proved. To describe the logic behaviour of a circuit node in terms of time, we adopt the notion of Timed Boolean Functions (TBFs) [9]. A TBF can be seen as a modified Boolean function, exhibiting all those properties that can model efficiently the behaviour of a circuit node for every time point. Manipulation of TBFs can be done by the TBF-Ordered Binary Decision Diagrams (TBF-OBDDs) [9], which have the inherent important property to solve the problem of temporal compatibility [7].

The rest of the paper is organized as follows. In section 2 the problem is formulated, while in Section 3 the mathematical model is given. The main principles of Timed Boolean Functions and TBF-OBDDs are presented in Section 4, while the procedure for the switching activity evaluation is given in Section 5. In Section 6 the experimental results prove the efficiency of the proposed method. Finally, the conclusions are presented in Section 7.

2   Problem Formulation

The power estimation problem of a combinational logic circuit, under real gate delay model can be stated as:

"Given the gate level description of a combinational circuit with n inputs and m outputs and the inertial delays of its gates, and, assuming that the time between two successive applied input vectors is greater or equal to the settling time of the circuit, estimate the average power consumption of the circuit for an input vector stream through the calculation of its average switching activity."

It is assumed that the combinational circuit is a part of a synchronous sequential circuit, which means that its inputs can switch synchronously with the clock, performing at most one transition at time $t=0$ during the clock period $[0,T)$. Moreover, an applied input signal is considered as an ideal step pulse without any voltage drops at circuit nodes, while its width is greater or equal to the period time T.

To clarify the proposed method and the introduced concepts, a running example is used throughout the whole paper.

Example: We assume a logic circuit with gate delays equal to one, shown in Figure 1. The logic behaviour of the node f can be described in time domain as follows:

$$f = F(x_1, x_2, t) = x_1\,(t-2)\,x_2\,(t-2)\,x_2\,(t-1).\tag{1}$$

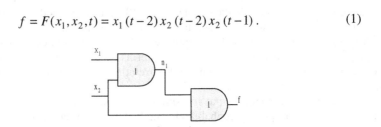

Fig. 1. A logic circuit with Unit Delay AND gates

The signal f may switch at two time instances, i.e. $t_1^f = 1$ and $t_2^f = 2$. The transition of the signal f at t_1, $t_1^f = 1$, depends on the transitions of the primary inputs x_1 and x_2 at time points $t_1^{x_1} = -1$, $t_1^{x_2} = -1$, and $t_1^{x_2} = 0$, while the transition of f at $t_2^f = 2$ depends on the transitions of the signals x_1 and x_2 at $t_2^{x_1} = 0$, $t_2^{x_2} = 0$, and $t_2^{x_2} = 1$. The corresponding logic functions of f derived by (1) at $t_1^f = 1$ and $t_2^f = 2$ are: $f_1 = F(x_1, x_2, x_3, 1) = x_1(-1)x_2(-1)x_3(0)$ and $f_2 = F(x_1, x_2, x_3, 2) = x_1(0)x_2(0)x_3$.

From the above example it is proved that the switching activity considering real-delay gate model is transformed to the switching activity estimation at multiple time instances. Also, the logic behavior of signal f at any time instance is described by the modified logic function of eq. (1). However, at each time instance the modified function is reduced to an ordinary Boolean function, where the Boolean variables are the corresponding logic values of the input signals at specific time points.

Having known the probabilistic properties (e.g. the switching activity) of its variables and manipulating the modified function efficiently the switching activity at any time point can be evaluated in similar manner to zero delay methods.

Having as starting point eq. (1), a new mathematical model, which describes the behavior of a logic signal in terms of time and signal correlation, should be introduced. We aim at the development of the new method, which reduces the power estimation problem with real delay model to a zero delay problem at certain switching time points. For that purpose, we introduce new formulas, which express parameters of real delay, that model power estimation problem in terms of zero delay parameters.

3   Mathematical Model

The behavior of a binary signal, x, at a time point, t, i.e. $x(t)$, is modelled as a random variable of a time homogeneous, Strict Sense Stationary, lag-one Markov stochastic process having two states, s, with $s \in S = \{0,1\}$.

The *transition probability*, $p_{kl}^x(t)$, expresses the probability of a signal x to perform a transition from the state k to the state l within two successive time points $(q-1)T$ and qT, where q is integer and T is the period of the input signals, and can be defined by:

$$p_{kl}^x(t) = p\big(x((q-1)T) = k \wedge x(qT) = l\big) \ \forall k, l \in S. \tag{2}$$

The *switching activity*, $E^x(qT)$, of a signal x at time instance qT is given by:

$$E^x(t) = p_{01}^x(t) + p_{10}^x(t). \tag{3}$$

The above stochastic process models the behavior of an input signal at times $t=0$, $t=T$, e.t.c., where the input signal performs a transition. However, as it has been shown in the example of Figure 1, the transition probabilities $p_{kl}^x(t)$ of an input signal x at several time points $t=\pm d$, $d \in \{1,2\}$, are needed. More specifically, the $p_{kl}^{x_2}(0)$, $p_{kl}^{x_2}(-1)$, and $p_{kl}^{x_2}(1)$ for the signal x_2 are needed. However, the transition probabilities of an input signal at any time point $t=\pm d$, are constant since the signal may perform transition at $t=0$ only.

We introduce the notion of transition probabilities of an input signal x in time intervals $(-T, 0)$ and $(0, T)$ as $p_{kk}^x(0^-)$ and $p_{ll}^x(0^+)$, respectively. Their corresponding values can be computed by the next lemma.

Lemma 1. The transition probability of a primary input signal, x, at time intervals $(-T, 0)$ and $(0, T)$ is expressed with respect to the transition probabilities at $t=0$ as:

$$p_{kk}^x\big(0^-\big) = p_{kk}^x(0) + p_{k(1-k)}^x(0) \ \forall k \in S, \tag{4}$$

$$p_{ll}^x\big(0^+\big) = p_{ll}^x(0) + p_{(1-l)l}^x(0) \ \forall l \in S, \tag{4.1}$$

$$p_{kl}^x(0^-) = p_{kl}^x(0^+) = 0 \ \ \forall k, l \in S \wedge k \neq l. \tag{4.2}$$

Proof: Due to lack of space the proof is omitted here.

Definition 1. A *Signal Transition Probability Set*, $P^x(t)$, of a signal x at a time instance t, is defined as the set of all transition probabilities $p_{kl}^x(t)$, where $k, l \in S$:

$$P^x(t) = \big\{p_{00}^x(t), p_{01}^x(t), p_{10}^x(t), p_{11}^x(t)\big\}. \tag{5}$$

The accuracy of the power estimation implies that the spatial correlation among the circuit signals should be considered. Let x_1 and x_2 be two signals. The corresponding stochastic machine has four states, which are the four combinations of the signal values of x_1 and x_2. Based on this stochastic machine, it has been proved in [5] that the spatial correlation can be captured by the *Transition Correlation Coefficient (TC)*. Assuming zero-delay model, the TC between two signals x_1 and x_2 is: [5]:

$$TC_{kl,\,mn}^{x_1,\,x_2} = \frac{p\big(x_1(t-1) = k \ \wedge x_1(t) = l \wedge x_2(t-1) = m \wedge x_2(t) = n\big)}{p\big(x_1(t-1) = k \ \wedge x_1(t) = l\big) \ \ p\big(x_2(t-1) = m \wedge x_2(t) = n\big)} \ . \tag{6}$$

Since we use a real delay gate model, the notion of the *TC* [5] should be generalized for capturing the spatial correlation of two signals for any two certain time instances. Thus, under a non-zero delay model, we define the *Generalized Transition Correlation Coefficient*.

Definition 2. The *Generalized Transition Correlation Coefficient*, $TC_{kl,mn}^{x_1,x_2}(t_1,t_2)$, between two signals x_1 and x_2, which perform a transition from the states k and m to l and n, at times t_1 and t_2, respectively, is defined as:

$$TC_{kl,mn}^{x_1,x_2}(t_1,t_2) = \frac{p\left(x_1(t_1-1)=k \ \wedge x_1(t_1)=l \wedge x_2(t_2-1)=m \wedge x_2(t_2)=n\right)}{p\left(x_1(t_1-1)=k \ \wedge x_1(t_1)=l\right) \ p\left(x_2(t_2-1)=m \wedge x_2(t_2)=n\right)} \quad (7)$$

where $k, l, m, n \in S$.

The spatial dependencies among three or more signals are captured by the pairwise *TCs*, approximately. For example, the *TC* of x_1, x_2 and x_3 can be expressed as:

$$TC_{kl,mn,pq}^{x_1,x_2,x_3}(t_1,\ t_2,\ t_3) = TC_{kl,mn}^{x_1,x_2}(t_1,t_2) \ TC_{kl,pq}^{x_1,x_3}(t_1,t_3) \ TC_{mn,pq}^{x_2,x_3}(t_2,t_3) \quad (8)$$

where $k, l, m, n, p, q \in S$.

Since we deal with three time instances, i.e. $t=0^-, t=0$, and $t=0^+$ (where $0^+ / 0^-$ denotes the time intervals $(-T,0) / (0,T)$), appropriate *TCs* between two signals x_1 and x_2 for capturing their spatiotemporal dependency should be determined.

Definition 3. The *Transition Correlation Coefficient Set*, $TC^{x_1,x_2}(t_1,t_2)$, between two signals x_1 and x_2 at time instances t_1 and t_2 is defined as the set of sixteen *TCs*:

$$TC^{x_1,x_2}(t_1,t_2) = \left\{TC_{00,00}^{x_1,x_2}(t_1,t_2),...,TC_{11,11}^{x_1,x_2}(t_1,t_2)\right\}. \quad (9)$$

Lemma 2. The spatiotemporal *TC* of two input signals x_1 and x_2, at $t_1, t_2 \in \left\{0^-,0,0^+\right\}$ are expressed in terms of their signal transition probability sets (i.e. eq. 5) and the associated *TC* set (i.e. eq. 9) at time points $t_1 = 0$ and $t_2 = 0$ as follows:

$$TC^{x_1,x_2}(t_1,t_2) = F\left(TC^{x_1,x_2}(0,0),P^{x_1}(0),P^{x_2}(0)\right) \quad (10)$$

and can be calculated by:

$$TC_{kk,mn}^{x_1,x_2}(0^-,0) = \frac{TC_{kk,mn}^{x_1,x_2}(0,0) \ p_{kk}^{x_1}(0) \ + \ TC_{k(1-k),mn}^{x_1,x_2}(0,0) \ p_{k(1-k)}^{x_1}(0)}{p_{kk}^{x_1}(0) \ p_{mn}^{x_2}(0) \ + \ p_{k(1-k)}^{x_1}(0) \ p_{mn}^{x_2}(0)} \quad (11)$$

$$TC_{ll,mn}^{x_1,x_2}(0^+,0) = \frac{TC_{ll,mn}^{x_1,x_2}(0,0) \ p_{ll}^{x_1}(0) \ + \ TC_{(1-l)l,mn}^{x_1,x_2}(0,0) \ p_{(1-l)l}^{x_1}(0)}{p_{ll}^{x_1}(0) \ p_{mn}^{x_2}(0) \ + \ p_{(1-l)l}^{x_1}(0) \ p_{mn}^{x_2}(0)} \quad (11.1)$$

$$TC_{kk,\,mm}^{x_1,\,x_2}\left(0^-,0^-\right)=\frac{TC_{kk,\,mm}^{x_1,\,x_2}(0,0)\;p_{kk}^{x_1}(0)\;p_{mm}^{x_2}(0)+\;TC_{kk,\,m(1-m)}^{x_1,\,x_2}(0,0)\;p_{kk}^{x_1}(0)\;p_{m(1-m)}^{x_2}(0)}{p_{kk}^{x_1}(0)\;p_{mm}^{x_2}(0)+p_{kk}^{x_1}(0)\;p_{m(1-m)}^{x_2}(0)+\;p_{k(1-k)}^{x_1}(0)\;p_{mm}^{x_2}(0)+\;p_{k(1-k)}^{x_1}(0)\;p_{m(1-m)}^{x_2}(0)}\;+$$

$$\frac{TC_{k(1-k),\,mm}^{x_1,\,x_2}(0,0)\;p_{k(1-k)}^{x_1}(0)\;p_{mm}^{x_2}(0)+TC_{k(1-k),\,m(1-m)}^{x_1,\,x_2}(0,0)\;p_{k(1-k)}^{x_1}(0)\;p_{m(1-m)}^{x_2}(0)}{p_{kk}^{x_1}(0)\;p_{mm}^{x_2}(0)+p_{kk}^{x_1}(0)\;p_{m(1-m)}^{x_2}(0)+\;p_{k(1-k)}^{x_1}(0)\;p_{mm}^{x_2}(0)+\;p_{k(1-k)}^{x_1}(0)\;p_{m(1-m)}^{x_2}(0)}\qquad(11.2)$$

$$TC_{ll,\,nn}^{x_1,\,x_2}\left(0^+,0^+\right)=\frac{TC_{ll,\,nn}^{x_1,\,x_2}(0,0)\;p_{ll}^{x_1}(0)\;p_{nn}^{x_2}(0)+\;TC_{ll,\,(1-n)n}^{x_1,\,x_2}(0,0)\;p_{ll}^{x_1}(0)\;p_{(1-n)n}^{x_2}(0)}{p_{ll}^{x_1}(0)\;p_{nn}^{x_2}(0)+p_{ll}^{x_1}(0)\;p_{(1-n)n}^{x_2}(0)+\;p_{(1-l)l}^{x_1}(0)\;p_{nn}^{x_2}(0)+\;p_{(1-l)l}^{x_1}(0)\;p_{(1-n)n}^{x_2}(0)}\;+$$

$$\frac{TC_{(1-l)l,\,nn}^{x_1,\,x_2}(0,0)\;p_{(1-l)l}^{x_1}(0)\;p_{nn}^{x_2}(0)+TC_{(1-l)l,\,(1-n)n}^{x_1,\,x_2}(0,0)\;p_{(1-l)l}^{x_1}(0)\;p_{(1-n)n}^{x_2}(0)}{p_{ll}^{x_1}(0)\;p_{nn}^{x_2}(0)+p_{ll}^{x_1}(0)\;p_{(1-n)n}^{x_2}(0)+\;p_{(1-l)l}^{x_1}(0)\;p_{nn}^{x_2}(0)+\;p_{(1-l)l}^{x_1}(0)\;p_{(1-n)n}^{x_2}(0)}\qquad(11.3)$$

$$TC_{kl,\,mn}^{x_1,\,x_2}(0^-,0)=TC_{kl,\,mn}^{x_1,\,x_2}(0^+,0)=0\quad\forall\;k,l\in S\;\wedge\;k\neq l,\qquad(11.4)$$

$$TC_{kl,\,mn}^{x_1,\,x_2}(0^-,0^-)=TC_{kl,\,mn}^{x_1,\,x_2}(0^+,0^+)=0\quad\forall\;k,l,m,n\in S\;\wedge\;k\neq l\;\wedge\;m\neq n\qquad(11.5)$$

where $k,\,l,\,m,\,n\in S$. Proof: Due to lack of space the proof is omitted here.

4   Timed Boolean Functions

As it has been mentioned, the glitch generation is strongly dependent on the time. Therefore, a modified Boolean function, which will describe the logic and timing behavior, is needed. This modified Boolean function, called *Timed Boolean Function* (TBF) and its mathematical foundation of TBFs was presented in [9]. Exploiting the timing properties of the input signals, a range interval $(0,T)$ can be partitioned in coarsest of sub-intervals, within each of which a TBF is an ordinary Boolean function. Each input signal with specified switching times can be represented by a TBF, using the unit step function. It is also has been proved in [9] that any binary signal with known switching times can be represented by a TBF using the unit step function. These inputs are modeled by TBFs and can be represented by a set of time intervals and the corresponding Boolean functions in each interval. Consequently, a TBF of an internal node can be regarded as a transformation of a set of intervals and Boolean functions of the inputs to another set of intervals and Boolean functions, considering the gate delays and their Boolean operations.

A gate operation is logically and temporally separable if the computation can be performed in two separate steps: i) delay the inputs and ii) perform Boolean operation on the delayed inputs. The TBF of the circuit output is a composition of the TBFs of its gates. Thus, the TBF of an output f is given by:

$$f(t)=F(t,x_1,....,x_n)=F\left(x_1(t-d_1),...,x_n(t-d_n)\right),\qquad(12)$$

where d_i is the delay of the path starting from the input x_i and terminated at f.

Definition 5. The Boolean variables of an input signal, $x_i(t)$, which is modeled as TBF, are defined by:

$$x_i(t) = x_i^- \text{ if } t \in (-T, 0), \tag{13}$$

$$x_i(t) = x_i^+ \text{ if } t \in (0, T). \tag{13.1}$$

Example: The corresponding TBFs of the nodes n_1 and f of the circuit shown in Figure 1 are:

$$n_1(t) = x_1(t-1)\, x_2(t-1), \tag{14}$$

$$f(t) = n_1(t-1)\, x_2(t-1) = x_1(t-2)\, x_2(t-2)\, x_2(t-1). \tag{14.1}$$

Considering the TBF of node f, we infer that there exist two valid switching time points, namely $t_1=1$ and $t_2=2$ and therefore, three time intervals, i.e. $(-\infty, 1)$, $(1, 2)$ and $(2, +\infty)$. All the TBF variables are positive for $t \in (2, +\infty)$ (i.e. x_1^+, x_2^+). Thus, within this time interval the signal f does not perform a transition and the TBF is reduced to $f = x_1^+ x_2^+$. Similarly, the associated Boolean functions in the transitionless time intervals $(-\infty, 1)$ and $(1, 2)$ are $f = x_1^- x_2^-$ and $f = x_1^- x_2^- x_2^+$, respectively.

In order to manipulate the TBFs efficiently the TBF-OBDDs has been presented in [9]. More specifically, a TBF-OBDD consists of the upper BDD called K-OBDD, and a set of ordinary OBDDs. The purpose of the K-OBDD is to represent the associative time intervals of the TBF. Any leaf node of the K-OBDD is a dummy node, which is replaced by the OBBD of the Boolean function of the corresponding interval. Thus, the OBDD of the ordinary Boolean function corresponding to the time interval (t_i, t_{i+1}), is the OBDD that replaces the leaf node of the right branch of the node K_i. Also, the OBDD of Boolean function of the time interval (t_{i-1}, t_i) corresponds to the OBDD that replaces the rightmost leaf node of the left branch of the node K_i (which is the same with the OBBD of the right branch of node K_{i-1}). It has been proved in [9] that the TBF-OBBDs are canonical and can be reduced and manipulated as ordinary OBDDs. The corresponding TBF-OBDD of node f of the example 1 is given below:

5   Switching Activity Evaluation

Generally, a signal f of the logic circuit is a Boolean function of a subset of the primary input signals, i.e. $f = F\big(x_1(t), x_2(t), ..., x_v(t),\big)$, where $v \le n$.

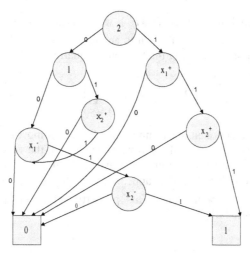

Fig. 2. The TBF-OBDD of circuit node f of Figure 1

Definition 4. We define as *Valid Time Points Set*, $\mathbf{T}^f = \{t_1^f, t_2^f, ..., t_r^f\}$, the transition time points of a signal f.

Determining the valid time points, the switching activity estimation problem is reduced to the estimation of $\mathbf{P}^f(t) \forall t \in \mathbf{T}^f$.

A circuit node, f, switches at time $t = t_i^f$, if the derivative with respect to time t of its TBF is equal to 1. Thus, the average switching activity, $E^f(t_i^f)$, is:

$$E^f(t_i^f) = p(\lim_{\varepsilon \to 0}\{f(t_i^f - \varepsilon) \oplus f(t_i^f + \varepsilon)\} = 1) \ . \tag{15}$$

Instead of performing the XORing between the Boolean functions corresponding to the time intervals (t_{i-1}^f, t_i^f) and (t_i^f, t_{i+1}^f), we can manipulate efficiently the TBF-OBDD, in order to evaluate the switching probability of node f at time point $t = t_i^f$. Taking into consideration the representation of the ordinary Boolean function on the TBF-OBDD, the evaluation of the transition $k \to l$ with $k, l \in S$ at time instance t_i^f can be done as described by the following steps (similar to the zero-delay procedure of [5]:

i) Find the set of paths, $\Pi_k^-(t = t_i^f)$, of the OBDD of the Boolean function corresponding to the time interval (t_{i-1}^f, t_i^f), which terminate at node k,

ii) Find the set of paths, $\Pi_l^+(t = t_i^f)$, of the OBDD of the Boolean function corresponding to the time interval (t_i^f, t_{i+1}^f), which terminate at node l,

iii) Combine each path of $\Pi_k^-(t=t_i^f)$ with all paths of $\Pi_i^+(t=t_i^f)$ and extract the switching behavior taking into account the temporal compatibility of each primary input signal using the following equation:

$$p_{kl}^f(t_i^f) = \sum_{\pi \in \Pi_k^-} \sum_{\pi' \in \Pi_i^+} \prod_{i=1}^v (p_{k_i l_i}^{x_i}(t_i^{x_i}) \, (\prod_{1 \le i < j \le v} TC_{k_i l_i, k_j l_j}^{x_i, x_j}(t_i^{x_i}, t_j^{x_j}))^{\frac{2}{v}}). \quad (16)$$

As it has been mentioned, each input signal x has at most two Boolean variables, namely the variables $x(0^-)=x^-$ and $x(0^+)=x^+$. The values of the variables x^- and x^+ denote the values of the signal x for the time intervals $(-T,0)$ and $(0,T)$, respectively.

In case of the variable x^- or x^+ is appeared in both paths π and π' then its value must be the same in both paths. This property solves the problem of temporal compatibility reported in [7].

6   Experimental Results

The proposed power estimation method is implemented by ANSI C language, whereas its efficiency is proved by a number of ISCAS'85 benchmark multilevel circuits. For the technology mapping step, a general library of primitive gates (i.e. AND, OR, e.t.c.) of up to 4 primary inputs, is used. For a signal x, we define as *Real Node Error* the quantity $Err(x)=|E_{eff}(x)-E'_{eff}(x)|/E_{eff}(x)$, where $E_{eff}(x)$ is the real effective switched capacitance of signal x and $E'_{eff}(x)$ is the estimated effective capacitance. The effective switched capacitance is calculated by the product of the switching activity $E(x)$ and the total capacitance load of the node x, $C_x=F_x C_g$, where F_x is the fanout of this node and $C_g = 0.05$ pF is a typical input gate capacitance.

For a combinatorial circuit with N signals and a specific input vector set V_j, we define as *Total Power Consumption* the quantity $Power(V_j)=\frac{1}{2} \cdot V_{dd}^2 \cdot f \cdot \sum_{i=1}^N E_{eff}(x_i)$,

as *Real Total Error* the quantity $Total\ Error(V_j)=|Power(V_j)-Power(V_j)'|/Power(V_j)$, as *Real Mean Error* the quantity $MeanError(V_j)=\frac{1}{N}\sum_{i=1}^N Err(x_i)$ and finally as *Real Maximum Error* the quantity $Max\ Error\ (V_j)=\max\{Err(x_1), Err(x_2),..., Err(x_N)\}$. If we choose M input vector sets (for a reliable comparison) the above formulas become:

$$Power^M = \sum_{j=1}^M Power(V_j), Total\ Error^M = \sum_{j=1}^M Power(V_j) - \sum_{j=1}^M Power(V_j)' / \sum_{j=1}^M Power(V_j),$$

$$Mean\ Error^M = \frac{1}{M}\sum_{j=1}^M Mean\ Error(V_j) \text{ and } Max\ Error^M = \max\{Mean\ Error(V_1),..,Mean\ Error(V_M)\}$$

For comparisons, three categories of input vectors are chosen: i) without spatial correlation (column NO), ii) with low-spatial correlation (column LOW) and iii) with high spatial correlation (column HIGH). For each category and circuit, $M=10$ input vector sets of 50000 vectors are generated.

We compare the proposed method and the method of [7] with Mentor's Graphics QUICKSIM II gate level simulator. The power consumption differences between each method and switch level simulator are depicted in Table 1 and 2. In particular, the columns TOTAL represent the *Total Error^M* of the total power dissipation for all of the 10 estimations, the columns MEAN is the *Mean Error^M* error, while the columns MAX contain the *Max Error^M* error for the 10 power estimations.

Table 1 shows the errors in power estimation (%) of the proposed method and proves the quality of the method. The average TOTAL error is about 0.07 % for NO spatial input correlation, 1.62 % for LOW spatial correlation, and 1.67 % for HIGH spatial correlation. The corresponding average MEAN error values are 0.88 %, 4.91 %, and 5.74 %, while the average MAX errors are 1.77 %, 7.65 %, and 8.85%. Table 2 shows the errors of the method of [7]. It can be seen that for NO spatial correlation column, the error values of [7] are comparable with the corresponding errors of the proposed method. In contrary, the average errors of the remaining two categories (i.e. LOW and HIGH spatial correlation) are large enough, that is, 7.60 % and 9.27 % for TOTAL power, 15.74 % and 18.83 % for MEAN power, and 24.96 % and 30.84 % for MAX power, respectively. It is concluded that the lack of the spatial correlations in the primary inputs increases the power estimation error (e.g. for HIGH correlation, the MAX error of circuit cm82 is more than 55%).

Table 1. Power estimation errors of the proposed method

Circuit	TOTAL			MEAN			MAX		
	NO	LOW	HIGH	NO	LOW	HIGH	NO	LOW	HIGH
9symml	0,012	0,406	0,007	0,549	2,537	3,068	0,940	3,025	3,582
C17	0,006	1,174	1,124	0,271	2,140	2,527	0,971	3,411	4,609
Cm163	0,019	1,968	2,260	0,870	6,376	7,409	1,592	9,670	11,227
Cm42	0,022	0,816	0,946	1,037	6,411	7,393	2,270	9,940	11,514
Cm82	0,012	3,422	3,357	0,265	4,886	5,581	0,480	8,592	9,354
Cm85	0,007	2,092	2,353	0,789	4,185	4,836	1,860	7,370	8,468
Cmb	0,205	0,131	0,350	2,265	5,193	6,394	4,512	8,491	9,696
Cu	0,336	0,080	0,000	1,134	2,268	2,634	1,625	2,827	3,349
Decod	0,104	1,886	2,177	1,577	9,341	10,885	1,973	15,865	18,346
F51m	0,058	1,814	1,616	0,597	3,446	3,811	1,157	4,766	5,105
Majority	0,017	0,339	0,029	0,482	3,230	4,390	1,503	4,820	6,424
Pm1	0,067	2,091	2,385	1,721	10,093	11,563	4,568	15,586	17,877
Rca4	0,108	6,072	6,852	0,325	6,767	7,576	0,583	11,476	12,940
x2	0,010	1,034	0,999	0,797	5,337	6,383	1,308	6,652	8,129
Z4ml	0,038	0,941	0,515	0,398	1,506	1,641	1,185	2,185	2,124
Average	**0,068**	**1,618**	**1,665**	**0,872**	**4,914**	**5,739**	**1,768**	**7,645**	**8,850**

Table 2. Power estimation errors of method [7]

Circuit	TOTAL			MEAN			MAX		
	NO	LOW	HIGH	NO	LOW	HIGH	NO	LOW	HIGH
9symml	3,205	0,616	1,685	2,657	7,818	3,068	4,007	10,611	15,707
C17	0,188	7,846	9,469	0,835	13,534	16,270	1,437	30,645	39,011
Cm163	0,097	5,824	7,081	1,354	15,045	18,590	2,219	21,615	27,256
cm42	0,006	1,986	2,405	1,455	16,029	19,398	3,148	24,035	29,262
Cm82	0,170	20,576	24,716	0,757	25,917	31,009	1,156	45,345	54,872
Cm85	0,079	8,814	10,615	1,521	12,256	15,213	5,324	22,205	26,182
Cmb	0,105	2,555	3,045	1,291	7,917	9,399	2,137	11,758	14,153
Cu	0,467	2,260	2,880	1,732	6,602	7,790	2,312	8,978	10,622
Decod	0,038	7,595	9,114	2,119	22,958	28,293	3,547	36,668	48,091
F51m	0,131	8,764	10,698	1,216	13,119	15,714	1,862	22,982	27,048
Majority	0,032	4,978	5,940	1,005	12,503	14,780	2,979	22,060	25,969
Pm1	0,041	5,631	7,028	2,208	24,772	32,438	5,181	34,826	44,208
Rca4	0,196	22,786	27,680	1,293	28,368	34,852	2,189	43,239	52,237
X2	0,041	5,815	6,988	1,196	19,366	23,846	1,939	24,148	29,069
Z4ml	0,169	7,972	9,744	1,156	9,911	11,834	1,943	15,287	18,831
Average	**0,331**	**7,601**	**9,273**	**1,453**	**15,741**	**18,833**	**2,759**	**24,960**	**30,835**

7   Conclusions

The proposed method constitutes an extension of the zero delay probabilistic method that presented in [5] and takes into account the first-order temporal correlations and the spatial correlations not only for the logic circuit structural dependencies but also for the data dependencies at the primary input signals as well. Since the proposed method is a global approach, our future work is to implement a method that propagates the primary input statistics and correlation coefficients through the logic network estimating efficiently the switching activity at any node and any valid time point.

References

1. J. Rabaey and M. Pedram, "*Low Power Design Methodologies,*" Kluwer Academic Publishers, 1996.
2. F. Najm, "*A Survey of Power Estimation Techniques in VLSI circuits (Invited paper),*" in IEEE Trans. On VLSI, vol 2, no 4, pp. 446-455, December 1995.
3. F. Najm, "*Transition Density: A new measure of activity in digital circuits,*" in IEEE Trans. On CAD, Vol. 12, No. 2, pp. 310-323, February 1995.
4. P. Schneider and U. Schlichmann, "*Decomposition of Boolean functions for low power based on a new power estimation technique,*" in Proc. of Int.Workshop on Low Power Design, pp. 123-128, NapaValley, CA, April 1994.
5. R. Marculescu, D. Marculescu, and M. Pedram "*Efficient Power estimation for highly correlated input streams,*" in Proc. of DAC. pp.628-634, 1995.

6. J. Monteiro, A. Ghosh, S. Devadas, K. Keutzer, and J. White, *"Estimation of average switching activity in combinatorial and sequential circuits"*, in IEEE Trans. on CAD, Vol. 16, No.1, pp. 121-127, January 1997.
7. J.C. Costa, J.C. Monteiro, and S. Devadas, *"Switching Activity Estimation using Limited Depth Reconvergent Path analysis"*, In Proc. of ISLPD, pp. 184-189, 1997.
8. K. Parker and E. McCluskey, *"Probabilistic Treatment of General Combinational Networks"*, in IEEE Trans. on Electronic Computers, c-24(6), pp. 668-670, 1975.
9. W. Lam and R.K. Brayton, *"Timed Boolean Functions: A Unified Formalism for Exact Timing Analysis"*, Kluwer Academic Publishers, 1994.

A Holistic Approach
to System Level Energy Optimization

Mary Jane Irwin, Mahmut Kandemir, N. Vijaykrishnan,
and Anand Sivasubramaniam

Department of Computer Science and Engineering
The Pennsylvania State University
University Park, PA 16802-6106
http://www.cse.psu.edu/~mdl

Abstract. Over the past few years, the design automation community
has expended a lot of effort in developing low power design methodologies. However, with the increasing software content in mobile environments and the proliferation of such devices in our day to day life, it
is essential to take a fresh holistic look at power optimization from an
integrated hardware and software perspective. This paper envisions the
tools and methodologies that will become necessary for performing such
optimizations. It also presents insights into the interaction and influence
of hardware and software optimizations on system energy.

1 Introduction

Energy has become an important design consideration, together with performance, in computer systems. While energy conscious design is obviously crucial
for battery driven mobile and embedded systems, it has also become important
for desktops and servers due to packaging and cooling requirements where power
consumption has grown from a few watts per chip to over a 100 watts. As a result,
there has been a great deal of interest recently (e.g., [8,9,10,11,12,13,14,15,16])
in examining optimizations for energy reduction from both the hardware and
software points of view.

From the hardware viewpoint, there are several complementary energy saving trends and techniques. These include the use of higher levels of integration
thereby clustering components into smaller and/or less energy consuming pack-
ages [17], the continuous scaling of supply voltages, and the use of hardware-
controlled clock-gating [7] that automatically shuts down portions of the chip
when not in use. Another important trend is the support of different operating modes, each consuming a different amount of energy at the cost of a loss
in performance. Some on-chip energy reduction operating modes are based on
scaling the supply voltage and/or clock frequency [18,19] under low load conditions. Others are based on the transitioning of unused hardware components
into energy-conserving modes under the direction of software control.

However, the power-aware computing community has long claimed that the
greatest energy savings benefits (other than supply voltage scaling) is to be obtained at the software and applications levels as illustrated in Figure 1. From the

D. Soudris, P. Pirsch, and E. Barke (Eds.): PATMOS 2000, LNCS 1918, pp. 88–107, 2000.

software viewpoint, new compiler, runtime, and application-directed techniques are being developed that target improvements in the energy-performance product or that selectively utilize as few hardware components as possible (without paying performance penalties) thereby allowing the remainder to be transitioned into energy-conserving modes [14]. Unlike hardware optimizations where the designer is usually faced with trading performance for reductions in energy, it is an open question as to whether the best performance-oriented compiler optimizations are the best from the energy point of view.

Another important consideration when tackling the energy problem is knowing the energy budget of one's system. Ensuring that the major energy consuming portions of the system are the ones being optimized will, of course, give the largest overall improvements. In fact, Amdahl's Law for performance can be modified to apply to energy. *'The performance benefits to be gained using some faster mode of execution is limited by the fraction of the time the faster mode can be used'* becomes *'the energy benefits to be gained by applying an energy saving optimization is limited by the fraction of the time that optimized component is used'*. As an example, if one is focused on achieving energy savings in the ALU and the ALU accounts for only 2% of the total energy budget, then the overall return will be very small indeed. Thus, it is important to know the energy budget of the system for the intended application environment. For example, Figure 2 shows the energy budget of the on-chip datapath and caches and the off-chip DRAM for two benchmark codes drawn from different application environments: a static compilation environment that targets array-dominated C codes (on the left) and a Java-based dynamic (runtime, on-demand) compilation environment. As we see on the left, overall energy of array based codes are dominated by instruction cache due to high frequency and good locality of instruction accesses. We observe that state-of-the-art compiler optimizations aggressively reduce the number of datapath operations, thereby causing the memory-bound instructions to be a significant portion of the energy budget. As compared to the array based application, the Java code is much more computation-intensive and datapath energy constitutes nearly 40% of the overall energy budget. This observation is interesting because dynamic compilation and other features of Java in general exercise the memory much more than statically-compiled C codes. However, memory-bound characteristics of the array-based domain dominate and mask the expected behavior. Thus, it is important for designers to understand the runtime environment as well as the application characteristics before focusing their efforts on optimizing specific components.

Although larger energy gains can be obtained by optimizations made at the software and application levels, hardware optimizations are still crucial. For example, we have found in our experiments that while certain widely used high-level compiler optimizations (e.g., loop permutation, loop unrolling, and loop tiling [6]) might optimize overall system energy, they can often increase on-chip energy, impacting chip packaging and cooling. This is due to a reduction of off-chip DRAM usage at the cost of an increased on-chip usage. This impact can be mitigated by optimizing the cache to reduce its energy (e.g., with block

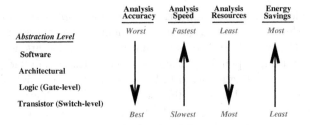

Fig. 1. Comparison of Energy Optimizations at Different Levels.

buffering [20] and cache subbanking [21] Also, to achieve the greatest energy savings, the designer must consider the interactions of hardware and software energy optimizations in the intended application environment. Obviously, compiler optimizations impact the energy gains of cache block buffering. In fact, optimizations made at the software level can be negated or improved by energy optimizations made at the hardware level, and vice versa.

When focusing on a set of energy optimizations it is important to be aware of changes in technology in order to ensure that the optimizations will be of benefit in the future. Supply voltage scaling, new interconnect materials, embedding additional components on-chip (e.g., eDRAM, RF), globally asynchronous-locally synchronous control styles all could have a significant impact on the relative benefits of a set of energy optimizations. We have looked at, in particular, how energy trade-offs in optimizations made in the memory system will be affected by the move from off-chip DRAM to eDRAM. While we concentrate here on reducing dynamic energy (the energy consumed when transistors change state), in the future power problems will be exacerbated by a dramatic increase in leakage power (currently less than 2% of the power equation) due to the scaling of supply voltages and, thus, threshold voltages.

The remainder of this paper details a variety of experiments we have done using energy modeling/simulation tools developed at Penn State in an attempt to address some of the issues raised above.

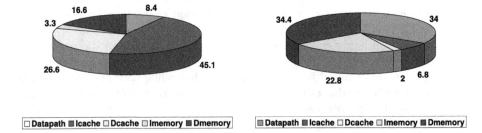

Fig. 2. Energy Breakdown Between Different Components. Left: Static Compilation of an Array-Dominated C Code. Right: Dynamic Compilation of a Java Code.

2   Tools

Tools for accurate power-performance prediction are essential for designing power-aware architectures, compilers, run-time support, communication protocols, and applications. Currently, there are tools to measure the power at either a very fine-grain (circuit or gate) level or coarse-grain (procedural or program) level. With fine-grain estimation, it is difficult or impossible to measure power usage in (future) billion transistor designs or for large programs. However, this is the most accurate approach to power estimation. On the other hand, coarse-grain measurements can only give gross estimates, but do so quite efficiently. Thus it is essential to provide a hierarchical spectrum of design tools for power estimation and optimization as shown in Figure 3. These tools can be used, as will be described shortly, to perform a series of 'what if' energy optimization experiments with various hardware and software design alternatives considering the computing system as a whole, rather than as a sum of parts (processor core, memory hierarchy, system level interconnect, etc.). In this section, we explain the design and use of *PowerMon*, a multi-granular power estimation and optimization tool currently being developed at Penn State.

Power estimation can be performed at the application, procedure, and instruction level granularity using *PowerMon* (see Figure 4). The monitoring capability of the Operating System (OS) along with energy measurement devices is utilized in developing *CoarseMon* to provide the coarse-grain estimates at the application and procedure levels. The energy hot-spots identified using this coarse-grain tool can, later, be studied in detail using a cycle-accurate instruction-level simulator, *FineMon*. This hierarchical approach provides an efficient mechanism for trading the simulation time and the estimation accuracy. Further, simulators for different types of processors such as scalar, pipelined, superscalar and VLIW can be plugged into *FineMon* to evaluate the influence of architectural choices on system energy. *FineMon* also provides the flexibility of choosing among traditional (direct-mapped, set-associative) and recently proposed power-efficient (way-prediction, sub-banking, isolated bit line, block buffering) cache architectures [22,20,32,34,31]. The accuracy and time can further be traded within *FineMon* based on the energy-models used.

Energy models for the datapath, control unit, system level interconnect, clock and memory components of the system can be either analytical or transition-sensitive. In transition-sensitive models, the switching activity of the input data is captured by the model. Analytical models, on the other hand, provide an energy consumption measure per access independent of the input data values. While transition-sensitive models are much more accurate they are also more time consuming to develop and incur longer running times.

2.1   CoarseMon

It is becoming increasingly apparent that the different hardware modules (processing unit, memory, disk, etc.) operating in an energy-constrained environment

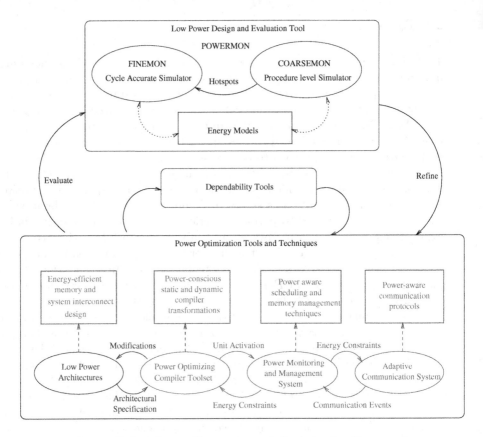

Fig. 3. Unified Energy Estimation and Optimization Framework.

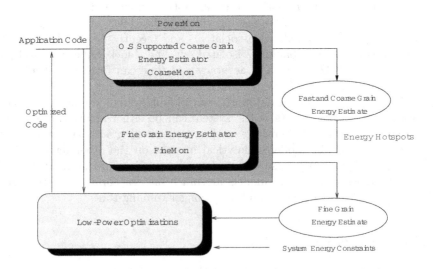

Fig. 4. *PowerMon:* Hierarchical Energy-Aware Design Framework.

should support several (at least more than one) modes of operation, each consuming a different amount of energy. When a module is not being exercised, the software can then selectively transition it to a lower energy consumption state/mode. Recognizing the importance of such capabilities, there has been recent interest in standardizing these different operating modes for each module in the form of the Advanced Configuration and Power Interface (ACPI) from both hardware and software vendors [24]. With time, one can expect different modules to support several modes (many of them already support different modes), and it is up to the software to effectively utilize these modes to lower the overall energy consumption of the system. However, response times are likely to suffer when there is a request to a module that has been transitioned to a lower energy consumption mode. Hence, the software has to employ intelligent heuristics to determine when to cause the transitions. Monitoring the system and application activity (based on the past and current behavior) to predict future usage can be valuable towards this goal as shown in several studies [36,35,29,25]. Further, the current energy usage/availability would be extremely useful for doing application-level adaptation (to the compiler generated 'smart' code or directly to the application); i.e., execute code that is more tailored for performance if there is adequate power, or vice-versa. Finally, monitoring and estimation is crucial to the design of the operating/runtime system itself and to develop the energy-delay aware services that are demanded from it.

Recognizing the importance of tools and techniques for energy monitoring/estimation, there have been prior studies looking at this issue [25,30]. Our coarse-grain monitoring tool, *CoarseMon*, attempts to reduce the overheads of the monitoring so that measurements can be taken more frequently and accurately. Instead of using an external device to measure the power, and interfacing with this device, we explore the use of energy counters that can be provided in the hardware. This is similar to the performance (and statistics) counters that many modules already support, except that they contain energy consumption information for the corresponding module since the last time they were read. Such counters can be built to monitor the signal switching activities that form the basis of FineMon's transition-sensitive energy models. Periodically (using the traditional timer-based mechanism), the OS reads off these counters and writes them into logs. Periodically, these logs can be flushed to the disk without becoming significantly intrusive on actual execution. The logs would contain program counter information (available on the stack during an interrupt) and the current energy counters in the last interval. Energy counter values can then be used to drive analytical energy models to estimate the consumption. Post-processing these logs, one can associate the energy consumption with the procedural level using the program counters (in the log) and the symbol table information of the compiled program. With such a design, monitoring would be relatively non-intrusive - only reading off counters within the timer interrupt mechanism (which would be called even in a non-monitored system). A further benefit is that there is no external device to interface with during the interrupts. As a result, the monitoring can be done more frequently for better accuracy.

CoarseMon can be used as a stand alone platform that can be used to drive energy-delay conscious application development and compilation, operating/runtime system, communication protocol and architectural design. In addition, the energy counters along with their associated run-time support software interface can be used to perform dynamic adaptation. For instance, based on current conditions, module state transitions can be initiated. Further, the application (or the compiled code) can use them to find out current conditions for dynamically changing the code to be executed.

2.2   FineMon

The power estimation tool, *FineMon*, is depicted in Figure 5. *FineMon* consists of a cycle accurate processor datapath simulator, a cache/bus simulator, energy models for the various components including clock and memory systems, and compiler/OS support tools. At each clock cycle, *FineMon* simulates the activities of all the components and calls corresponding power estimation interfaces. It continues the simulation until the predefined program halt instruction is fetched. In order to support 'what if' architectural level experimentation, the datapath is specified only to the RTL level so that many different architectural alternatives can be quickly evaluated. In order to keep the simulator technology independent, register transfer language (RTL) power estimation interfaces have been developed for all the components. These interfaces utilize the technology dependent energy models.

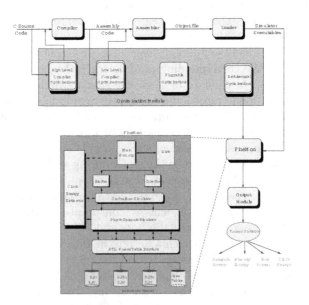

Fig. 5. *FineMon* Energy Estimation Framework.

A prototype of *FineMon* has already been developed for two processors. One was based on the ISA of the Hitachi SH-DSP, a merged DSP/RISC processor with on-chip code and data RAM. This prototype was validated by comparing its results against measurements made by Hitachi using gate-level power simulation [27,28]. A more recent prototype of *FineMon*, *SimplePower* — a single-issue five-stage pipelined architecture based on the SimpleScalar instruction set architecture (ISA) [26] with a cache-based memory hierarchy — has been developed and used to perform experiments in architectural and compiler optimizations to reduce energy consumption [14].

3   Hardware versus Software

Hardware and software techniques to reduce energy consumption have become an essential part of current system designs. In this section, we seek answers for the following questions:

- What is the impact of current performance-oriented software optimizations (that primarily aim at maximizing data locality and enhancing parallelism [6]) on energy? How do they affect the energy consumption of different system components (memory system, datapath, etc.)?
- What are the relative gains obtained using software and hardware optimization techniques? How can one exploit the interaction between these optimizations to reduce energy further?
- Is the most efficient code from the performance perspective the same as that for the energy viewpoint? If not, why?
- How does the impact of these optimizations get affected as a result of anticipated technological improvements in the future?

Of course, answering all these questions completely in a such a short article is not possible. However, we believe that any progress made in answering them will pave the way for our understanding of impacts and interactions of hardware and software optimizations.

3.1   Impact of Software Optimizations on System Energy

In this section, we evaluate the impact of three widely used high-level compiler optimizations on a simple matrix multiply code. The optimizations considered are as follows:

Linear Loop Transformation: The linear loop transformations attempt to improve cache performance, instruction scheduling, and iteration-level parallelism by modifying the traversal order of the iteration space of the loop nest. The simplest form of loop transformation, called loop interchange [6], can improve data locality (cache utilization) by changing the order of the loops.

Loop Tiling: Another important technique used to improve cache performance is blocking, or tiling [37]. When it is used for cache locality, arrays that are too big to fit in the cache are broken up into smaller pieces (to fit in the cache) and the nested loop in question is restructured accordingly. In the extreme case, loop tiling can double the number of loops in the nest.

Loop Unrolling: This optimization unrolls a given loop, thereby reducing loop overhead and increasing the amount of computation per iteration.

Energy Consumptions. We evaluated the energy consumptions for the matrix multiply code for different cache topologies (configurations) and program versions (each corresponding to different combinations of three optimizations mentioned above). The first observation we made is that all optimizations except loop unrolling increase the core power. This is due to the fact that the optimized versions generally have more complex loop structures; that, in turn, means extra branches and more complex subscript and loop bound calculations. Loop unrolling is an exception, as it reduces loop control overhead and enables better loop scheduling.

When considering the memory power, on the other hand, we made the following observations. First, with the increasing cache size and/or associativity, tiling performs better than pure linear loop transformations and unrolling. Unlike those optimizations, tiling exploits locality in all loop nest dimensions; increasing associativity helps to eliminate conflict misses between different array tiles. Second, in the original (unoptimized) code, the memory power is 5 to 47 times larger than the core power. However, after some optimizations, this picture changes. In particular, beyond a 2K, 2-way set associative cache (i.e., higher associativities or larger caches), the core and memory powers become comparable when some optimizations are applied. For example, when tiling is applied for a 2K, 4-way associative cache, the memory energy is 0.0764 J, which is smaller than the core energy, 0.0837 J. Similarly, for the most optimized version (that uses all three optimizations), the core and memory energy consumptions are very close for a 4K, 4-way set associative cache. This shows that when we apply optimizations, we reduce the memory energy significantly making the contribution of the core energy more important. Since we expect these optimizations (in particular, loop tiling) to be applied frequently by optimizing compilers, reducing core power using additional techniques might become very important. Overall, the power optimizations should not focus only on memory, but need to consider the overall system power. In fact, the choice of best optimization for this example depends strongly on the underlying cache topology. For instance, when we consider the total energy consumed in the system, for a 4K, 2-way cache, the version that uses only loop permutation and unrolling performs best. Whereas for an 8K, 8-way cache, the most optimized version (that uses all three optimizations) outperforms the rest. In fact, given a search domain for optimizations and a target cache topology, an optimizing compiler can decide which optimizations will be most suitable.

Cache Miss Rates versus Energy Consumptions. We now investigate the correlation between cache miss rate and energy consumption. Figure 6 gives the miss rates for some selected cases. This subsection will make some correlations

Version		Miss Rates			
	$\downarrow\rightarrow$	1-way	2-way	4-way	8-way
original	1K	0.1117	0.1020	0.1013	0.1013
	2K	0.0918	0.0989	0.1013	0.1013
	4K	0.0737	0.0330	0.0245	0.0150
	8K	0.0680	0.0214	0.0117	0.0117
linear transformed	1K	0.0278	0.0119	0.0113	0.0104
	2K	0.0185	0.0107	0.0099	0.0099
	4K	0.0135	0.0100	0.0099	0.0099
	8K	0.0118	0.0099	0.0099	0.0099
unrolled	1K	0.0678	0.0384	0.0359	0.0359
	2K	0.0479	0.0362	0.0359	0.0359
	4K	0.0358	0.0198	0.0145	0.0173
	8K	0.0294	0.0135	0.0077	0.0077
tiled	1K	0.0180	0.0055	0.0039	0.0039
	2K	0.0105	0.0028	0.0016	0.0016
	4K	0.0046	0.0016	0.0012	0.0013
	8K	0.0027	0.0008	0.0007	0.0006

Fig. 6. Miss Rates for the Matrix Multiply Code.

between miss rates and energy consumptions. Let us first consider the miss rates and energy consumption of the original (unoptimized) code. When we move from one cache configuration to another, we have a similar reduction rate for energy as that for miss rate. For instance, going from 1K, 1-way to 1K, 2-way reduces the miss rate by a factor of 1.10 and reduces the energy by the same factor. As another example, when we move from 1K, 1-way to 4K, 8-way, we reduce the miss rate by a factor of 7.45, and the corresponding energy reduction is a factor of 7.20. These results show that the gain in energy obtained by increasing associativity is not offset, in general, by the increasing complexity of the cache topology. As long as a larger or higher-associative cache reduces miss rates significantly (for a given code), we might prefer it, as the negative impact of the additional complexity is not excessive. However, we note that when moving from one cache configuration to another, if there is not a significant change in miss rate (as was the case in our experiments when going from 1K, 4-way to 1K, 8-way), we incur an energy increase. This can be expected as, everything else being equal, a more complex cache consumes more power (due to more complex matching logic).

Next, we investigate the impact of various optimizations for a fixed cache (and memory) topology. The following three measures are used to capture the correlation between the miss rates and energy consumption of the original and optimized versions.

$$\text{Improvement}_m = \frac{\text{Miss rate of the original code}}{\text{Miss rate of the optimized code}},$$

$$\text{Improvement}_e = \frac{\text{Memory energy consumption of the original code}}{\text{Memory energy consumption of the optimized code}},$$

$$\text{Improvement}_t = \frac{\text{Total energy consumption of the original code}}{\text{Total energy consumption of the optimized code}}.$$

In the following discussion, we consider four different cache configurations: 1K, 1-way; 2K, 4-way; 4K, 2-way; and 8K, 8-way. Given a cache configuration, the following table shows how these three measures vary when we move from the original (unoptimized) version to an optimized (tiled) version of the matrix multiply code.

	1K, 1-way	2K, 4-way	4K, 2-way	8K, 8-way
Improvement$_m$	6.21	63.31	20.63	19.50
Improvement$_e$	2.13	18.77	5.75	2.88
Improvement$_t$	1.96	9.27	3.08	1.47

We see that in spite of very large reductions in miss rates as a result of tiling, the reduction in energy consumption is not as high. Nevertheless, it still follows the miss rate. We made the same observation in different benchmark codes as well. We have found that Improvement$_e$ is smaller than Improvement$_m$ by a factor of 2 - 15. Including the core (datapath) power makes the situation worse for tiling (from the energy point of view), as this optimization increases the core energy consumption. Therefore, compiler writers for energy-aware systems can expect an overall energy reduction as a result of tiling, but not as much as the reduction in the miss rate. Thus, optimizing compilers that estimate the miss rate (before and after tiling) statically at compile time can also be used to estimate an approximate value for the energy variation. The following table gives the same improvement measures for the loop unrolled version of the matrix multiply code.

	1K, 1-way	2K, 4-way	4K, 2-way	8K, 8-way
Improvement$_m$	1.65	2.82	1.67	1.52
Improvement$_e$	2.07	3.53	2.07	1.83
Improvement$_t$	2.03	3.37	1.97	1.68

The overall picture here is totally different. First, Improvement$_e$ is larger than Improvement$_m$, which proves that loop unrolling is a very useful transformation from the energy point of view. Including the core power makes only a small difference, as this optimization reduces the core power as well. We should mention that our other experiments (not presented here due to lack of space) yielded similar results. We now look at the loop transformed version of the same code:

	1K, 1-way	2K, 4-way	4K, 2-way	8K, 8-way
Improvement$_m$	4.02	10.23	3.30	1.18
Improvement$_e$	3.42	8.51	2.74	0.99
Improvement$_t$	3.17	6.84	2.32	0.94

Here, Improvement$_e$ closely follows Improvement$_m$. Including the core energy brings the energy improvement down further, as in this example, the loop optimization results in extra operations for the core. In the experiments with other cache configurations, we observed similar trends: Improvement$_e$ generally follows Improvement$_m$; but it is slightly lower. And, Improvement$_t$ is smaller than Improvement$_e$ by a factor of 1.05 to 1.80.

We can conclude that the energy variations do not necessarily follow miss rate variations in the optimized array-dominated codes. More correlations between energy behavior and performance metrics can be found in [5].

3.2   Relative Impact of Hardware and Software Optimizations on Memory Energy

In this section, we focus specifically on memory system energy due to data accesses and illustrate how software and hardware optimizations effect this energy.

Hardware Optimizations. A host of hardware optimizations have been proposed to reduce the energy consumption. In this section, we focus on two cache optimizations, namely, block buffering [20] and cache subbanking [21]. Note that none of these optimizations cause a noticeable negative impact on performance. In the block buffering scheme, the previously accessed cache line is buffered for subsequent accesses [20]. If the data within the same cache line is accessed on the next data request, only the buffer needs to be accessed. This avoids the unnecessary and more energy consuming access to the entire cache data and tag array. Multiple block buffers can be thought of as a small sized Level 0 cache. In the cache subbanking optimization, which is also known as column multiplexing [21], the data array of the cache is divided into several subbanks and only the subbank where the desired data is located is accessed. This optimization reduces the per access energy consumption.

We studied the energy consumed by the matrix multiply code in the data cache with different configurations of block buffers and subbanks (the number of block buffers being either 2, 4 or 8 and the number of sub-banks varying from 1 to 4) for a 4K cache with various associativities. This result showed that increasing the number of sub-banks from one to two provides an energy saving of 45% for the data cache accesses. An additional 22% saving is obtained by increasing the number of sub-banks to 4. It must be observed that the savings are not linear as one may expect. This is because the energy cost of the tag arrays remains constant, while there being a small increase in energy due to additional sub-bank decoding. We found that for block buffering adding a single block buffer reduced the energy by up to 50%. This reduction is achieved by capturing the locality of the buffered cache line, thereby avoiding accesses to the entire data array. However, access patterns in many applications can be regular and repeating across a varied number of different cache blocks. In order to capture this effect, we varied the number of block buffers to two, four, and eight as well. We observed that, for our matrix multiply benchmark, an additional 17% (as compared to a single buffer) energy saving can be achieved using four buffers.

We also found that using a combination of eight block buffers and four subbanks, the energy consumed in 4K (16K) data cache could be reduced on an average by 88% (89%). Thus, such hardware techniques can reduce the energy consumed by processors with on-chip caches. However, if we consider the entire memory system including the off-chip memory energy consumption, the energy savings from these techniques amount to only 4% (15%) when using a 4K (16K) data cache. Thus, it may be necessary to investigate optimizations at the software level to supplement these optimizations.

Combined Optimizations for Memory Energy. It was found that when a combination of different software (loop tiling, loop unrolling, and linear loop transformations) and hardware (block buffering and subbanking) optimizations is applied, tiling performs the best among the three individual compiler optimizations applied in terms of memory system energy across different cache configurations. Since, we mentioned earlier that tiling increases the cache energy consumption, subbanking and block buffering are of particular importance here. For the tiled code, moving from a base data cache configuration to one with eight block buffers and four subbanks reduces the overall memory system energy by around 10%. Thus, it is important to use a combination of hardware and software optimizations in designing an energy-efficient system.

Further, we observed that the linear loop transformed codes exploited the block buffers better than the original code and other optimizations. For example, when using two (eight) block buffers in a 4K 2-way cache, the block buffer hit rate was 69% (82%) as compared to the 55% (72%) for the unoptimized matrix multiply code. Thus, it is also important to choose the software optimizations such that they provide the maximum benefits from the available hardware optimizations.

Overall, we observe that even performance based compiler optimizations provide a significantly higher energy savings as opposed to those gained using the pure hardware optimizations considered. However, a closer observation reveals that hardware optimization become more critical for on-chip cache energy reduction when executing optimized codes. We refer the reader to [3] for more discussion on this topic.

3.3   Impact of Software Optimizations on Instruction Cache

Aggressive compiler optimizations that enhance locality of data accesses tend to increase the energy spent due to instruction accesses (as many of these optimizations reduce the instruction reuse). For studying this impact on instruction energy, we used four different motion estimation codes (Full Search, 3Step-Logarithmic Search, Hierarchical Motion Estimation, and Parallel Hierarchical One-Dimensional Search (PHODS))[33]. These codes also show the importance of choosing appropriate algorithms (i.e., application design) for energy savings. For instance, among the different algorithms employed to perform the motion estimation, the most data-intensive full search code consumes about 8 times more

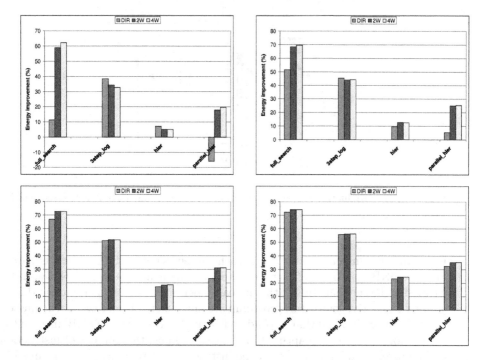

Fig. 7. Energy Reduction (%) due to High-Level Compiler Optimizations for Data Accesses Using Different Cache Configurations (from Top to Bottom, Cache Size of 8KB, 16KB, 32KB, and 64KB).

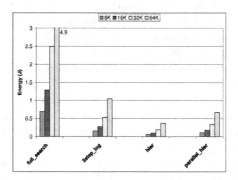

Fig. 8. Energy (J) Consumption due to Instruction Accesses for Two-Way Set-Associative Caches.

energy for data accesses than the most energy-efficient PHODS algorithm when using an 8K direct-mapped data cache.

Further, we observed that, for the direct-mapped data caches, the energy expended during data accesses reduces when cache size is increased from 8KB to 16KB. But, this trend changes with further increase in cache size. This behavior

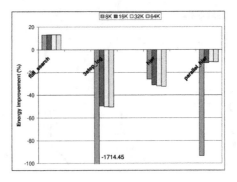

Fig. 9. Energy Reduction (%) for Instruction Accesses Using Two-Way Set-Associative Cache.

is due to the significant reduction in cache misses when cache size increases from 8KB to 16KB resulting in fewer energy-expensive memory accesses. However, for cache sizes larger than 16KB, the increased per-access cache energy cost (due to a larger capacitive load) starts to dominate any benefits from fewer cache misses.

It was also observed that beyond an instruction cache size of 8KB, most of the instruction accesses are captured in the cache. Thus, the number of instruction cache misses is small and most of the instruction related energy is consumed in accessing the instruction cache. Further, it is observed that the energy cost for instruction accesses is comparable to the energy consumed by data accesses for most configurations. This observation is important as most of the state-of-the-art compiler optimizations currently target only improving data accesses.

Next, we tried to apply linear loop transformations, loop unrolling and loop tiling to the motion estimation codes. In optimizing the motion estimation codes, the compiler could not find any opportunities to apply tiling due to imperfectly-nested nature of the loops in these codes. In two of the codes, however, it successfully applied loop unrolling with an unroll factor of 5 and 6. When we analyzed the resulting optimized C codes, we observed that in all of them, there was an expansion in static code size as compared to the original. This is mainly due to loop unrolling and scalar replacement exercised by the compiler to improve cache and register performance.

Figure 7 shows the change in energy consumption due to data accesses after applying the high-level optimizations. It is observed that the energy reduction is most significant for the full search algorithm that is most data-intensive. This reduction is due to the significant decrease in number of data accesses as a result of improved locality. For example, scalar replacement converts memory references to register accesses. However, this also leads to an increase in dynamic instruction count. We can also see from Figure 7 that, except for one case, high-level compiler optimizations improve the data energy consumption for all motion estimation codes in all configurations. The average data energy reduction over all studied cache sizes is 30.9% for direct-mapped caches, 39.4% for 2-way caches and 39.8% for 4-way caches. Our experiments also show that in `hier` and `parallel_hier`, after the optimizations, there is an increase in the number of conflict misses (as we do not use array padding). In particular, with `parallel_hier`, when the cache size is very small and cache is direct-mapped, these conflict misses offset any benefits that would otherwise be obtained from improved data locality, thereby degrading the performance from the energy perspective. Increasing the associativity eliminates this conflict miss problem.

It can be observed from Figure 9 that the energy consumed by instruction accesses increases on an average by 466%, 30% and 32% for the `3step_log`, `hier` and `parallel_hier` optimized codes, respectively. The main reason for this increase is the aggressive use of scalar replacement in these codes. While this optimization helps caches and registers to exploit temporal data locality, the use of scalar replacement in the inner loops of a given nest structure leads to significant increase in the dynamic instruction count. For example, in the optimized version of `hier`, dynamic instruction count increased to 62 million from 46 million. In contrast, the energy consumed by instruction accesses for `full_search` decreases by 13%. The data access pattern for `full_search` is more regular as compared to the other algorithms. Consequently, the MIPSpro optimizer was less aggressive with scalar replacement. Further, the application of loop unrolling on `full_search` reduced the number of branch instructions.

The overall impact of the optimizations considering both the instruction and data accesses was also studied. It was observed that the optimizations decrease the energy consumption by 26% for `full_search` on the average. However, due to the detrimental impact on energy consumed by instruction accesses, the overall energy consumption increased by approximately 153%, 11% and 43% for `3step_log`, `hier` and `parallel_hier`, respectively.

3.4   Technological Trends

We now investigate the relative magnitudes of the core power and the memory power for a specific optimization: loop tiling. Figure 10 shows the memory energy for different values of E_m (energy cost per access) for four different cache organizations. Note that $E_m = 4.95 \times 10^{-9}$J is a reasonable value for today's technology and is based on the Cypress SRAM CY7C1326-133. The lowest value that we experiment with in this section (4.95×10^{-11}) corresponds to the magnitude of energy per first-level on-chip cache access with current technology. E_m can be reduced through better process technology, reduction in physical distance between memory and core (or using new memory implementation techniques). Considering the fact that large amounts of storage capacity are coming closer to the CPU, we expect to see lower E_m values in the future. This can make the energy consumed in the core larger than the energy consumed in memory. Even for $E_m = 4.95 \times 10^{-9}$, in a 4K, 2-way cache, the two energy values (core and memory) are the same. Given the fact that optimizations such as tiling are very popular and used by commercial compilers extensively, we predict that research (hardware and software) on reducing the core power will become even more important. We refer the reader to [4] for a thorough discussion of the impact of compiler optimizations with varying energy cost per access values.

Confi-guration	Memory Energy (J)							
	4.95×10^{-11}	2.475×10^{-10}	4.95×10^{-10}	2.475×10^{-9}	4.95×10^{-9}	2.475×10^{-8}	4.95×10^{-8}	2.475×10^{-7}
1K, 1-way	0.0164	0.0462	0.0836	0.3821	0.7553	3.7408	7.4727	37.3280
1K, 4-way	0.0090	0.0154	0.0234	0.0872	0.1671	0.8056	1.6038	7.9892
4K, 1-way	0.0194	0.0270	0.0364	0.1119	0.2062	0.9611	1.9047	9.4533
4K, 2-way	0.0183	0.0210	0.0243	0.0507	0.0837	0.3477	0.6778	3.3183

Fig. 10. Impact of Different E_m Values on Total Memory System Energy Consumption for Tiled Matrix Multiply.

4   Future Challenges

Software content is continuing to form increasing portions of energy-constrained systems. Thus, it is of utmost importance to develop a closely intertwined monitoring and optimizing mechanism involving the OS, compiler and communication software to provide an integrated approach to optimizing the overall system power. In particular, we see potential for the following areas:

- *Fast and Accurate Energy Models:* It remains extremely important to develop accurate and fast energy models for different system components. Such models can be utilized in power estimation and optimization tools (such as cycle-accurate energy simulators and profilers), and can also be employed in an optimizing compiler framework that specifically targets power. Since an optimizing compiler may need to estimate energy for a given code many times during compilation, such models should be efficient. In addition, such models need to provide accurate information so that they can guide high-level and low-level compiler optimizations.
- *Energy-Aware Compilation Framework:* It is important to design and implement compilation frameworks for high-quality power-aware code generation. Such a framework should take into account the power constraints known at compile time as well as the power constraints that change dynamically during the run time. Among the important optimization problems are minimizing memory requirements, improving data locality and optimizing data decomposition in multiple memory spaces during static (compile time) power-aware compilation, and minimizing bus switching activities. It is also important to consider dynamic situations where the compiler does not know the possible ranges of power constraints at compile time. In such cases, the compiler can obtain dynamic power constraint information from the operating system and can dynamically change the run time activity for reducing power consumption.
- *Power-Aware Operating Systems:* Operating system can play a major role in power reduction by providing feedback to the compiler, architecture and communication subsystems regarding dynamic system condition. It can be used for both coarse-level and fine-level power monitoring and management. We anticipate scheduling, synchronization and memory management techniques to play a major role in minimizing overall system energy. Already,

there are pointers in literature [30] that illustrate the promising potential of such techniques.

- *Power-Conscious Communication System:* With increasing mobility of power-aware systems, the need for addressing energy optimizations for wireless communication is becoming critical. It is also predicted that the RF components associated with communication will dominate the energy budget of future mobile devices. A coordinated effort between different layers of the OS and the communication protocol layers seems to be essential.
- *Unified Optimizations for Energy:* So far, majority of the efforts focussed on specifically hardware or software. However, the improvements in both areas indicate the limitations of these techniques and suggest a unified approach that involves both hardware and software. We envision a system in which the software is aware of the low power features of hardware components, and dynamically adapts itself or the hardware to optimize energy. Similarly, the hardware can provide a feedback mechanism to the software that enables the latter to initiate dynamic energy optimizations.

5   Conclusions

The goal of this study is to investigate the interaction and influence of hardware and software optimizations on system energy. Towards this goal, we evaluate three widely used high-level compiler optimizations from energy perspective considering a variety of cache configurations including conventional direct-mapped and associative caches as well as new energy-efficient subbanking and block buffering designs.

Our results show that, as far as reducing the overall system energy is concerned, software optimizations are more effective. However, they have an important negative effect: they increase the energy consumption on datapath (core) and instruction cache. Consequently, hardware-based energy optimizations can be used to mitigate that effect. This preliminary study identifies developing hierarchical, fast, and accurate energy models as an important area of future research.

References

1. J. Bunda, W. C. Athas, and D. Fussell. Evaluating power implication of CMOS microprocessor design decisions. In Proc. the 1994 International Workshop on Low Power Design, April 1994.
2. R. Y. Chen, R. M. Owens, and M. J. Irwin. Validation of an architectural level power analysis technique. In Proc. the 35th Design Automation Conference, June 1998.
3. G. Esakkimuthu, N. Vijaykrishnan, M. Kandemir, and M. J. Irwin. Memory system energy: Influence of hardware-software optimizations. In Proc. ACM/IEEE International Symposium on Low Power Electronics and Design, Rapallo/Portofino Coast, Italy, July, 2000.

4. M. Kandemir, N. Vijaykrishnan, M. J. Irwin, and H. S. Kim. Towards energy-aware iteration space tiling. In Proc. the Workshop on Languages, Compilers, and Tools for Embedded Systems, Vancouver, B.C., June, 2000.

5. M. Kandemir, N. Vijaykrishnan, M. J. Irwin, and W. Ye. Influence of Compiler Optimizations on System Power. Submitted to IEEE Transactions on VLSI, March 2000.

6. M. Wolfe. High Performance Compilers for Parallel Computing, Addison Wesley, CA, 1996.

7. W. Ye, N. Vijaykrishnan, M. Kandemir, and M. J. Irwin. The design and use of SimplePower: a cycle-accurate energy estimation tool. In Proc. the 37th Design Automation Conference, Los Angeles, CA, June 5–9, 2000.

8. G. Albera and R. I. Bahar. Power and performance tradeoffs using various cache configurations. In Proc. *Power Driven Micro-architecture Workshop,* in conjunction with *ISCA '98,* Barcelona, Spain, June 1998.

9. D. H. Albonesi. Selective cache ways: On-demand cache resource allocation. In Proc. *the 32nd International Symposium on Microarchitecture,* pp. 248–259, November 1999.

10. F. Balasa, F. Catthoor, and H. De Man. Exact evaluation of memory area for multi-dimensional processing systems. In Proc. *the IEEE International Conference on Computer Aided Design,* Santa Clara, CA, pages 669–672, November 1993.

11. D. Brooks, V. Tiwari, and M. Martonosi. Wattch: A framework for architectural-level power analysis and optimizations. In Proc *the 27th International Symposium on Computer Architecture,* Vancouver, British Columbia, June 2000.

12. R. Gonzales and M. Horowitz. Energy dissipation in general purpose processors. *IEEE Journal of Solid-State Circuits,* 31(9):1277–1283, Sept 1996.

13. M. K. Gowan, L. L. Biro, and D. B. Jackson. Power considerations in the desing of the Alpha 21264 microprocessor. In Proc. *the Design Automation Conference,* San Francisco, CA, 1998.

14. N. Vijaykrishnan, M. Kandemir, M. J. Irwin, H. Y. Kim, and W. Ye. Energy-driven integrated hardware-software optimizations using SimplePower. In Proc. *the International Symposium on Computer Architecture,* Vancouver, British Columbia, June 2000.

15. K. Roy and M. C. Johnson. Software design for low power. *Low Power Design in Deep Sub-micron Electronics,* Kluwer Academic Press, October 1996, Edt. J. Mermet and W. Nebel, pp. 433–459.

16. M. J. Irwin and N. Vijaykrishnan. Low-power design: From soup to nuts. Tutorial Notes, *ISCA,* 2000.

17. V. Zyuban and P. Kogge. Inherently lower-power high-performance superscalar architectures, submitted to *IEEE Transactions on Computers.*

18. http://www.transmeta.com/articles/

19. http://www.intel.com/pressroom/archive/releases/mp042400.htm

20. J. Kin et al. The filter cache: An energy efficient memory structure. In Proc. *International Symposium on Microarchitecture,* December 1997.

21. C.-L. Su and A. M. Despain. Cache design trade-offs for power and performance optimization: A case study, In Proc. *International Symposium on Low Power Electronics and Design,* pp. 63–68, 1995.

22. M. B. Kamble and K. Ghose. Analytical energy dissipation models for low power caches. In Proc. *International Symposium on Low Power Electronics and Design,* pages 143–148, 1997.

23. K. Itoh, K. Sasaki, and Y. Nakagome. Trends in low-power ram circuit technologies. *Proceedings of the IEEE,* pages 524 –543, Vol. 83. No. 4, April 1995.

24. Advanced configuration and power interface specification. Intel, Microsoft, and Toshiba, Revision 1.0b, Feb 2, 1999.
25. L. Benini, A. Bogliolo, S. Cavallucci, and B. Ricco. Monitoring system activity for os directed dynamic power management. In *Proceedings of the International Symposium on Low Power Electronics and Design*, pages 185–190, 1998.
26. D. Burger and T. Austin. The simplescalar tool set, version 2.0. Technical report, Computer Sciences Department, University of Wisconsin, June, 1997.
27. R. Y. Chen, R. M. Owens, and M. J. Irwin. Architectural level power estimation and design experiments. *To appear in ACM Transactions on Design Automation of Electronic Systems.*
28. R. Y. Chen, R. M. Owens, and M. J. Irwin. Validation of an architectural level power analysis technique. In *Proceedings of the 35th Design Automation Conference*, pages 242–245, June 1998.
29. F. Douglis, P. Krishnan, and B. Marsh. Thwarting the power-hungry disk. In *Proceedings of the 1994 Winter USENIX Conference*, pages 293–306, January 1994.
30. J. Flinn and M. Satyanarayanan. Powerscope: A tool for profiling the energy usage of mobile applications. In *Proceedings of the 2nd IEEE Workshop on Mobile Computing Systems and Applications*, 1999.
31. J. Hezavei, N. Vijaykrishnan, and M. J. Irwin. A comparative study of power efficient SRAM designs. In *to appear in Proc. of Great Lakes Symposium on VLSI*, 2000.
32. K. Inoue, T.Ishihara, and K. Murakami. Way-predicting set-associative cache for high performance and low energy consumption. In *Proceedings of the International Symposium on Low Power Electronics and Design*, pages 273–275, 1999.
33. M. J. Irwin and N. Vijaykrishnan. Energy issues in multimedia systems. In *Proc. of Workshop on Signal Processing System*, pages 24–33, October 1999.
34. K. Itoh, K. Sasaki, and Y. Nakagome. Trends in low-power ram circuit technologies. *Proceedings of IEEE*, 83(4):524 –543, April 1995.
35. K. Li, R. Kumpf, P. Horton, and T. Anderson. A quantitative analysis of disk drive power management in portable computers. In *Proceedings of the 1994 Winter USENIX Conference*, pages 279–292, January 1994.
36. J. R. Lorch and A. J. Smith. Software strategies for portable computer energy management. *IEEE Personal Communications*, pages 60–73, June 1998.
37. M. Wolf and M. Lam. A data locality optimizing algorithm. In *Proceedings of ACM SIGPLAN 91 Conference Programming Language Design and Implementation*, pages 30–44, June 1991.

Early Power Estimation
for System-on-Chip Designs

M. Lajolo[1], L. Lavagno[2], M. Sonza Reorda[3], and M. Violante[3]

[1]NEC C&C Research Labs, Princeton, NJ, USA
lajolo@ccrl.nj.nec.com
[2]DIEGM, Università di Udine, Udine, Italy
lavagno@uniud.it
[3]Dipartimento di Automatica e Informatica, Politecnico di Torino, Torino, Italy
http://www.cad.polito.it

Abstract. Reduction of chip packaging and cooling costs for deep sub-micron System-On-Chip (SOC) designs is an emerging issue. We present a simulation-based methodology able to realistically model the complex environment in which a SOC design operates in order to provide early and accurate power consumption estimation. We show that a rich functional test bench provided by a designer with a deep knowledge of a complex system is very often not appropriate for power analysis and can lead to power estimation errors of some orders of magnitude. To address this issue, we propose an automatic input sequence generation approach based on a heuristic algorithm able to upgrade a set of test vectors provided by the designer. The obtained sequence closely reflects the worst-case power consumption for the chip and allows looking at how the chip is going to work over time.

1 Introduction

In the last years, new technologies allowed to integrate entire systems on a single chip, thus causing the appearance of new electronic devices, called System-on-Chips (SOCs). SOC products represent a real challenge not just from the manufacturing point of view, but even when design issues are concerned.

To cope with SOC design requirements, researchers developed co-design environments, whose main characteristic is to allow the designer to quickly evaluate the costs and benefits of different architectures, including both hardware and software components. To perform design space exploration, efficient and accurate analysis tools are required. In particular, power consumption is a major design issue and thus it mandates the availability of effective power estimation tools. Moreover, it is known that power analysis and optimization during the early design phases, starting from the system level, can lead to large power savings [1], [2], [3]. As a consequence, several efforts have been devoted to develop methodologies for system-level power estimation.

Early works on low-power design techniques have mostly focussed on estimating and optimizing power consumption in the individual SOC components (application-specific hardware, embedded software, memory hierarchy, buses, etc.) separately.

D. Soudris, P. Pirsch, and E. Barke (Eds.): PATMOS 2000, LNCS 1918, pp. 108–117, 2000.

Various power estimation and minimization techniques for hardware at the transistor, logic, architecture, and algorithm levels have been developed in the recent years, and are summarized in [1], [2], [3], [4], [5].

Recently, researchers have started investigating system-level trade-off and optimizations whose effects transcend the individual component boundaries. Techniques for synthesis of multiprocessor system architectures and heterogeneous distributed HW/SW architectures for real-time specifications were presented in [6], [7]. In [8] and [9], separate execution of an instruction set simulator (ISS) based software power estimator, and a gate-level hardware power estimator were used to drive exploration of tradeoffs in an embedded processor with memory hierarchy, and to study HW/SW partitioning tradeoffs. The traces for the ISS and hardware power estimator were obtained from timing-independent system-level behavioral simulation. To cope with the heterogeneous components SOCs usually embed, in [10] a tool is proposed based on the concurrent and synchronized execution of multiple power estimators that analyze different parts of the SOC, driven by a system-level simulation master.

To assist designers in defining a suitable input streams for power estimation purposes, we developed an algorithm that improves an initial input sequence (either provided by designers or randomly generated) so that it activates all the functions of the system while trying to maximize the power it consumes. As a result, being the sequence able to exhaustively activate the whole system, more accurate power figures can be obtained.

To generate suitable input sequences for SOCs where hardware and software tasks are mixed together, we need a system representation that abstracts architectural details. We achieve this goal by developing our algorithm in a co-design environment. The benefits that stem from this solution are twofold:

1. by abstracting the behavior from the architecture, we deal with a high-level system description that we can simulate with a very low cost in terms of CPU time;
2. the sequences we compute are reusable, i.e., the same sequence can be used during power estimation at every level of abstraction. This allows us to use the developed test sequences, for example, for evaluating the power consumption of both the algorithm and the architecture implementing it.

A prototype of the proposed input sequence generation technique has been implemented using the POLIS [11] co-design tool. It is based on a heuristic algorithm that automatically generates an input sequence able to exercise as much as possible of the specification, by interacting with a simulator executing a specification of the system under analysis.

The remainder of the paper is organized as follows. In Section 2 we motivate our work. In Section 3 we describe the system representation we exploited. Section 4 describes the optimization we developed, while Section 5 reports some preliminary experimental results assessing the effectiveness of the proposed approach. Finally, Section 6 draws some conclusions.

2   Motivations

In a SOC design, the entire system is implemented using a single chip module. Due to this fact, the power budget requirements will be dictated by the most power consuming modules present on the chip and this will determine the cooling requirements of the entire chip. This is radically different with respect to a multi-chip implementation in which each chip component can a priori have different power budget and cooling requirements. In order to cope with this scenario it is necessary to provide early and accurate power estimation capabilities in the system level design methodology.

It has already been demonstrated [10] that power consumption in hardware and software cannot be addressed separately due to the effects of system resources that they share (such as the cache and the buses). The most promising solution is the use of high-level hardware/software co-simulation in order to address power analysis in hardware and software concurrently.

The approach that is presented in this paper is based on enriching a co-simulation based power estimation methodology with a tool able to generate vectors in order to improve the coverage of the functional specification.

It is important to underline that the input sequence generator deals with the entire design and not with single modules separately. The reason behind this choice is that if the test generator can deal with the entire design, the resulting tests will only include vectors that are possible in normal functional modes. On the other hand, if test generation is performed on single modules, the generated vectors may contain illegal sequences that cannot be applied to the module from the primary inputs of the design.

Moreover, since the system-level power estimation techniques proposed so far are essentially simulation-based ones, they require the system to be simulated under typical input sequences provided by designers. The definition of a proper set of input stimuli is a very time consuming and difficult task, since all the details of the design must be understood for generating suitable input sequences. The right trade-off between designer's time and power estimation accuracy is often difficult to find, and this often results in power figures that underestimate the actual power consumption. Moreover, in the generation of typical input sequences the designer may be *biased* by his knowledge of the desired system or module behavior, so that he often fails in identifying input sequences really able to activate possible critical points in the description.

3   System Representation

In POLIS the system is represented as a network of interacting Co-design Finite State Machines (CFSMs). CFSMs extend Finite State Machines with arithmetic computations without side effects on each transition edge. The communication edges between CFSMs are events, which may or may not carry values. A CFSM can execute a transition only when an input event has *occurred*.

A CFSM network operates in a *Globally Asynchronous Locally Synchronous* fashion, where each CFSM has its own *clock*, modeling the fact that different resources (e.g., HW or SW) can operate at widely different speeds. CFSMs communicate via non-blocking depth-one buffers. Initially there is no relation between local clocks and physical time, that is defined later by a process called *architectural mapping*.

This involves allocating individual CFSMs to computation resources and assigning a scheduling policy to shared resources. CFSMs implemented in hardware have local clocks that coincide with the hardware clocking. CFSMs implemented in software have local clocks with a variable period, that depends both on the execution delay of the code implementing each transition and on the chosen scheduling policy (e.g., allowing task preemption).

For each CFSM in the system description, a Control Flow Graph representation (Fig. 1) is computed, called S-Graph.

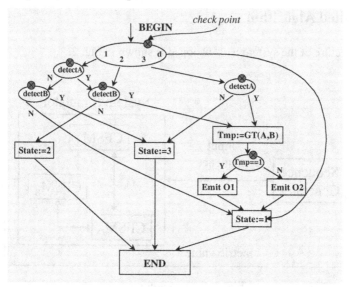

Fig. 1. An example of S-Graph

An S-Graph has a straightforward and efficient implementation as sequential code on a processor. In the C code that POLIS is able to generate from the S-Graph representation, each statement is almost in a 1-to-1 correspondence with a node in the S-Graph. Thus the model can be simulated as native object code very efficiently.

Our goal is to compute input sequences that exercise as much as possible of the system specification, while trying to maximize the power consumption; we thus adopt statement coverage as a metric of the activity in the specification. Moreover, as far as power estimation is concerned, we characterize each statement with a value representing the cost in terms of energy required to execute the statement.

In order to compute the adopted metric and to gather information during simulation of an input sequence, we instrument the simulation model by inserting:

1. *trace points* associated to each statement in the S-Graph. A trace point is a fragment of code that records the number of times the statement has been executed;
2. *check points* associated to each test in the S-Graph. A check point is a fragment of code that records the number of times the test has been executed.

During simulations, we use *trace points* to evaluate statement coverage, while we *check points* to direct the search towards repeated traversal of test nodes that have still uncovered outgoing branches.

In our simulation model, we assume that a new vector is placed on the system inputs if and only if the response to the previous input has already been computed and the system is in a steady state.

4   Adopted Algorithm

The architecture of the system we developed is shown in Fig. 2.

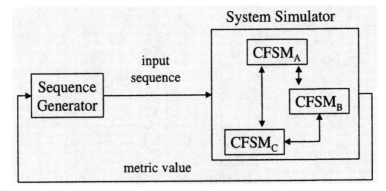

Fig. 2. System architecture

A Sequence Generator computes some input vectors and sends them to the System Simulator. The model specification is then simulated and the value of the adopted metric is computed. Finally, this value is sent back to the Sequence Generator. By exploiting this feedback architecture, the Sequence Generator can grade input vectors according to the associated metric figures.

```
sequence hill_climber( sequence initial_sequence )
{ sequence current_solution = initial_sequence;
  int iter, weight, new_weight;
  weight = evaluate( current_solution );
  for( iter = 0; iter < MAX_ITER; iter++ )
  { modify( current_solution );
    new_weight = evaluate( current_solution )
    if( new_weight > weigth )
      weight = new_weight;
    else
      revert_modification( current_solution );
  }
  return( current_solution );
}
```

Fig. 3. The optimization algorithm

Given a metric to measure the goodness of a sequence of input vectors, we adopted a heuristic algorithm to implement the Sequence Generator, whose goal is to find a sequence that maximizes the value of the adopted metric. In order to reuse the information, e.g., input sequence, already provided by designers we adopted a random mutation hill climber, whose pseudo-code is reported in Fig. 3. The algorithm randomly modifies the initial sequence provided by designers and evaluates it. A sequence is accepted if and only if it improves the adopted metric.

We apply a sequence of vectors to the system inputs. Each vector is a set of events that are concurrently applied to the system input at a given time. We coded the sequence as a matrix of bits, where SEQUENCE_LENGTH is the number of rows in the matrix and thus it represents the number of vectors to be applied on the system inputs. Conversely, N_INPUTS is the number of system inputs and thus the number of columns in the matrix. The number of bits use to represent an input event e is selected as follows:

1. 1 bit if e is an input event without value
2. $\log_2 n$ bits, where n is the number of different values associated to the event e, if e is an input event with value.

Given and initial solution, we randomly modify it and accept the new sequence if and only if it improves the metric we adopted.

The metric we use in this paper is defined as follows:

$$f(S) = K_1 \cdot \left(C_1 \cdot \sum_{i=0}^{N} OP_i + C_2 \cdot \sum_{j=0}^{M} \sum_{i=0}^{Nj} (1 - OP_i) \cdot NT_j \right) + K_2 \cdot Power(S) \qquad (1)$$

Where:
1. S is the input sequence to be evaluated;
2. N is the number of trace points;
3. M is the number of check points;
4. N_j is the number of trace points associated to checkpoint j;
5. OP_i is equal to 1 if the trace point i has been traversed, 0 otherwise;
6. NT_j is the number of times the check point associated to the test j has been executed during the simulation of the input sequence S;
7. $Power(S)$ is the power consumption of the sequence S;
8. C_1, C_2, K_1 and K_2 are constants.

The metric is intended to maximize the number of statements covered by the input sequence, while maximizing the power consumption. This is motivated by observing that a fair measure of the system power consumption can be attained only if the input sequence is capable of effectively activating the entire system. An input sequence that fails in activating a subset of the system can lead to power figures that do not reflect actual power consumption.

The first part of (1) measures how many statements the sequence S traverses and tends to favor sequences that execute those tests whose outgoing branches have not yet been covered. In order to preserve the already covered statements while trying to cover new ones the first part must dominate the second (in the experiments we used C_1=1,000 and C_2=10). Conversely, the second part of (1) is intended to take into account the power consumed by the application of S to the system. In particular, this term tries to favor those vectors increasing the power consumption.

Given the previously reported considerations, the number of covered statements prevails over the power consumption and therefore $K_1 \gg K_2$.

5   Experimental Results

We implemented a prototype of the proposed algorithm, called *Hill Climber Test Bench Generator* (HC-TBG), in C language.

Using this prototype, we performed a set of experiments, whose purpose was to assess the effectiveness of the proposed approach; the preliminary experiments have been run on a set of small benchmarks. All the results have been gathered on a Sun UltraSparc 5/360 running at 360 MHz and equipped with 256 Mbytes of RAM.

We considered three control-dominated benchmarks: a belt control system, a traffic light controller and a dashboard system, whose characteristics are reported in Table 1, in terms of number of CFSMs and number of statements for each CFSM.

Table 2 reports the results gathered with our algorithm. We have compared the metric figures our algorithm attains with the ones attained by random sequences and, when available, with functional vectors provided by designers.

In Table 2, the column *Benchmark* reports the system under analysis, *Vec* reports the number of vectors in the input sequence, while *CPU* reports the time spent for running HC-TBG. The remaining columns report the statement coverage (*S*) and the energy consumption (*E*) attained by respectively HC-TBG, Random and Functional generated sequences.

Table 1. Benchmarks characteristics

Benchmark	CFSM	Statements [#]
Belt Controller	BELT_CONTROLLER	31
	TIMER	25
Traffic Light Controller	CONTROLLER	66
	TIMER	13
Dashboard	BELT	25
	DISPLAY	73
	FRC	26
	FUEL	35
	ODOMETER	18
	TIMER	75
	SPEEDOMETER3	17
	SPEEDOMETER4	17

As shown in Table 2, HC-TBG sequences are far more effective than random generated one, and better than the functional ones.

To better investigate the effectiveness of the approach we propose, we carried out a second set of experiments on the Dashboard benchmark. In particular, we selected two partitioning and synthesized the corresponding hardware/software systems. We then simulated the systems with the HC-TBG and functional sequences already adopted in the previous experiment. Table 3 summarizes the attained results, where the energy breakdown is reported. In all cases HC-TBG is able to provide results at least comparable with those obtained with functional vectors. For 3 CFSMs out of 8 (implemented either in software or in hardware), HC-TBG attains an energy estimation that is 2 or 3 orders of magnitude higher than that attained by functional vectors, thus prompting the importance of having input sequences able to activating most of the system functionality. For the remaining CFSMs the two approaches attain comparable results; in these cases we can thus conclude that the functional vectors provided by designers are already able to produce good power estimations.

Table 2. Statement coverage and Energy results

Benchmark	Vec [#]	CPU [s]	HC-TBG E [#]	HC-TBG S [%]	Random E [#]	Random S [%]	Functional E [#]	Functional S [%]
Belt Controller	1,000	408	2,945,741	89.3	2,433,689	52.7	n.a	n.a.
Traffic Light Controller	1,000	441	3,602,133	94.9	1,819,476	83.5	n.a.	n.a.
Dashboard	1,000	12,696	20,311,073	80.4	16,384,701	72.7	14,061,184	71.7

Table 3. The synthesized version of Dashboard

CFSM	Partitioning 1 Impl.	Partitioning 1 HC-TBG [μJ]	Partitioning 1 Functional [μJ]	Partitioning 2 Impl.	Partitioning 2 HC-TBG [μJ]	Partitioning 2 Functional [μJ]
BELT	SW	1131.9	2.1	HW	$5.1 \cdot 10^{-4}$	$9.1 \cdot 10^{-5}$
DISPLAY	SW	6.1	8.7	SW	6.2	8.7
FRC	SW	3736.3	3736.3	SW	3736.3	3736.3
FUEL	SW	2004.4	1930.6	SW	2008.9	1930.6
ODOMETER	SW	2033.3	41.9	HW	$1.7 \cdot 10^{-2}$	$9.8 \cdot 10^{-4}$
TIMER	SW	1951.6	1951.6	SW	1951.6	1951.6
SPEEDOMETER3	SW	1671.2	34.7	HW	$3.1 \cdot 10^{-2}$	$1.1 \cdot 10^{-3}$
SPEEDOMETER4	SW	1667.8	168.4	HW	$3.1 \cdot 10^{-2}$	$3.7 \cdot 10^{-3}$
TOTAL		**12,198.2**	**7,874.3**		**7,703.0**	**7,627.2**

The energy figures for Partitioning 1 are also compared in Fig. 4.

6 Conclusions

This paper proposed an algorithm for computing input sequences intended for simulation-based system-level power estimation of SOC design.

The approach is able to upgrade test vectors given by designers with ad-hoc vectors generated by a heuristic algorithm able to cover much more extensively the system-level specification with particular emphasis dedicated to the most consuming parts.

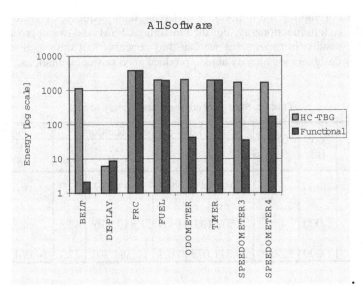

Fig. 4. Comparing energy figures

The algorithm can be exploited since the early design phases; it indeed deals with the system behavior only, while its architecture is neglected. Moreover, the sequences it produces can be exploited in the following design phases, when a more detailed description of the system is available, and thus providing more accurate power estimation figures.

We have presented experimental results that show a difference of 2 or 3 orders of magnitude on a reasonably complex case study, which confirms the usefulness of the methodology.

The methodology proposed can be very useful in order to model much more extensively the environment in which the system operates by taking into account more input sequences with respect to the ones that can be thought by the designer.

Moreover, an automatic test bench generation approach can also be useful in order to predict the power dissipated in the chip during the SOC manufacturing test (the test at the end of the production of the chip), where the activity produced in the chip, and hence the chip power consumption, can be much higher than during normal operation.

References

1. A. R. Chandrakasan, R. W. Brodersen, Low Power Digital CMOS Design, Kluwer Academic Publishers, 1995
2. J. Rabaey, M. Pedram, Low Power Design Methodologies, Kluwer Academic Publishers, 1996
3. L. Benini, G. De Micheli, Dynamic Power Management: Design Techniques and CAD Tools, Kluwer Academic Publishers, 1997
4. J. Monteiro, S. Devadas, Computer-Aided Design techniques for Low Power Sequential Logic Circuits, Kluwer Academic Publishers, 1996

5. E. Macii, M. Pedram, F. Somenzi, High-level power modeling, estimation and optimization, Proc. Design Automation Conference, pp. 504-511, 1997
6. D. Kirkovski, M. Potkonjak, System-level synthesis of low-power hard real-time systems, Proc. Design Automation Conference, pp. 697-702, 1997
7. B. Dave, G. Lakshminarayana, N. H. Jha, COSYN: Hardware-software co-synthesis of embedded systems, Proc. Design Automation Conference, pp. 703-708, 1997
8. Y. Li, J. Henkel, A framework for estimating and minimizing energy dissipation of embedded HW/SW systems, Proc. Design Automation Conference, pp. 188-193, 1998
9. J. Henkel, A power hardware/software partitioning approach for core-based embedded systems, Proc. Design Automation Conference, pp. 122-127, 1999
10. M. Lajolo, L. Lavagno, A. Raghunathan, S. Dey, Efficient Power Co-estimation Techniques for System-on-Chip Design, Proc. Design Automation and Test in Europe, pp. 27-34, 2000
11. F. Balarin et al., Hardware-Software Co-design of Embedded Systems: The POLIS Approach, Kluwer Academic Publishers, 1997
12. H. Hsieh, A. Sangiovanni-Vincentelli, et al. Synchronous equivalence for embedded systems: a tool for design exploration, Proc. ICCAD 99.

Design-Space Exploration of Low Power Coarse Grained Reconfigurable Datapath Array Architectures

R. Hartenstein, Th. Hoffmann, and U. Nageldinger

Computer Structures Group (Rechnerstrukturen), Informatik

University of Kaiserslautern, D-67653 Kaiserslautern, Germany

hartenst@rhrk.uni-kl.de - http://xputers.informatik.uni-kl.de - Fax: +49 631 205 2640

Abstract. Coarse-grain reconfigurable architectures promise to be more adequate for computational tasks due to their better efficiency and higher speed. Since the coarse granularity implies also a reduction of flexibility, a universal architecture seems to be hardly feasible, especially under consideration of low power applications like mobile communication. Based on the KressArray architecture family, a design-space exploration system is being implemented, which supports the designer in finding an appropriate architecture featuring an optimized performance / power trade-off for a given application domain. By comparative analysis of the results of a number of different experimental application-to-array mappings, the explorer system derives architectural suggestions. This paper proposes the application of the exploration approach for low power KressArrays. Hereby, both the interconnect power dissipation and the operator activity is taken into account.

1   Introduction

Many of today's application areas, e.g. mobile communication, require very high performance as well as a certain flexibility. It has shown, that the classic ways of realizing the associated algorithms, ASIC implementation and microprocessors, are often not adequate, as microprocessors cannot provide the performance, while ASIC implementations lack flexibility.

As a third way of implementation, reconfigurable computing has gained importance in the recent years. It is expected, that in order to obtain sufficient flexibility for high production volume most future SoC implementations need some percentage of reconfigurable circuitry [1]. Reconfigurable computing even has the potential to question the dominance of the microprocessor [2].

Early approaches in reconfigurable computing were based on Field-Programmable Gate Arrays (FPGAs), which provide programmability at bit-level. These solutions soon turned out to have some disadvantages for computing applications, as complex operators have to be composed from bit-level logic blocks. This leads to the following drawbacks among others:

- A large amount of configuration data is needed, which makes fast configuration hard and increases power dissipation during configuration.
- As the operators are made of several logic blocks, regularity is mostly lost and an extra routing-overhead occurs. This extra routing also increases the power dissipation of the circuit during run-time.

D. Soudris, P. Pirsch, and E. Barke (Eds.): PATMOS 2000, LNCS 1918, pp. 118-128, 2000.

To encounter the disadvantages of FPGA-based solutions, coarse-grain reconfigurable architectures have been developed for computational applications [3], [4], [5], [6], [7], [8], [9], [10]. These devices are capable of implementing high-level operators in their processing elements, featuring multiple-bit wide datapaths. Coarse grain reconfigurable architectures avoid several drawbacks of FPGAs. Featuring relatively few powerful operators instead of many logic blocks, coarse grain architectures need much less configuration data. Also, as the processing elements can be implemented in an optimized way, both the expected performance and the power dissipation are lower than for FPGAs, as shown in [11]. However, for the use of a coarse granularity some problems still have to be solved. to cope with the reduced flexibility compared to FPGA-based solutions:

- Processing elements of coarse-grain architectures are more "expensive" than the logic blocks of an FPGA. While it is possible for FPGA mappings, that a number of logic blocks is unused or cannot be reached by the routing resources, especially the latter situation is quite annoying for coarse-grain architectures due to the fewer processing elements of higher area consumption.

- While the multi-bit datapath applies well to operators from high-level languages, operations working on smaller word-lengths and especially bit manipulation operations are either weakly supported, or need a sophisticated architecture. If such operations occur, like e.g. in data compression algorithms, they may result in a complex implementation requiring several processing elements, or, in difficult architectural requirements.

- Although the power consumption caused by the routing resources can be expected to be lower than for FPGAs [11], the problem of a careful architectural design of the interconnect network, which provides both low power and adequate flexibility remains.

Due to these problems, the selection of architectural properties like datapath width, routing resources and operator repertory is a general problem in the design of coarse-grain architectures. Thus, a design space exploration is done in many cases, to determine suitable architectural properties. As the requirements to the architecture are mostly dependent on the set of typical applications to be mapped onto it, previous efforts use normally a set of example applications, e.g. DES encryption, DCT, or FIR filters. Examples for such approaches are published in [9] and [10]. According to these methods,

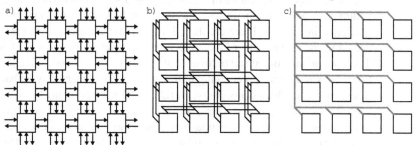

Fig. 1. Three levels of interconnect for KressArray architectures: a) nearest neighbor links, b) backbuses in each row or column (for more examples see figure 2), c) serial global bus.

coarse-grain architectures are often optimized for a specific application area, like DSP or multimedia applications.

While most of these explorations focus on performance or area, power considerations are often thrown over the wall from architectural exploration to physical design. But in regard to interconnect networks, Zhang et al. have presented a comparison and analysis of reconfigurable interconnect architectures using energy as a metric in [12], giving general results for DSP applications. However, we feel that optimized architectures can also be found for more specific application domains, which are defined by a number of sample applications.

To find a suitable architecture for a given domain, the KressArray Xplorer framework is currently being implemented. The framework uses the KressArray [7] [8] architecture family as basis for an interactive exploration process. When a suitable architecture has been found, the mapping of the application is provided directly. The designer is supported during the exploration by suggestions of the system how the current architecture may be enhanced. This paper proposes the application of this framework for power aware architecture exploration, with the discussion of related issues.

The rest of this paper is structured as follows: To give an overview on the topic, the next two sections briefly sketch the KressArray architecture family and the design space for the exploration process published elsewhere [13]. In section 4, our general approach for an interactive design space exploration for an application domain is presented. After this, a short overview on the KressArray Xplorer framework is given. The next section outlines our approach for the generation of design suggestions, which can be used to incorporate the models for power estimation presented in the following section. Finally the paper is concluded.

2   The KressArray Architecture Family

The KressArray family is based on the original mesh-connected (no extra routing areas see figure 2 d, e) KressArray-1 (aka rDPA) architecture published elsewhere [8]. An architecture of the KressArray family is a regular array of coarse grain reconfigurable DataPath Units (rDPUs), each featuring a multiple-bit datapath and providing a set of coarse grain operators. The original KressArray-1 architecture provided a datapath of 32 bits and all integer operators of C, the proposed system can handle also other datapath widths and operator repertories. The different types of communication resources are illustrated in figure 1. There are three levels of interconnect: First, a rDPU can be connected via nearest neighbor links to its four neighbors to the north, east, south and west. There are unidirectional and bidirectional links. The data transfer direction of the bidirectional ones is determined at configuration time. Second, there may be backbuses in each row or column, which connect several rDPUs. These buses may be segmented, forming several independent buses. Third, all rDPUs are connected by one single global bus which allows only serial data transfers. This type of connection makes only sense for coarse grain architectures with a relatively low number of elements. However, a global bus effectively avoids the situation, that a mapping fails due to lack of routing resources. The rDPUs themselves can serve as pure routing elements, as an operator, or as an oper ator with additional routing paths going through. Some more communication architecture examples are shown in figure 2.

The number and type of the routing resources as well as the operator repertory are subject of change during the exploration process. Typically, a trade-off has to be found between the estimated silicon area, the performance, and the power dissipation of the architecture, where both performance and power dissipation will typically depend on the application to be implemented.

3   The KressArray Design Space

The KressArray structure defines an architecture class rather than a single architecture. The class members differ mainly by the available communication resources and the operator repertory. Both issues have obviously a considerable impact on the performance of the architecture. In the following, we define the design space for KressArray architectures based on the introduction given in section 2.

The following aspects of a KressArray architecture are subject to the exploration process and can be modified by the tools of the exploration framework:

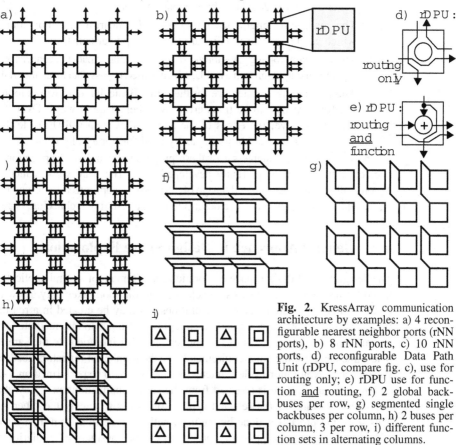

Fig. 2. KressArray communication architecture by examples: a) 4 reconfigurable nearest neighbor ports (rNN ports), b) 8 rNN ports, c) 10 rNN ports, d) reconfigurable Data Path Unit (rDPU, compare fig. c), use for routing only; e) rDPU use for function and routing, f) 2 global backbuses per row, g) segmented single backbuses per column, h) 2 buses per column, 3 per row, i) different function sets in alternating columns.

- The size of the array.
- The operator repertory of the rDPUs.
- The available repertory of nearest neighbor connections. The numbers of horizontal and vertical connections can be specified individually for each side and in any combination of unidirectional or bidirectional links.
- The torus structure of the array. This can be specified separately for each nearest neighbor connection. The possible options are no torus structure or torus connection to the same, next or previous row or column respectively.
- The available repertory of row and column buses. Here, the number of buses is specified as well as properties for each single bus: The number of segments, the maximal number of writers, and the length of the first segment, which allows buses having the same length but spanning different parts of the array.
- Areas with different rDPU functionality. For example, a complex operator may be available only in specific parts of the array. This allows also the inclusion of special devices in the architecture, like embedded memories. The operator repertory can be set for arbitrary areas of the array, using generic patterns described by few parameters.
- The maximum length of routing paths for nearest neighbor connections, which can be used to satisfy hard timing or power constraints.
- The number of routing paths through a rDPU. A routing path is a connection from an input to an output through a rDPU, which is used to pass data to another rDPU.
- The interfacing architecture for the array. Basically, data words to and from the KressArray can be transferred by either of three ways: Over the serial global bus, over the edges of the array, or over an rDPU inside the array, where the latter possibility is mostly used for library elements.

In order to find a suitable set of these properties for a given application domain, an interactive framework is currently developed [13]. The framework, called KressArray Xplorer, allows the user a guided design of a KressArray optimized for a specified problem. At the end of the design process, a description of the resulting architecture is generated.

4   General Approach to Design Space Exploration

The design flow of design space exploration for a domain of several applications is illustrated by figure 3. In most cases a cycle through the loop takes only a few minutes, so that a number of alternative architectural designs may be created in a reasonable time. First, all applications are compiled into a representation in an intermediate format, which contains the expression trees of the applications. All intermediate files are analyzed to determine basic architecture requirements like the number of operators needed. The main design space exploration cycle is interactive and meant to be performed on a single application. This application is selected from the set of applications in a way, that the optimized architecture for the selected application will also satisfy the other applications. This selection process is done by the user, with a suggestion from the system. For low power exploration, two application properties can be considered: Regularity and the esti mated power consumption without regarding the component of the routing architecture. An approach to measure the regularity of an application has been published in [14]. The power estimation can be derived from scheduled data flow graphs using a methodology described in [15] and [16].

The exploration itself is an interactive process, which is supported by suggestions to the designer, how the current architecture could be modified. The application is first mapped onto the current architecture. The generated mapping is then analyzed and statistic data is generated. Based on this data, suggestions for architecture improvement are created using a fuzzy-logic based approach described below. The designer may then chose to apply the suggestion, propose a different modification, return to a previous design version, or end the exploration. Some modifications allow the new architecture to be used directly for the next mapping step, while others will require a re-evaluation of the basic architecture requirements and/or a re-selection of the application for the exploration. Especially, a change of the operator repertory for the rDPUs requires the replacement of subtrees in the dataflow graph, thus effecting the number of required rDPUs in the array, complexity of all applications, and the power consumption. Thus, for a change of the operator repertory, a re-evaluation is required.

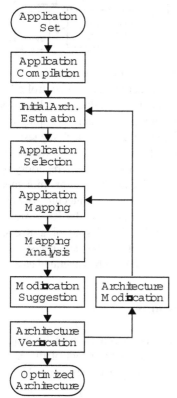

Fig. 3. Global approach for domain-specific architecture optimization.

After the exploration cycle has ended, the final architecture has to be verified by mapping the remaining applications onto it. On the one hand, this step produces mappings of all applications to be used for implementation, while on the other hand, it is checked if the architecture will satisfy the requirements of the whole application domain.

5   The KressArray Xplorer

This section will give a brief description of the components of the KressArray Xplorer, which has been published elsewhere [13]. An overview on the Xplorer is shown in figure 4. The framework is based on a design system, which can handle multiple KressArray architectures within a short time.

It consists of a compiler for the high-level language ALE-X, a scheduler for performance estimation, and a simulated-annealing based mapper. This system works on an intermediate file format, which contains the net list of the application, delay parameters for performance estimation, the architecture description, and the mapping information. The latter is added by the mapper at the end of the synthesis process. An architecture estimator determines the minimum architecture requirements in terms of operator numbers and suggests the application with the expected

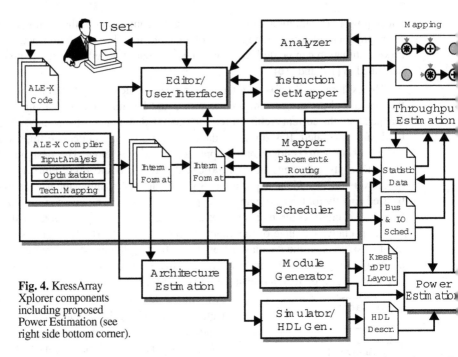

Fig. 4. KressArray Xplorer components including proposed Power Estimation (see right side bottom corner).

worst requirements to power and routing resources to the user through an interactiv graphical user interface. This interface is also generally used to control all other tool Further, it contains two interactive editors, an architecture editor, which allows to chang the architecture independently by the design suggestions, and a mapping editor, whic allows to fine-tune the result of the mapper. An analyzer generates suggestions for arch tecture improvements by information gathered directly from the mapping and oth sources. An instruction mapper allows the change of the operator repertory by exchang ing complex operators in the expression tree with multi-operator implementations. simulator allows both simulation of the application on the architecture as well as gener tion of a behavioral HDL (currently Verilog) description of the KressArray. Finally, Module Generator (planned) should generate the final layout of a KressArray ce Throughput estimations are generated from Scheduler results and statistical data. Fro Module Generator parameters also an area estimation can be generated easily.

6   Generation of Design Suggestions

In this chapter, we will give a short overview on our approach to analysis and generatic of design suggestions, which is performed by the analyzer tool of the Xplorer fram work. The problem of the generation of feedback on a given design can be split into se eral subproblems:

- Analysis of the current architecture and gathering of information. This includes the combination of data to gain derived information.
- Generation of suggestions from this information.

- Ranking of the suggestions after their importance.

In our approach, the basis for the information gathering step is the mapping produced by the design system. Primary data gathered includes the usage of buses and nearest-neighbor connections in percent, the number of serial bus connections, the estimated number of execution cycles, and others. This primary data can be used to derive secondary information, like the estimated power dissipation, based on the mapping and the actual usage of routing resources by the application.

To allow the gathering of a variety of information from the mapping, the analyzer tool is implemented using a plug-in-based approach, which provides also the flexibility to extend the system. The plug-ins are controlled by the analyzer tool, which holds also data structures for the analysis results.

The design suggestions themselves are then generated using an approximate reasoning approach based on fuzzy logic [17], [18]. The knowledge how to guide the exploration process is expressed in implication rules, like known from expert systems. The fuzzy approach has been chosen to allow the generation of suggestions based on inexact information, as during the exploration process, it is assumed the designer will apply a 'fast' mapping (by tuning the parameters of the simulated annealing), which will probably result in a mapping with a lower quality.

7   Power Estimation for the Exploration Process

In this section, we propose an approach to be used to generate a power estimation. Given an application and a KressArray architecture, onto which this application has been mapped, the total power consumption is composed from the power consumed by the operators and the power consumed by the interconnect network. We will now discuss how to estimate these in a way adequate for the exploration process. Note, that for the exploration process, a very accurate estimation of the power consumption is not necessary. Instead, we will propose simplified measures, which allow a relative quantitative comparison of power dissipation of different architectures.

The operator component is determined by the operator repertory (taken from the intermediate form seen in figure 4), by the implementation of the operators (to be derived from Module Generator parameters), by the configuration (indicating which operator has been selected) and by the switching activity (to be extracted from the intermediate form (or from the HDL description) and the Scheduler results: see figure 4).

In our Xplorer framework, we assume the operator repertory to be organized in several sets, which can be switched by the instruction set mapper mentioned in section 5, and that according power models for the final implementation of those sets are available from the library of the Module Generator (figure 4). With these preliminaries, the power consumption for a given application can be estimated by techniques like those presented in [15], [16], since the required data flow graph can be extracted from the intermediate format (s. figure 4). The resulting estimation does not consider the routing architecture or an actual mapping of the application at all (however, see next paragraph). Though it can be used to distinguish applications in order to select the one for the exploration process described in section 4. This selection takes place at the beginning and when the operator set has been changed.

The power consumption caused by the routing network depends also on the actual use of the routing resources during run-time. However, for our purposes, we can employ a more relaxed metric for the different routing resources of KressArrays. Instead of the power, we use the energy of the routing resources for measure, neglecting the actual usage during execution. Generally, for the energy consumption E of an interconnect structure, the following estimation can be used (a modified version of the model in [12] for a net con starting from an operator op in the mapping:

$$E(con) = (\ C_{wire} + C_{switch} + Fanout_{op} \bullet C_{load}) \bullet V^2 \tag{1}$$

where C_{wire} and C_{switch} are the according capacitances of the wire segments and switches for each routing resource, $Fanout_{op}$ is the fan-out of the operator resembling the source of the net, C_{load} is the load capacitance of an operator input, and V is the supply voltage. For our purposes, we can simplify this equation to:

$$E(con) = K_L \bullet L + K_S \bullet NR \bullet S + Fanout_{op} \bullet K_{load} \bullet NR \tag{2}$$

Where K_L is the energy per wire segment, L is the number of segments used, K_S is the energy per switch, NR is the number of all routing resources meeting at an rDPU which determines the width of the required switch, S is the number of switches, and K_{load} is the energy per fan-out connection. The values K_L, K_S, and K_{load} can be assumed to be constant and known for each interconnect type. Then, we get the following estimates for the energy for each routing resource:

Global Bus. For this type of connection, the capacity of the whole bus structure has to be considered, as this bus is not switched, but operates in a serial manner. The capacity is dependent of the array size and the actual layout of the bus structure. For relative measure between different architectures, we can assume a general layout like in figure 1c. If the Sizes of the array in x and y-direction are denoted as AS_x and AS_y respectively, the number of bus segments is approximated by:

$L = AS_x \bullet (\ AS_y + 1\ )$, which we simplify to $L = AS_x \bullet AS_y$ to keep the measure independent from the aspect ratio of the array. Thus, we get:

$$E_{global}(con) = K_L \bullet AS_x \bullet AS_y + Fanout_{op} \bullet K_{load} \bullet NR \tag{3}$$

for a global bus connection con with source operator op.

Row/Column backbuses. For those connections, both source and one or more sinks lie on one bus segment, which in itself is not switched. Thus, we get:

$$E_{backbus}(con) = K_L \bullet Segment(op) + Fanout_{op} \bullet K_{load} \bullet NR \tag{4}$$

with $Segment(op)$ denoting the length of the bus segment holding the source operation op.

Nearest Neighbor Connects. Using the nearest neighbor connects, a connection from one source operator to several sinks can be implemented, resulting in a path composed of several subpaths (cf. figure 5). There is a subpath for each sink composed of a sequence of length-1 connections, whereby different subpaths do not share such segments and each rDPU lying on this sequence inflicts additional switching energy. We get the following estimation for a path made up of $Fanout_{op}$ subpaths, each of which with the according subpath length SPL:

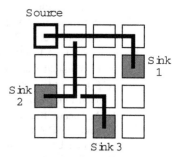

Fig. 5. Path composed of three subpaths

$$E_{NN}(con) = \sum_{i=1}^{Fanout_{op}} (SPL_i \cdot K_L + (SPL_i - 1) \cdot K_L \cdot NR + K_{load} \cdot NR)$$

According to the software architecture of the framework outlined in section 5, the implementation of these estimation functions can easily done as plug-ins, which use available data from the intermediate file, given the hardware-parameters are provided.

8   Conclusions

An interactive approach for the design-space exploration of mesh-based reconfigurable architectures from the KressArray family has been presented. An according framework called KressArray Xplorer is based on a design system which allows the specification of the input language in a high-level language. During the exploration process, which is based on iterative refinement of the current architecture, the designer is supported by suggestions on how the current solution may be improved. To apply the framework for exploration of low power architectures, according models have been proposed, which can easily be integrated into the framework.

References

1. J. Rabaey. "Low-Power Silicon Architectures for Wireless Communications"; Embedded Tutorial, ASP-DAC 2000, Yokohama, Japan, Jan. 2000.

2. R. Hartenstein (invited paper): The Microprocessor is no more General Purpose: why Future Reconfigurable Platforms will win; Int'l Conf. on Innovative Systems in Silicon, ISIS'97, Austin, Texas, USA, Oct 1997

3. E. Mirsky, A. DeHon: „MATRIX: A Reconfigurable Computing Architecture with Configurable Instruction Distribution and Deployable Resources", Proc. FPGAs for Custom Computing Machines, pp. 157-166, IEEE CS Press, Los Alamitos, CA, U.S.A., 1996.

4. A. Marshall et al.: A Reconfigurable Arithmetic Array for Multimedia Applications; FPGA'99, Int'l Symposium on Field Programmable Gate Arrays, Monterey, CA, U.S.A., Febr. 21 - 23, 1999

5. C. Ebeling, D. Cronquist, P. Franklin: „RaPiD: Reconfigurable Pipelined Datapath", Int'l Workshop on Field Programmable Logic and Applications, FPL'96, Darmstadt, Germany, Sept 1996.

6. R. A. Bittner, P. M. Athanas and M. D. Musgrove: „Colt: An Experiment in Wormhole Run-

time Reconfiguration"; SPIE Photonics East `96, Boston, MA, USA, November 1996.

7. R. Kress et al.: A Datapath Synthesis System for the Reconfigurable Datapath Architecture; Asia and South Pacific Design Automation Conf. (ASP-DAC'95), Chiba, Japan, Aug. 29 - Sept. 1, 199.

8. R. Kress: „A Fast Reconfigurable ALUs for Xputers", Ph.D. thesis, Univ. Kaiserslautern, 1996.

9. E. Waingold et al.: „Baring it all to Software: Raw Machines", IEEE Computer 30, pp. 86-93

10. S. C. Goldstein, H. Schmit, et al.: „PipeRench: A Coprocessor for Streaming Multimedia Acceleration"; Int'l Symposium on Computer Architecture 1999, Atlanta, GA, USA, May 1999.

11. A. Abnous, K. Seno, Y. Ichikawa, M. Wan and J. Rabaey: Evaluation of a Low-Power Reconfigurable DSP-Architecture; Proceedings of the Reconfigurable Architectures Workshop, Orlando Florida, USA, March 1998.

12. H. Zhang, M. Wan, V. George and J. Rabaey: Interconnect Architecture Exploration for Low-Energy Reconfigurable Single-Chip DSPs; Proceedings of the WVLSI, Orlando, Florida, USA, April 1999

13. R. Hartenstein, M. Herz, Th. Hoffmann, U. Nageldinger: KressArray Xplorer: A New CAD Environment to Optimize Reconfigurable Datapath Array Architectures; 5th Asia and South Pacific Design Automation Conference 2000, ASP-DAC 2000, Yokohama, Japan, January 25-28, 2000

14. L. Guerra, M. Potkonjak and J. Rabaey: „System-Level Design Guidance Using Algorithm Properties"; IEEE VLSI Signal Processing Workshop, 1994.

15. L. Kruse, E. Schmidt, G. Jochens, A. Stammermann and W. Nebel: Lower Bounds on the Power Consumption in Scheduled Data Flow Graphs with Resource Constraints; Proceedings of the European Design and Test Conference DATE 2000.

16. L. Kruse, E. Schmidt, G. Jochens and W. Nebel: Lower and Upper Bounds on the Switching Activity in Scheduled Data Flow Graphs; Proceedings of ISPLED 1999.

17. W. Pedrycz: „Fuzzy Modelling - Paradigms and Practice"; Kluwer Academic Publishers, 1996.

18. B.R. Gaines: „Foundations of Fuzzy Reasoning"; Int'l Journal of Man-Machine Studies, Vol. 8 1976.

Internal Power Dissipation Modeling and Minimization for Submicronic CMOS Design

P. Maurine, M. Rezzoug, and D. Auvergne

LIRMM, UMR CNRS/Université de Montpellier II, (C5506),
161 rue Ada, 34392 Montpellier, France
auvergne@lirmm.fr

Abstract. Based on a concept of equivalent capacitance, previously developed, we present a novel analytical linear representation of internal power dissipation components in CMOS structures. An extension to gates is proposed using an equivalent inverter representation, deduced from the evaluation of an equivalent transistor for serial transistors arrays. Validation of this model is given by comparing the calculated results to the simulated values (using foundries model card), with different design conditions, implemented in 0.25μm and 0.18μm CMOS processes. Application is given to delay and power optimisation of buffer and path.

1   Introduction

Careful study of the power dissipation in submicron CMOS structures shows that the internal component may have contributions greater than necessary to control the different gates. It becomes then essential to get an accurate and design oriented power estimation of static CMOS family for the design and optimisation of high performance circuits. As a result many authors developed accurate models of power dissipation dedicated to submicron technologies. In [1] a short circuit power consumption formula is derived from a piece wise linear representation of the short circuit current. A complex modelling of the output waveform permits an accurate evaluation of the short circuit dissipation in [2]. A macro-model of short circuit power consumption has been deduced from a detailed delay analysis in [3].

Unfortunately, these models are too complicated to define low power design criteria at cell level. We present here a design-oriented model of the internal power dissipation component based on a previously developed concept of equivalent capacitance. The defined target is to allow direct comparison between the components commonly considered as significant for CMOS circuits: the external dynamic component, associated to the gate output capacitance charge and discharge, and the internal dynamic component due to the short circuit occurring between N and P blocks and to the overshoot discharge resulting from the input to output coupling.

In order to define low power design criteria, at cell level, applicable to buffer design, we propose a design-oriented modelling of the short circuit power component.

D. Soudris, P. Pirsch, and E. Barke (Eds.): PATMOS 2000, LNCS 1918, pp. 129-138, 2000.

The major focus here will be to define the propitious design conditions minimizing the internal power component with respect to the external (capacitive) one. The internal power consumption model is described in section 2. A linear representation of this model is proposed in section 3 where we give validations of this approach targeting 0.25μm and 0.18μm process. In section 4 we model the serial transistor array through an equivalent transistor that we define. This allows to represent each gate by an equivalent inverter for which we can predict the internal power consumption . In section 5, we define sizing criteria for power minimisation. Application to buffer sizing and path optimisation is given in section 6. Conclusion is drawn in section 7.

2   Power Consumption Model Description

Using the equivalent capacitance concept proposed in [16], we can express the internal power consumption as follow :

$$P_{INT} = \eta . f .(C_{SC} + C_{OV}).V_{DD}^2 \tag{1}$$

where η, f, and V_{DD} are respectively the activity rate, the switching frequency and the supply voltage; C_{SC} and C_{OV} are the equivalent capacitances that would generate the same power dissipation as the short circuit while reported on the output node. The evaluation of C_{OV} is done using the expression proposed in [3]. The short circuit equivalent capacitance C_{SC} is expressed as :

$$C_{SC} = \frac{1}{V_{DD}} \int_{t_0}^{t_1} I_{SC}(t).dt \tag{2}$$

To evaluate C_{sc} we can, as in [2], directly perform the integration between t_0 and t_1 which are respectively the beginning and the end of the short circuit. We can also [4] use symmetrical properties of the short circuit current (Fig. 1). This allows to perform the integration only between t_0 and t_{SP} which corresponds to the maximum short circuit current occurrence. We can also as in [1] assume that the short circuit current presents a linear variation between t_0 and t_{SP}; therefore the integration is reduced to the evaluation of a triangular surface. The height and the triangle basis being respectively $I_{SC\text{-}MAX} = I_{SC}(t_{SP})$ and $\Delta t_{SC} = 2.(t_{SP}\text{-}t_0)$ which represents the short circuit duration. This leads to the following expression of the short circuit equivalent capacitance :

$$C_{SCLH} = \frac{(t_{SP} - t_{OVLH})}{V_{DD}} . I_{SC\text{-}MAX} \tag{3}$$

Where $t_{OVLH} = t_0$ corresponds to the end of the overshoot discharge [3]. Thus, the main difficulty here is to accurately evaluate the values of t_{SP} and $I_{SC\text{-}MAX}$. In [3], these values have been calculated from the switching delays of the structure considering that the maximum current occurs when the operating mode of the short circuiting transistor evolves from linear to saturated mode. If this hypothesis has been sufficient for 0.7μm process it appears that with deep submicron (0.25μm and less) the position of the maximum current appears in linear mode and is modulated by desaturation effects of carrier speed [4]. With such considerations, t_{SP} is evaluated from a temporal derivation of the short circuit current. To perform this derivation we assume, according to [6], that around t_{SP} the variation of the output voltage is linear (Eq. 4, 5), as a consequence the output slope duration is proportional to the step response t_{HLS} of the inverter. Under this assumption the drain to source and gate to source voltage values necessary to determine the current evolution can be easily calculated.

$$V_{GSP}(t_{SP}) = V_{DD} \cdot \left(1 - \frac{t_{SP}}{\tau_{INLH}}\right) \quad (4) \qquad V_{DSP}(t_{SP}) = V_{DD} \cdot \left(\frac{t_{SP} - t_{OVLH}}{2.t_{HLS}}\right) \quad (5)$$

Canceling the derivative of the current expression with respect to time gives directly the expression of t_{SP} as:

$$t_{SP} = \frac{1}{2} \cdot \left(1 - v_{TPL} + \frac{t_{OVLH}}{\tau_{INLH}}\right) \cdot \tau_{INLH} \quad (6)$$

With this value of t_{SP} the equivalent short circuit capacitance is deduced from (3), considering a linear variation of the voltages around t_{SP} (Eq. 4, 5):

$$C_{SCLH} = \frac{(t_{SP} - t_{OVLH})}{V_{DD}} \cdot \frac{\mu_0.C_{OX}.W_P / L_{GEO}}{1 + \theta.(V_{GSP}(t_{SP}) - v_{TPL})} * (V_{GSP}(t_{SP}) - v_{TPL})V_{DSP}(t_{SP}) \quad (7)$$

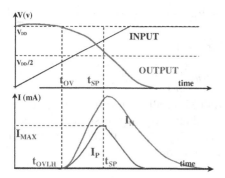

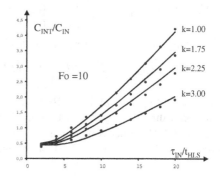

Fig. 1. Evolution of the output voltage and current in a CMOS inverter; the short circuit current shape is triangular in shape

Fig. 2. Simulated and calculated values of C_{INT} for an inverter ($W_N=1\mu m$) design in 0.18μm process and with different configuration ratio values k

These equations have been used to calculate the equivalent internal capacitance values of inverters designed in 0.25μm, and 0.18μm process with various conditions of control, loads, sizes and configuration ratio. Validations have been obtained by comparing these values to values simulated from HSPICE simulations using the Kang's method [7,8]. As shown in Fig. 2 the agreement between simulated and calculated values over the considered design range is good enough to validate the proposed method. Currently observed discrepancies are lower than 13%.

3   Linear Representation

3.1   Linearization

The expression (7) is a product of two terms. The first one is the short circuit duration and the second one is the maximum short circuit duration. Thus in the aim of simplifying (7), let us consider the short circuit duration. At the first order the overshoot duration is given by;

$$t_{OVLH} = v_{TNL} \cdot \tau_{INLH} \tag{8}$$

Therefore the short circuit duration can be expressed as a linear function of the input slope duration τ_{INLH}:

$$\Delta t_{SC} = 2.(t_{SP} - t_{OVLH}) \approx (1 - v_{TPL} - v_{TPN}).\tau_{INLH} \tag{9}$$

Let us now consider the maximum short circuit current, neglecting the small variation of the gate to source voltage with respect to τ_{INLH} (10), we note that the maximum short circuit current is proportional to the transistor width. Its evolution with the input slope can represented by a linear function of τ_{INLH} / t_{HLS} (11).

$$V_{GSP}(t_{SP}) = V_{DD} \cdot \left(1 - \frac{t_{SP}}{\tau_{INLH}}\right) \approx \frac{V_{DD}}{2}.(1 - v_{TNL} + v_{TPL}) \tag{10}$$

$$V_{DSP}(t_{SP}) = V_{DD} \cdot \left(\frac{t_{SP} - t_{OVLH}}{2.t_{HLS}}\right) V_{DSP}(t_{SP}) \approx V_{DD} \cdot \frac{(1 - v_{TN} - v_{TP}).\tau_{INLH}}{4.t_{HLS}} \tag{11}$$

Where t_{HLS} is the inverter response to a step input [6] :

$$C_{SCLH} = \frac{(1 - v_{TPL} - v_{TPN}).\tau_{INLH}}{2.V_{DD}} \cdot \left(a_{LH}.\frac{\tau_{INLH}}{t_{HLS}} + b_{LH}\right) W_P \tag{12}$$

And a_{LH} and b_{LH} are single coefficients determined for all the inverters of the library. They can directly be obtained from (7) or calibrated on Hspice simulations. For a 0.25μm CMOS process we obtain $a_{HL} = 7.5.10^3$, $b_{HL} = 3.75.10^2$, for an input falling edge, $a_{LH} = 2.25.10^3$, $b_{LH} = 3.10^2$ for an input rising edge.

3.2   Validation

Validation of the linear representation given by Eq. 12, is illustrated in Fig. 3. In this figure we compare for a 0.18μm CMOS process, simulated and calculated values of the equivalent capacitance representing the total internal power dissipation component. As shown we obtain a good agreement between calculated and simulated values over all the considered design range. Commonly, discrepancies are lower than 15 %.

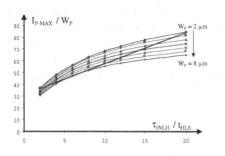

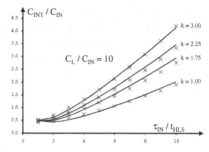

Fig. 3. Simulated values of I_{P-MAX}/W_P of an inverter ($W_N=4$ μm) loaded by $C_L=10.C_{IN}$

Fig. 4. Simulated (lines) and calculated (with the linear representation) values of the internal power dissipation of the inverter defined in Fig. 2.

3.3   Comparison with Previous Work

In Table 1 we give a more complete validation in comparing previously proposed formulas [4,9,10] to the macro-model presented here.

Table 1. Comparison of internal power dissipation formulae with Spice simulation results for an inverter ($C_{IN}=20.7$fF loaded by 145fF $L_{GEO}=0.25$μm) for various input slope conditions

τ_{IN}/t_{HLS}	C_{INT}/C_{IN}				
	[4] Eq. 10	[9] Eq. 11	[10] Eq. 12	Eq. 12	HSPICE
2	1,14	0,94	0.75	0,63	0,69
4	2,29	1,88	1,21	1,22	1,36
6	3,43	2,82	1,98	1,99	2,25
8	4,57	3,77	3,07	2,93	3,24
10	5,72	4,71	4,46	4,03	4,33
12	6,86	5,65	6,16	5,31	5,48
16	9,15	7,53	10,48	8,36	7,80
20	11,43	9,41	16,05	12,06	10,34

Eq. 10 of Veendrick has been updated for submicron process and represents the zero-load model. Eq. 11 of [9] presented by Sakurai and based on the α-power model uses the zero load assumption, Eq. 12 from [10] is also based on the α-power model but does not consider the overshoot component. As shown in the Table 1 the proposed expression (Eq. 11) including linear representation of overshoot and short circuit components is still accurate for a large range of input slew.

4   Extension to Gates

The main difference between an inverter and a logic gate is the presence of a serial array of transistors in the N or the P block. To evaluate the internal power dissipated in CMOS gates, we represent each gate by an equivalent inverter defined for each possible switching condition. As an example let us consider a Nand2 where two switching conditions may be defined depending on the input control data (Fig. 5). Considering the top input of the serial array as the controlling one, it is clear that its driving gate strength is reduced by the voltage drop across the bottom transistor working in the linear mode (Fig. 6).

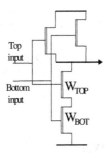

Fig. 5. Two input Nand structure

This defines two reduction factors, Red_{LIN} and Red_{SAT} for falling and rising input controlling edges respectively, for which the top transistor is working in linear or saturated mode, as:

$$Red_{LIN} = 1 + \frac{\mu_{EFF} \cdot C_{OX}}{L} \cdot W_{TOP} \cdot R \cdot V_{OUT}(t_{SP}) \quad (13) \qquad Red_{SAT} = 1 + K_{NSAT} \cdot W_{TOP} \cdot R \quad (14)$$

where R is the drain source resistance of a unit transistor (Width=1μm), K_{NSAT} is the conduction coefficient of the N transistor working in saturation. Considering the bottom transistor, due to the voltage level degradation occurring on the conducting top N transistor, its working mode is the same than that of the N transistor of an inverter working under a lower supply voltage $V_{DD}^{*} = V_{DD} - V_{TN}$ (Fig. 7). In this condition the overshoot duration increases and the short circuit duration will be lower than that of an inverter. Consequently, the equivalent short circuit value will be modified through the supply voltage reduction ($V_{DD} - V_{TN}$ instead of V_{DD}).

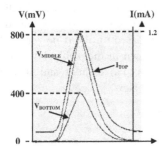

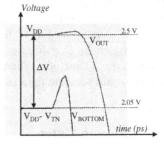

Fig. 6. Resistive comportment of the bottom transistor of a Nand3 gate when only the top input has switched

Fig. 7. Voltage drop across the top transistor of a Nand2 gate when the bottom input is switching

For the Nand2 it is then possible to express the equivalent short circuit capacitance for a rising input as:

$$C_{SCLH} = \frac{1}{2.V_{DD}}\left(1 - v_{TPL} - \frac{V_{TNL}}{V_{DD} - V_{TNL}}\right)\left(a_{ND2-LH} \cdot \frac{\tau_{INLH}}{red_{SAT} \cdot t_{HLS}} + b_{ND2-LH}\right).W_P \quad (15)$$

with an equivalent expression for the falling edge. Here V_{TN} is the voltage drop across the transistor Top, a_{ND-LH} and b_{ND-LH} are process parameters. As for inverters, we compare in Table 2 simulated and calculated values of the internal capacitances. As shown the observed discrepancy is kept lower than 15%.

Table 2. Comparisons between simulated and calculated internal capacitance for a Nand2 (W_N = 4μm Wp=12μm C_L=198 fF L_{GEO}=0.25μm) only the bottom input has switched

τ_{IN}/t_{HLS}	C_{INT}/C_{IN}		$\|\Delta C_{INT}\|/C_{IN}$
	HSPICE	MODELE	
4	0.95	1.09	14%
6	1.88	1.87	1%
8	2.95	2.89	2%
10	4.12	4.15	1%
12	5.33	5.64	6%
15	7.24	8.33	15%

5   Sizing Criteria for Power Minimization

5.1   Buffer Sizing

Since we are able to reduce each gate into an equivalent inverter, minimizing the power dissipated along a combinatorial path results in optimizing an inverter chain. Thus our concern here is to find the sizing of each inverter which minimizes the power. To get facilities let us define the smallest inverter chain that is possible to optimize for power. Although the external power dissipated by an inverter is imposed by its input gate capacitance, its internal contribution depends on the preceding stage (the input slew) and the following stage (the load). That is a real input output slew control.

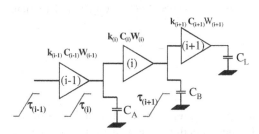

Fig. 8. Smallest inverter chain to be optimized for power

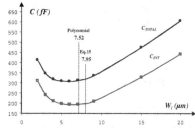

Fig. 9. Illustration of the variations, with respect to the size of the drive (i), of the internal and total power components necessary to drive a composite load

Usual design alternatives are expressed with respect to load and fan-out factors, therefore optimizing the smallest inverter chain results in finding the optimal sizing of the stage (i) which determines the fan-out factors between stages (i-1) and (i) minimizing the power dissipation. This is illustrated in Fig. 9 which represents the variation of internal and total power dissipation on the preceding array, versus the size of inverter (i).

C_{INT} represents the equivalent capacitance relative to the internal power dissipation on the array, C_{TOTAL} is the sum of this capacitance and of the total array input capacitance. As shown out of the minimum area bad selection of the buffer (to small or too large) will result in an unnecessary extra power dissipation.

Using Eq. 10, it is quite easy to evaluate the internal power component of the array given in Fig. 9 and to search for the condition on the intermediate inverter size minimizing this component. The summation of all the contributions results in a 6th order polynomial expression that can only be solved graphically or numerically. Neglecting in Eq. 10 the $a_{LH,HL}$, contributions results in a 3rd order expression which can be solved as:

$$C_i^3 = \frac{C_{(i+1)}^2 C_{(i-1)}}{2}\left(1+\frac{C_B}{C_{(i+1)}}\right) - \frac{1}{27}\left(\frac{C_A}{2}+\frac{C_{(i-1)}}{\alpha}\right)^3 \qquad \alpha = \frac{(1-v_{TN}-v_{TP})}{2.V_{DD}.C_{OX}.L_{geo}}\left(b_{LH}.R_\mu + b_{LH}\right)$$

(16)   (17)

The validation of Eq. (16) and (17) has been done by comparing the optimal values of the buffer input capacitance ($C_{(i)}$) obtained from Hspice to the values predicted by the model. As shown in Fig. 10, we obtain a good agreement between simulated and calculated buffer size.

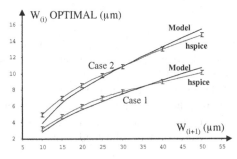

Fig. 10. Comparisons between calculated and simulated of W(i) optimal values for two cases (L_{GEO}=0.25μm) : Case1: W(i-1) =1μm $k_{(i)}$ = 2.23 C_L/C(i+1)=5 C_A=C_B=0 Case2: W(i-1)=1μm, k(i)=3 CL/C(i+1)=10 C_A=10fF C_B=2.C(i+1)

5.2   Path Sizing

The optimization criteria for delay and power proposed in [11] and in Eq. 16, respectively, are mostly non linear. In these conditions global optimisation of a combinatorial path is unpractical for logical depth value greater than 5. We propose here to apply a local optimisation in which processing for the selected path from output to input we optimize each element with its real load but driven by a reference inverter. This procedure has been shown [11] efficient and fast enough to be applied to this optimisation problem.

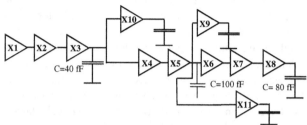

Fig. 11. Circuit used to illustrate the optimisation procedure: X_i represents the inverters (gates) to be sized, the capacitance values on the different nodes represent the parasitic loads

Results are quasi optimum, delay constraints being managed by sizing the global reference. For illustration we applied this procedure to the example given in Fig. 11. First we implement an initial solution where all the transistors have the minimum size. Then we identify the critical path (X_1 to X_8 in this example). For a given fan out factor of the last stage of this critical path, we size all the stages of this path, processing backward according to the following rule: the sizing for minimizing the power is obtained in the same way applying the sizing Eq. 16.

Non critical paths are sized under the same criteria by varying the C_{REF} value in order to satisfy the delay defined by the critical path. This procedure has been applied to the example shown in Fig. 11. The results obtained have been compared to those obtain using a regular sizing (Fan-out =cste) and a uniform sizing (all transistors are identical), when imposing a delay constraint equal to 460 ps. Those comparisons are summarized in Table 3 where we report the values of the critical path delay, the total power dissipation, the power delay product (PDP), the power delay surface product (PDPS), the total transistor width used as an indicator of the total active area and the slope out.

Table 3. Comparison between our, the uniform and the regular sizing (Fo = cste)

SIZING	REGULAR	UNIFORM	OUR METHOD
Delay (ps)	450	471	456
Total Power (mw)	200	224	187
ΣW (µm)	31	30	23
P.D.P. (fJ)	88.6	101	85.3
P.D.P.S (pJ.µm)	2.7	3	2
Slope out (ps)	114	239	153

6   Conclusion

Based on a concept of equivalent capacitance we demonstrate the possibility in characterizing the internal power dissipation of CMOS inverters through accurate design oriented expressions. Clear indication of input controlling slopes, output loads and gate structure may help designers in defining fan out control as a design strategy for minimizing internal power dissipation. Extension to general gates as been proposed through an algorithm for gate reduction to an equivalent inverter.

The equivalent capacitance concept we used gives facilities in comparing directly the different power dissipation components in terms of fan-out factors that can be obtained at the circuit level and used to drive optimisation alternatives.

Validations have been performed by comparing calculated and simulated (HSPICE) values (0.25-0.18µm CMOS process) on a wide range of input control slew. The interest of this model is in defining transistor resizing rules for power optimisation. Application has been given to power optimisation under a delay constraint

References

1. A. Hirata, H. Onodera, K. Tamaru « Estimation of Propagation Delay Considering Short-Circuit For Static C.M.O.S. » IEEE Transactions on Circuits and Systems –I :Fundamental Theory and Applications, Vol. 45, n°. 11, pp.304-310, March 1998.
2. L. Bisdounis, S. Nikolaidis, O. Koufopavlou " Propagation Delay and Short-Circuit Power Dissipation Modelling of the C.M.O.S. Inverter " IEEE Transactions on Circuits and Systems –I :Fundamental Theory and Applications, Vol. 45, n°3, March 1998.
3. S. Turgis, D. Auvergne "A novel macromodel for power estimation for CMOS structures" IEEE Trans. On CAD of integrated circuits and systems vol.17, n°11, pp.1090-1098, nov.98.
4. H.J.M. Veendrick, "Short circuit power dissipation of static CMOS circuitry and its impact on the design of buffer circuits", IEEE J. Solid State Circuits, vol. SC-19, pp.468-473, Aug. 1984.
5. Kai-Yap Toh, Ping-Keung Ko, R. G. Meyer " An Engineering Model for Short-Channel MOS Devices " IEEE Journal of Solid-states Circuits, Vol. 23, n°. 4, August 1988.
6. J. Daga, D. Auvergne " A Comprehensive Delay Macro-Model of Submicron CMOS Logics" IEEE Journal of Solid States Circuits, vol 34, n°1, pp.42-55, January 1999.
7. S.M.Kang: "Accurate simulation of power dissipation in VLSI circuits", IEEE J. Solid State Circuits, vol. SC-21, pp. 889-891, oct.1986.
8. G. Y. Yacoub, W. H. Ku : "An accurate simulation technique for short circuit dissipation based on current component isolation" Proceedings of ISCAS'89 pp. 1157, 1161.
9. T. Sakurai, R. Newton "Alpha-power law MOSFET model and its applications to CMOS inverter delay and other formulas", IEEE Journal of Solid States Circuits, vol.25, n°2, pp.584-593, April 1990.
10. S.R. Vemuru, N. Scheinberg" Short circuit power dissipation for CMOS logic gates", IEEE Trans. On Circuits and systems Part 1, vol.41, n°11, pp.762-764, Nov. 1994.
11. D.Auvergne, N.Azemard, V.Bonzon, D.Deschacht, M.Robert " Formal sizing rules of CMOS circuits " EDAC Amsterdam February 25-28, 1991

Impact of Voltage Scaling
on Glitch Power Consumption

Henrik Eriksson and Per Larsson-Edefors

Electronic Devices, Department of Physics and Measurement Technology,
Linköpings universitet, SE-581 83 Linköping, Sweden.
{hener,perla}@ifm.liu.se

Abstract. To be able to predict the importance of glitches in future
deep-submicron processes with lowered supply and threshold voltages,
a study has been conducted on designs, which experience glitching, at
supply voltages in the range from 3.5 V to 1.0 V. The results show that
the dynamic power consumption caused by glitches will, in comparison to
the dynamic power consumption of transitions, be at least as important
in the future as it is today.

1   Introduction

Glitches are unnecessary signal transitions which do not contribute any information or functionality. The glitches can be divided into two different groups: generated and propagated. A generated glitch can occur if the input signals to a gate are skewed. If a generated glitch occurs at the input of a gate, the glitch may propagate through the gate; in that case we have a propagated glitch.

The number of glitches in a circuit depends mainly on the logic depth, gate fan-outs and how well the delays in the circuit are balanced. One obvious way to reduce glitching is to introduce pipelining, which would reduce the logic depth at the cost of power from pipeline registers. In circuits with large logic depths, the power consumption caused by glitches can be severe. In a non-pipelined 16×16-bit array multiplier, 75% of the switching in the circuit are glitches [1].

At almost all levels of abstraction, from the circuit level to the behavioral, techniques have been suggested to reduce the power consumption caused by glitches. At the circuit level, one popular way of reducing the power consumption is path balancing, where gates are resized and buffers are inserted to equalize the delays to the gates [2]. Restructuring multiplexer networks and clocking of control signals are techniques that can be used at the register-transfer level [3].

In future processes, the supply voltage has to be scaled even more than today to accommodate the demands for a lower power consumption. Other driving forces for supply voltage reduction are reduced channel lengths and reliability of gate dielectrics [4]. To retain the performance of the circuits, the threshold voltage, V_T, has to be scaled accordingly. However, a 100 mV decrease in V_T will increase the leakage current 10 times (@85°C). Therefore, the scaling of V_T is done at a slower pace and it might stop at a V_T of approximately 0.2 V [5].

D. Soudris, P. Pirsch, and E. Barke (Eds.): PATMOS 2000, LNCS 1918, pp. 139–148, 2000.

In this paper we will examine the swing distribution and dynamic power consumption of glitches when the supply voltage is lowered. Two different scenarios are considered; in the first scenario the threshold voltage is kept constant when the supply voltage is lowered. In the second scenario, the threshold voltage is scaled proportionally to the supply-voltage scaling.

2   Simulations

To see what is happening to the glitches when the supply voltage is lowered, some circuits, which experience a lot of glitching, i.e. adders and multipliers, have been simulated in *HSpice*™ and the behavior of the glitches has been studied.

In order to keep track of the glitches in a circuit, a *C* program, which detects glitches in the transient file from *HSpice*™, has been used. The output from the *C* program has been processed and analyzed using *Matlab*.

2.1   Circuit Selection

A large number of glitches are needed in the simulated circuits in order to make the analysis more valid. Both adders and multipliers are known to experience a lot of glitching [1,6]; therefore, one 8-bit adder and two array multipliers of different sizes have been implemented in layout and extracted netlists have been used in the simulations. The AMS 0.35 μm process has been used in the implementations and all transistors in the designs are minimum sized.

The 8-bit adder is an ordinary ripple-carry adder (RCA8) and it is implemented in static CMOS, in this case the mirror adder [6]. The multipliers are one 4×4-bit and one 8×8-bit array multiplier; both have been implemented as carry-propagate multipliers [6].

2.2   The Glitch-Detection Program

The program, which detects and calculates the power consumption of glitches and transitions, is written in *C*. The user of the program has to specify in which nodes the program should search for glitches and transitions. In the power calculations, only the dynamic power consumption is considered, i.e. short-circuit and leakage power consumptions are neglected. Neglecting the leakage current is still valid today, but in future processes, the leakage power will increase its importance significantly.

In our circuit analysis, we have chosen to study the nodes in which the transitions can have rail-to-rail swing, i.e. nodes that are situated between an NMOS and a PMOS net. The intermediate nodes between transistors inside an NMOS or a PMOS net have been ignored to reduce the simulated data.

The program uses the transient file and the capacitance table from *HSpice*™ to find and compute the power consumption of glitches and transitions. The power consumption of a glitch (corresponding to two transitions) is calculated as

$$P = f_{clk} \cdot C \cdot V_{DD} \cdot \Delta V \tag{1}$$

where f_{clk} is the clock frequency, C is the capacitance for the node as given by *HSpice*™, V_{DD} is the supply voltage, and ΔV is the swing of the node.

The program keeps track of the nodes specified in the setup and checks if the node voltage has changed more than a predetermined glitch amplitude value, e.g. 10%. If the voltage becomes larger than the glitch amplitude value we either have a transition or a glitch. If the voltage level returns to the threshold within a clock period we have a glitch, otherwise we have a transition. After we have registered a glitch we have the possibility to register another glitch or a transition. In Fig. 1, a glitch followed by a transition in a node of the simulated 8×8-bit multiplier is shown, $V_{DD} = 2.8$ V.

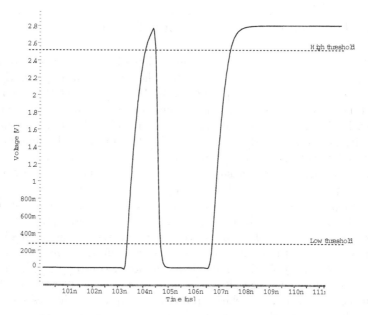

Fig. 1. A glitch and a transition during the same clock cycle, $f_{clk} = 100$ MHz

The program outputs the start and stop time and the maximum amplitude of all glitches it has found, together with the node to which the glitch belongs. We also get information from the program about how much of the power consumption originates from glitches and how much originates from transitions.

2.3   Simulation Strategy

To be able to make some predictions of the importance of glitching in the future, the circuits have been simulated under two different conditions. In the first case, we have lowered the supply voltage from 3.5 V down to 1.0 V without changing the threshold voltage. In the second case, the threshold voltage is scaled proportionally to the supply voltage. For example, at 3.3 V, the threshold voltage is 0.38 V for the NMOS transistor which is scaled to 0.17 V at V_{DD} 1.5 V. At a supply voltage of 1.0 V, the scaled V_T becomes 0.12 V which is an unrealistic value. The minimum usable V_T is approximately 0.2 V at room temperature. If a lower V_T is used, the leakage becomes intolerable [5]. However, the two simulation conditions (constant V_T and scaled V_T, respectively) can be used as limits for predicting the future importance of glitches. The V_T scaling will certainly lie somewhere within these limits, but it is hard to predict exactly where.

Two-hundred random test-vectors have been fed to the simulated circuits and the supply voltage has been decreased in steps of 0.1 V. The simulation and processing time for the 8×8-bit multiplier has been 10 days on a Sun Ultra 10, 333 MHz.

3   Discussion and Simulation Results

The output data from the C program have been processed and plotted using $Matlab^{TM}$. In Fig. 2, we have plotted the power consumption of glitches and transitions for different supply voltages. We have also plotted the ratio between the power consumption of glitches and transitions. In the left column we have the RCA8 and the 4×4-multiplier and in the right column we have the 8×8-multiplier. The dotted lines are constant V_T, and the solid lines are scaled V_T.

As expected, the power consumption of transitions, plots (c) and (d), falls off with the square of the supply voltage. The glitches, on the other hand, which are in plots (a) and (b), show a somewhat different behavior.

In plots (e) and (f), we have plotted the relative power consumption of glitches compared with the total dynamic power consumption. We can see that approximately 40% of the power consumption stems from glitches in the 8×8-bit multiplier. For the RCA8 and the 4×4-bit multiplier, the figures are 15% and 10% respectively. In the multipliers, the power consumption of glitches goes up for lower supply voltages; it can be hard to spot in the plot though. In the RCA8, on the other hand, the relative power consumption of glitches goes down for low supply voltages. This is of course in the case where a constant V_T is used. If V_T is scaled proportionally, the glitch power consumption falls off at exactly the same rate as the power consumption of transitions.

In an attempt of trying to understand why the glitches show different behavior for the two scenarios and also between different structures, we have plotted the voltage swing distribution of glitches at different supply voltages in Fig. 3. On the x-axis, we have the voltage swing relative to V_{DD} and on the y-axis we have the supply voltage. The number of glitches is plotted in the z-direction. The voltage swing has been divided into 0.05 wide bins to improve the readability.

The first two things that one observes are that almost all glitches have full swing and that there are no glitches with an amplitude lower than 10% of V_{DD}. The low-swing glitches are missing because of the glitch amplitude value set in the glitch-detection program. If the glitch amplitude value is changed to a lower value we would get a similar peak as for full-swing glitches. The drawback with

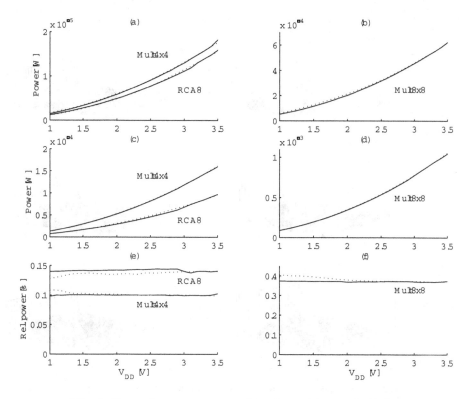

Fig. 2. Dynamic power consumption of transitions and glitches

a lower threshold voltage is that we might fail to detect glitches whose peak is close to one of the supply rails.

In circuits where there are many short signal paths and few longer ones, the vast majority of glitches are generated in the gates from which they are output, whereas very few glitches are the result of mere propagation. There are simply very few paths to propagate glitches through. If the logic depth, or rather the number of long paths, is larger, the propagated glitches will consequently be much more common. In circuits with large gate fan-outs the number of glitches that are propagated may even increase exponentially. At some point, for a certain size and a certain structure of the circuit, the propagated glitches may very well dominate the total number of glitches. For the circuits considered in this paper, we have the 4×4-bit multiplier, where the ratio between propagated and generated glitches is larger than in the RCA8 circuit, and the 8×8-bit multiplier, where the ratio has grown even larger.

A generated glitch is a function of the difference in arrival times of the input signals to the gate producing the glitch. A propagated glitch, on the other hand, is a function of the gate transfer characteristics. Let us now consider the CMOS gate transfer function: Any such transfer function tends to make glitches with

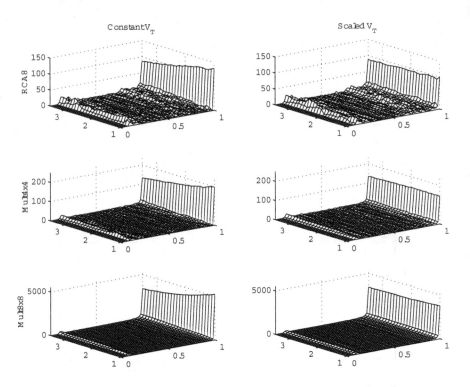

Fig. 3. Voltage-swing distribution of glitches at different supply voltages

low swing even smaller and large-swing glitches even larger. However, the sharper the knee of the function, the more pronounced this effect is. As shown in Fig. 4 reducing V_{DD}, while keeping V_T constant, makes the transfer function sharper. Reducing V_{DD}, while proportionally reducing V_T, will obviously keep the shape of the transfer function intact.

The implication of the ratio between the number of propagated and generated glitches is that the bigger this ratio is for a circuit, the more the circuit is depending on the transfer characteristics of the gates. There are, relatively speaking, very few glitches having medium-range swing inside circuits, which contain a large number of long paths or where the average gate fan-out is fairly large. Consequently, for the 8×8-bit multiplier, where there are many long paths, the propagated glitches are dominating. Thus, this circuit depends heavily on the transfer functions of the gates, i.e. we have relatively few medium-range glitches in this circuit. With the same line of discussion, the RCA8 circuit will have fairly many glitches with medium-range swing, at least in comparison to the multipliers. This is clearly illustrated in Fig 3.

Also, the dependence of the CMOS gate transfer function on V_{DD} and V_T can be observed in Fig. 3. In all three graphs showing V_{DD} reduction at constant V_T, the number of medium-range glitches is decreasing with reduced V_{DD}. This is due to the fact that the slope of the transfer function is getting steeper with

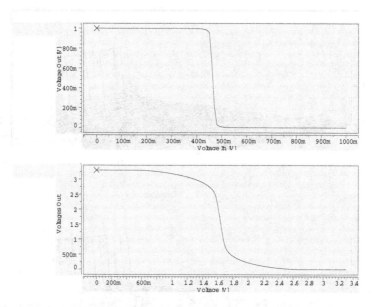

Fig. 4. Voltage-transfer functions of an inverter at V_{DD}=1.0 V and V_{DD}=3.3 V, constant V_T=0.38 V

reduced V_{DD}. With the same line of reasoning, for the simulations using V_{DD} reduction with scaled V_T, there is no redistribution of glitches in terms of voltage swing, since the transfer function stays constant in shape.

The next thing we can observe is that if V_T is kept constant, the number of full-swing glitches increases when V_{DD} is lowered. To explain this, we use Sakurai's alpha-power model [7]. From the model we get the following expressions

$$t_p = \left(\frac{1}{2} - \frac{1 - \frac{V_T}{V_{DD}}}{1 + \alpha} \right) t_T + \frac{C_L V_{DD}}{2 I_{D0}} \tag{2}$$

$$t_T = \frac{C_L V_{DD}}{I_{D0}} \left(\frac{0.9}{0.8} + \frac{V_{D0}}{0.8 V_{DD}} \ln \frac{10 V_{D0}}{e V_{DD}} \right) \tag{3}$$

$$V_{D0} = \left(\frac{V_{DD} - V_T}{V_{DD,ref} - V_T} \right)^{\alpha/2} V_{D0,ref} \tag{4}$$

where t_p and t_T are the propagation and transition times respectively, and Eq. 4 is used to recalculate the drain-saturation voltage to a different V_{DD}. If Eq. 2 and Eq. 3 are combined, we get the following expression for the propagation time

$$t_p = \frac{C_L V_{DD}}{I_{D0}} \left(\left(\frac{1}{2} - \frac{1 - \frac{V_T}{V_{DD}}}{1 + \alpha} \right) \cdot \underbrace{\left(\frac{0.9}{0.8} + \frac{V_{D0}}{0.8 V_{DD}} \ln \frac{10 V_{D0}}{e V_{DD}} \right)}_{k_1} + \frac{1}{2} \right) \tag{5}$$

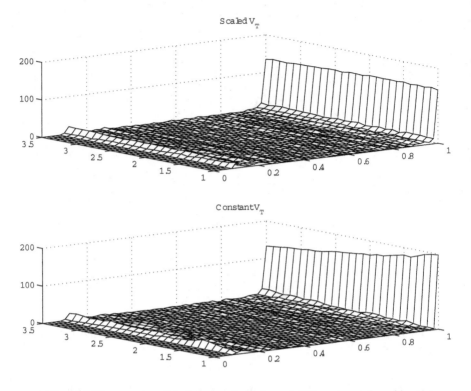

Fig. 5. Voltage-swing distribution of glitches at different supply voltages

Now, let us assume that we have a gate, e.g. an inverter, loaded by a capacitance, C_L, and also make the assumption that the capacitance is discharged by the constant current, I. We have the following expression for the voltage swing of the output node

$$I = C_L \frac{\Delta V}{\Delta t} \Rightarrow \Delta V = \frac{I \Delta t}{C_L} \qquad (6)$$

where Δt is the duration time of the input signal which causes the discharge of the node. If a glitch appears at the input of the gate, it must be due to differences in propagation delay and thus, we can model its duration time as $k_2 \cdot t_p$. If we assume that we have a typical short-channel device, i.e. $\alpha \approx 1$ and that expression k_1 in Eq. 5 is constant, i.e. independent of V_{DD}, we get the following expression for the voltage swing

$$\Delta V = \frac{I_{D0} \cdot k_2 \frac{C_L V_{DD}}{I_{D0}} \left(\left(\frac{1}{2} - \frac{1 - \frac{V_T}{V_{DD}}}{1 + \alpha} \right) \cdot k_1 + \frac{1}{2} \right)}{C_L} = \frac{k_2}{2} (k_1 V_T + V_{DD}) \qquad (7)$$

If we now calculate the relative voltage swing with constant V_T at two different supply voltages: $V_{DD1} = V_{DD}$ and $V_{DD2} = V_{DD}/S$, where S is the scaling factor, we get the following results

$$\frac{\Delta V_1}{V_{DD1}} = \frac{\frac{k_2}{2}(k_1 V_T + V_{DD1})}{V_{DD1}} = \frac{\frac{k_2}{2}(k_1 V_T + V_{DD})}{V_{DD}} \tag{8}$$

$$\frac{\Delta V_2}{V_{DD2}} = \frac{\frac{k_2}{2}(k_1 V_T + V_{DD2})}{V_{DD2}} = \frac{\frac{k_2}{2}(k_1 V_T + V_{DD}/S)}{V_{DD}/S} = \frac{\frac{k_2}{2}(k_1 V_T S + V_{DD})}{V_{DD}} \tag{9}$$

Eqs. 8 and 9 show that if the supply voltage is scaled down and the threshold voltage is kept constant, the swing of the node increases. A recalculation of Eq. 9 with a proportionally scaled threshold voltage, $V_{T2} = V_T/S$ gives

$$\frac{\Delta V_2}{V_{DD2}} = \frac{\frac{k_2}{2}(k_1 V_{T2} + V_{DD2})}{V_{DD2}} = \frac{\frac{k_2}{2}(k_1 \frac{V_T}{S} + \frac{V_{DD}}{S})}{\frac{V_{DD}}{S}} = \frac{\frac{k_2}{2}(k_1 V_T + V_{DD})}{V_{DD}} \tag{10}$$

which gives as a result that the swing of the node is constant if V_T is scaled proportionally. Despite all rough approximations, both these statements agree with the results from the simulations in Fig. 3. A magnified plot of the glitch distribution of the 4×4-bit multiplier is shown in Fig. 5.

3.1 Verification Using 0.13 μm Process Parameters

To evaluate the simulation results, the circuits have also been simulated using the 0.13 μm parameters from the Device Research Group at Berkeley [8]. The supply voltage was 1.5 V and the NMOS threshold voltage 0.24 V giving a V_T-V_{DD} ratio of 0.16 which is between a constant V_T, ratio = 0.25, and a proportionally scaled V_T, ratio = 0.12 of the 0.35 μm process. We use the number of full-swing glitches as a measuring device in the evaluation. From the simulations we get the results in Tab. 1. Since the values of the 0.13 μm process are in between our predicted limits, they do not contradict our results.

Table 1. Number of full-swing glitches

Circuit	Const. V_T, 0.35 μm	Sim., 0.13 μm	Scaled V_T, 0.35 μm	Pred., 0.13 μm	Error
RCA8	122	94	86	97	-3.1%
Mult4x4	167	142	139	147	-3.5%
Mult8x8	4543	3542	3435	3767	-6.0%

4   Conclusion

The power consumption and voltage-swing distribution of glitches have been studied for two types of circuits; the ripple-carry adder and the array multiplier. The main reason for the study was to see if the importance of glitches will increase or decrease in future processes when the supply voltage is scaled down further.

Two different scenarios have been considered, one with constant V_T and one with V_T scaled proportionally to V_{DD}. Neither of these two scenarios will predict the future V_T scaling, but the truth will certainly lie within the limits of the simulations of the two.

When V_T is scaled proportionally to the supply voltage, the relative power consumption of glitches stays almost constant. Furthermore, the voltage-swing distribution remains the same during V_{DD} scaling. That is, if V_T is scaled proportionally, the conditions for glitches will be the same in the future as it is today. However, as mentioned earlier, such V_T scaling is impossible due to leakage.

In the other scenario, where V_T is kept constant, the relative power consumption of glitches increases by some percent for the multipliers and decreases by some percent for the ripple-carry adder. The voltage-swing distribution changes in this scenario. The number of full-swing glitches increases when the supply voltage is lowered. This is the cause of the small relative increase in glitch power for the multipliers.

Under the assumption that the leakage power can be kept at a reasonable level in future processes, the overall conclusion drawn from this study is that the power consumption of glitches will at least be at the same relative level as today.

References

1. J. Leijten, J. van Meerbergen, and J. Jess, "Analysis and Reduction of Glitches in Synchronous Networks," in *Proceedings of the 1995 European Design and Test Conference*, 1995, pp. 398–403.
2. S. Kim and S.-Y. Hwang, "Efficient Algorithm for Glitch Power Reduction," *Electronics Letters*, vol. 35, no. 13, pp. 1040–1041, June 1999.
3. A. Raghunathan, S. Dey, and N. K. Jha, "Register Transfer Level Power Optimization with Emphasis on Glitch Analysis and Reduction," *IEEE Transactions on Computer-Aided Design of Integrated Circuits and Systems*, vol. 18, no. 8, pp. 1114–1131, Aug. 1999.
4. Semiconductor Industry Association, *International Technology Roadmap for Semiconductors: 1999 edition*, Austin, TX:International SEMATECH, 1999.
5. B. Davari, "CMOS Technology: Present and Future," in *Digest of Technical Papers, 1999 Symposium on VLSI Circuits*, 1999, pp. 5–10.
6. J. M. Rabaey, *Digital Integrated Circuits, A Design Perspective*, Electronics and VLSI Series. Prentice Hall, 1996.
7. T. Sakurai and A. R. Newton, "Alpha-Power Law MOSFET Model and its Applications to CMOS Inverter Delay and Other Formulas," *IEEE Journal of Solid-State Circuits*, vol. 25, no. 2, pp. 584–594, Apr. 1990.
8. BSIM Homepage, *http://www-device.EECS.Berkeley.EDU/~bsim3/*, Device Research Group of the Dept. of EE and CS, University of California, Berkeley, 2000.

Degradation Delay Model Extension to CMOS Gates

Jorge Juan-Chico[1], Manuel J. Bellido[1], Paulino Ruiz-de-Clavijo[1],
Antonio J. Acosta[2], and Manuel Valencia[1]

Instituto de Microelectrónica de Sevilla. Centro Nacional de Microelectrónica.
Edificio CICA, Avda. Reina Mercedes s/n, 41012-Sevilla, Spain.
{jjchico, bellido, acojim, manolov}@imse.cnm.es
[1]also with Dpto. de Tecnología Electrónica. Universidad de Sevilla.
[2]also with Dpto. de Electrónica y Electromagnetismo. Universidad de Sevilla.
Tlf. +34-955 05 66 66. FAX: +34-955 05 66 86

Abstract. This contribution extends the *Degradation Delay Model* (DDM), previously developed for CMOS inverters, to simple logic gates. A gate-level approach is followed. At a first stage, all input collisions producing degradation are studied and classified. Then, an exhaustive model is proposed, which defines a set of parameters for each particular collision. This way, a full and accurate description of the degradation effect is obtained (compared to HSPICE) at the cost of storing a rather high number of parameters. To solve that, a simplified model is also proposed maintaining similar accuracy but with a reduced number of parameters and a simplified characterization process. Finally, the complexity of both models is compared.

1   Introduction

As digital circuits become larger and faster, better analysis tools are required. It means that logic simulators must be able to handle bigger circuitry in a more and more accurate way. Simulating larger circuits is aided by the evolution of computer systems capabilities, and accuracy is improved by providing more realistic delay models.

Currently, there exist accurate delay models which take account of most modern issues [1, 2, 3, 4]: low voltage operation, sub-micron and deep sub-micron devices, transition wave-form, etc. Besides these effects there are also dynamic situations which might be handled by the delay model. The most important dynamic effects are the so-called *input collisions* [5]: a gate behavior when two or more input transitions happen close in time may be quite different from the response to an isolate input transition. Of all these input collisions, there is a special interest in the *glitch collisions*, which are those that may cause an output glitch. Being able to handle these glitch collisions is important since they are more and more likely to happen in current fast circuits, and will help us to determine race conditions and truly power consumption due to glitches [6, 7]. This is also strongly related to the modeling of the *inertial effect* [8], which determines when a glitch is filtered, and to the triggering of metastable behavior in latches [9, 10, 11, 12]. Other authors have treated the problem of glitches, either partially or not very accurately [5, 6, 7, 13].

D. Soudris, P. Pirsch, and E. Barke (Eds.): PATMOS 2000, LNCS 1918, pp. 149-158, 2000.

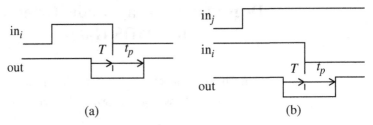

Fig. 1. Quantification of delay degradation: a) degradation due to a narrow pulse, b) degradatio due to a glitch collision.

In a previous work [14, 15] we have studies the problem from a more general poi of view, called the *Delay Degradation Effect*, showing its importance and proposing very accurate model for the CMOS inverter. The model obtained is called *Degradatio Delay Model* (DDM).

In the present paper we extent the model to simple gates (<N>AND, <N>OR) fror the viewpoint of a gate-level modeling, looking for an external characterization suite to standard cell characterization. In Sect. 2 we summarize the basic aspects of th DDM. Then we will make the extension to gates, studying the types of glitch collision and defining an *exhaustive model* for degradation at the gate level in Sect. 3. From th characterization results in section Sect. 4, we will derive a *simplified model*, whic accuracy and complexity is compared to the exhaustive one. Finally, we derive som conclusions.

2   Degradation Delay Model (DDM)

The degradation effect consists in the reduction of the propagation delay of an inpu transition to a gate, when this input transition takes place close in time to a previou input transition. This effect includes the propagation of narrow pulses and fast puls trains, and the delay produced by glitch collisions. This reduction in the delay can b expressed with an attenuating factor applied to the *normal propagation delay*, t_{p0} which is the delay for a single, isolated transition without taking account of the degra dation effect:

$$t_p = t_{p0}\left(1 - e^{-\frac{T-T_0}{\tau}}\right),$$

where T is the time elapsed since the last output transition, and determines how muc degradation applies to the current transition, and T_0 and τ are the *degradation param eters*, which are determined by fitting to electrical simulation data. For a given inpu transition, degradation will depend on the value of T, which express the internal sta of the gate when the transition arrives, caused by previous transitions (Fig. 1). Parame ters t_{p0}, T_0 and τ, in turn, depend on multiple factors: input transition time (τ_{in}

output load (C_L), supply voltage (V_{DD}) and gate's geometry (W_N and W_P). For the normal propagation delay, t_{p0}, good models can be found in the literature [2] and any of them can be used here. In [14] we obtained expressions for T_0 and τ as a function of these parameters:

$$\tau_x V_{DD} = a_x + b_x \frac{C_L}{W_y}$$

$$T_{0x} = \left(\frac{1}{2} - c_x \frac{V_{Ty}}{V_{DD}}\right) \tau_{in} \qquad (2)$$

where the pair (x, y) is (f, N) or (r, P) to distinguish falling from rising output transitions respectively. V_{TN} and V_{TP} are the MOS transistors thresholds. The parameters a, b and c are obtained in order to fit simulation data and characterize the process.

3   Degradation Delay Model at the Gate Level

In this section we will extent the DDM to simple gates (<N>AND, <N>OR) by performing three steps:

1. Reformulate (2) at the gate level, when no information about the gate's internal structure is available. Gate-level degradation parameters are defined in this step.

2. Finding out which distinct cases may lay to delay degradation. These are the *glitch collisions* or *degraded collisions*.

3. Defining a set of parameters for each *glitch collision*.

Due to point 3, the model defined this way may contain many parameters, with a particular set for each glitch collision case. Thus, this model will be referred to as *gate-level exhaustive model for delay degradation*. The purpose of this model is to be able to reproduce the propagation of each glitch collision with maximum accuracy.

3.1   DDM Reformulation at the Gate Level

To rewrite (2) we join together in a single new gate-level parameter the old ones and those *internal* parameters, not visible at the gate level. In other words, a becomes A, b_x/W_y becomes B and $c_x V_{ty}$ becomes C. This way, (2) is rewritten as

$$\tau V_{DD} = A + BC_L$$

$$T_0 = \left(\frac{1}{2} - \frac{C}{V_{DD}}\right) \tau_{in} \qquad (3)$$

A gives the value of τ when $C_L = 0$, and is strongly related to the gate's internal

Table 1. Glitch collisions characteristics for NOR and NAND gates. "i" is the index of the input changing alone or in second place. "j" is the index of the input changing in first place.

Type of collision	Input evolution		Final output transition	
	NOR	NAND	NOR	NAND
Type 1	i: 0-1-0 rest: 0	i: 1-0-1 rest: 1	rising (r)	falling (f)
Type 2	j: 1-0 i: 0-1 rest: 0	j: 0-1 i: 1-0 rest: 1	falling (f)	rising (r)

output capacitance; B depends on the geometry (or equivalent geometry) of the gate and C is related to some "effective" gate threshold. A single value of A, B and C will be calculated for each glitch collision.

3.2   Glitch Collisions

In a simple gate we can distinguish two types of glitch collisions, depending on how and to which values inputs change. To be able to talk in a general sense we will call S the *sensitizing logic value*, or the logic value of the inputs which makes the output of the gate *sensible* to other inputs. It is "0" for (N)OR gates and "1" for (N)AND gates. The opposite value will be noted as $\bar{S}$ (*non-sensitizing logic value*).

When in a simple gate all inputs are equal to S, the output value is S for non inverting gates and $\bar{S}$ for inverting gates. For any other input vector, the value of the output is the opposite. In the following we will consider inverting gates since a similar discussion can be applied to the non-inverting case. Using this, two types of glitch collisions can be defined

- Type 1: Initially, have value S and the output is $\bar{S}$. The output *may* change if an input changes, and a glitch may occur only if the same input changes again to value S. This type corresponds to a positive pulse in one input of a NOR gate or a negative pulse in one input of a NAND gate. Only one input is involved in this type of glitch collision and then, n possible collisions of type 1 exist for a n-input simple gate.

- Type 2: In this case, every input except one (the j-th) have value S and the output is also S. The output *may* change only if input j changes to S, and an output glitch may occur if any input (the i-th) changes to $\bar{S}$. This way, any input pair (even if $i = j$) may produce a glitch collision of type 2, resulting in n^2 possibilities.

We use *collision-i* to refer to type-1 collisions with i-th input changing, and *collision-* to refer to a type-2 collision with input i-th changing after input j-th. In Table 1 we have summarized the properties of both types of collisions for NOR and NAND gates.

Table 2. Vector/matrix form of gate-level degradation parameter for an INVETER and two-inputs NOR and NAND gates.

Type of gate	Parameter A	Parameter B	Parameter C
NOR2	$\tilde{A}_r = \begin{bmatrix} A_{r1} & A_{r2} \end{bmatrix}$ $\tilde{A}_f = \begin{bmatrix} A_{f11} & A_{f12} \\ A_{f21} & A_{f22} \end{bmatrix}$	$\tilde{B}_r = \begin{bmatrix} B_{r1} & B_{r2} \end{bmatrix}$ $\tilde{B}_f = \begin{bmatrix} B_{f11} & B_{f12} \\ B_{f21} & B_{f22} \end{bmatrix}$	$\tilde{C}_r = \begin{bmatrix} C_{r1} & C_{r2} \end{bmatrix}$ $\tilde{C}_f = \begin{bmatrix} C_{f11} & C_{f12} \\ C_{f21} & C_{f22} \end{bmatrix}$
NAND2	$\tilde{A}_r = \begin{bmatrix} A_{r11} & A_{r12} \\ A_{r21} & A_{r22} \end{bmatrix}$ $\tilde{A}_f = \begin{bmatrix} A_{f1} & A_{f2} \end{bmatrix}$	$\tilde{B}_r = \begin{bmatrix} B_{r11} & B_{r12} \\ B_{r21} & B_{r22} \end{bmatrix}$ $\tilde{B}_f = \begin{bmatrix} B_{f1} & B_{f2} \end{bmatrix}$	$\tilde{C}_r = \begin{bmatrix} C_{r11} & C_{r12} \\ C_{r21} & C_{r22} \end{bmatrix}$ $\tilde{C}_f = \begin{bmatrix} C_{f1} & C_{f2} \end{bmatrix}$
INV	$\tilde{A}_r = A_r$ $\tilde{A}_f = A_f$	$\tilde{B}_r = B_r$ $\tilde{B}_f = B_f$	$\tilde{C}_r = C_r$ $\tilde{C}_f = C_f$

3.3   Exhaustive Model for Gate-Level Delay Degradation

The total number of collisions for a n-input gate including type-1 and type-2 is

$$n + n^2 = n(n+1) \ . \tag{4}$$

Any of such collisions may be studied like an inverter under a narrow pulse input. Equations (1) and (3) can be applied to each case and a particular set of (A, B, C) parameters obtained for each collision. In this sense, if we make Δ to represent any of τ, T_0, A, B or C, we can refer to any single value with a notation like this:

- Δ_{Si}: value of parameter Δ for collision-i.

- $\Delta_{\bar{S}ij}$: value of parameter Δ for collision-ij.

These parameters can be expressed in vector/matrix notation like this:

$$\tilde{\Delta}_S = [\Delta_{S1}, \Delta_{S2}, ..., \Delta_{Sn}]$$

$$\tilde{\Delta}_{\bar{S}} = \begin{bmatrix} \Delta_{\bar{S}11} & \cdots & \Delta_{\bar{S}1n} \\ \cdots & \cdots & \cdots \\ \Delta_{\bar{S}n1} & \cdots & \Delta_{\bar{S}nn} \end{bmatrix} \ . \tag{5}$$

In Table 2 we show the vector/matrix form or parameters A, B and C for gates NOR2, NAND2 and INVERTER. Using (5), the expressions in (3) can also be written in vector/matrix form:

$$\tilde{\tau}_S V_{DD} = \tilde{A}_S + \tilde{B}_S C_L \qquad\qquad \tilde{\tau}_{\bar{S}} V_{DD} = \tilde{A}_{\bar{S}} + \tilde{B}_{\bar{S}} C_L$$

$$\tilde{T}_{0S} = \left(\frac{1}{2}\tilde{U}_n - \frac{\tilde{C}_S}{V_{DD}}\right)\tau_{in} \qquad \tilde{T}_{0\bar{S}} = \left(\frac{1}{2}\tilde{U}_{nn} - \frac{\tilde{C}_{\bar{S}}}{V_{DD}}\right)\tau_{in} \tag{6}$$

where $\tilde{U}_n$ and $\tilde{U}_{nn}$ are n-dimensional all-1's vector and matrix respectively.

4 Results

To obtain the whole set of parameter for a gate we use a characterization process which consists in two tasks:

1. Obtain t_p vs. T curves (see eq. 1) using an electrical simulator like HSPICE. For each curve, a value of τ and T_0 is obtained by fitting the simulation data to (1).

2. Task 1 is done repeatedly using different values of C_L and τ_{in}. The resulting τ and T_0 data is fitted to (3) and a value of A, B and C obtained.

The two phases are carried out for each glitch collision. The whole process in order to fully characterize a gate is quite complex. For example, the exhaustive characterization of a NAND4 gate requires performing about 8000 transient analysis. To make such complexity affordable, we have developed an automatic characterization tool which handles the whole characterization process, from launching the electrical simulator which performs the transient analysis, to make the curve fitting tasks. Using this tool, it is quite straight forward to study a wide set of gates.

Qualitatively, the results obtained for all gates analyzed are quite similar in the sense that simulation data can be easily fitted to (1) and (3), validating the degradation model. An example can be seen in Fig. 2. Gates ranging from 1 to 4 inputs have been analyzed. As an example, we present the results for a NAND4 and a NOR4 gates in Table 3. NAND4 data is also in graphical form in Fig. 3, and serves as example since all gates give quite similar qualitative results.

5 Simplified Model

It can be easily observed in Fig. 2 how A, B and C are almost independent of the first changing input (j) in type-2 collisions. It means that in practice, the degradation effect does not depend on which input triggered the last output transition, only on when that output transition took place. In other words, it depends on the state of the gate, but not on which input put the gate on that state. This makes that degradation parameters of the form $\Delta_{\bar{S}ij}$ to be very similar for different values of j.

Based on this result we propose a *simplified degradation model* for gates, in which we consider a single value of the parameter regardless the value of j. It means substituting each row in the matrices of Table 3 for a single value. This single value is partic

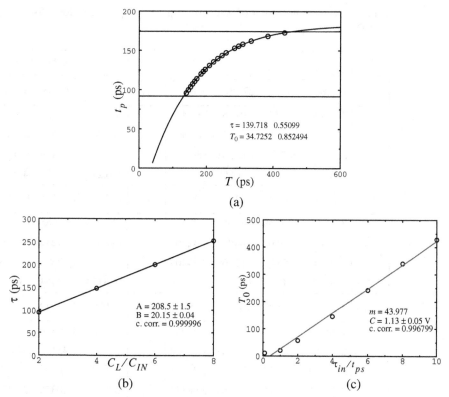

Fig. 2. Example of simulation data fitting to degradation model: a) t_p vs. T, b) τ vs. C_L, c) T_0 vs. τ_{in}.

ular one taken from each row $(\Delta_{\bar{S}ik})$ and is noted $\Delta_{\bar{S}i}$. It is

$$\Delta_{\bar{S}ij} = \Delta_{\bar{S}i} = \Delta_{\bar{S}ik} \qquad \forall (i, j) . \tag{7}$$

Any value of k with $1 \le k \le n$ is possible. Our criterion is to take an intermediate value of the form

$$k = \text{int}\left(\frac{n+1}{2}\right) . \tag{8}$$

This way, each matrix in Table 3 is reduced to a single column, which can be written like a vector. The resulting simplified set of parameter for NOR4 and NAND4 gates of the previous example are shown in Table 4. The number of glitch collisions that we need to take into account is reduced to $2n$.

The values of the parameter for different j are so similar that the simplified model is almost as accurate as the exhaustive model, but the number of parameters is greatly reduced, as well as the characterization process complexity. In Table 5 we compare the

Table 3. Vector/matrix form of gate-level degradation parameter for a four-inputs NOR and NAND gates.

	NOR4					NAND4			
$\tilde{A}_r$	112.819	145.08	275.101	568.706	$\tilde{A}_f$	341.335	363.03	432.19	533.097
$\tilde{A}_f$	788.806	804.331	780.062	786.426	$\tilde{A}_r$	364.451	356.81	359.536	357.584
	824.225	824.258	823.485	824.397		374.961	364.568	365.183	365.746
	860.778	847.25	852.561	850.086		395.57	391.429	390.884	388.101
	875.267	876.37	881.897	878.463		436.244	432.208	421.57	416.158
$\tilde{B}_r$	2.71788	2.62542	2.41312	1.83907	$\tilde{B}_f$	15.2991	15.4685	15.3365	14.7835
$\tilde{B}_f$	7.32507	7.21159	7.30652	7.29638	$\tilde{B}_r$	14.7053	14.5088	14.4525	14.5096
	7.43454	7.45502	7.44032	7.42662		15.2026	15.4239	15.4003	15.4015
	7.49901	7.5641	7.52869	7.54409		15.6956	15.7685	15.7861	15.833
	7.60508	7.60983	7.58054	7.61039		16.3134	16.2464	16.3738	16.4578
$\tilde{C}_r$	1.56364	1.47036	1.39764	1.29989	$\tilde{C}_f$	1.49791	1.39779	1.27071	1.04927
$\tilde{C}_f$	1.80267	1.76748	1.69145	1.67959	$\tilde{C}_r$	1.97685	1.89809	1.8573	1.84559
	2.14557	2.09964	2.05788	2.02964		2.49992	2.43175	2.40956	2.39455
	2.42609	2.37594	2.3378	2.31878		2.90296	2.90767	2.752	2.74911
	2.74211	2.70625	2.67864	2.68137		3.2206	3.20356	3.1773	3.15793

Table 4. Vector form of simplified gate-level degradation parameter for a four-inputs NOR and NAND gates.

	NOR4					NAND4			
$\tilde{A}_r$	112.819	145.08	275.101	568.706	$\tilde{A}_f$	341.335	363.03	432.19	533.097
$\tilde{A}_f$	804.331	824.258	847.25	876.37	$\tilde{A}_r$	356.81	364.568	391.429	432.208
$\tilde{B}_r$	2.71788	2.62542	2.41312	1.83907	$\tilde{B}_f$	15.2991	15.4685	15.3365	14.7835
$\tilde{B}_f$	7.21159	7.45502	7.5641	7.60983	$\tilde{B}_r$	14.5088	15.4239	15.7685	16.2464
$\tilde{C}_r$	1.56364	1.47036	1.39764	1.29989	$\tilde{C}_f$	1.49791	1.39779	1.27071	1.04927
$\tilde{C}_f$	1.76748	2.09964	2.37594	2.70625	$\tilde{C}_r$	1.89809	2.43175	2.90767	3.20356

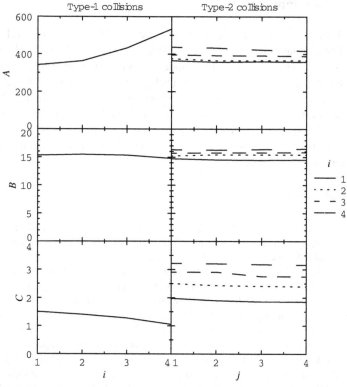

Fig. 3. Graphical representation of gate-level degradation parameter for a NAND4 gate. i is the changing input in type-1 collisions. j and i are the first and second changing inputs respectively in type-2 collisions. The graphs show the variation of degradation parameters with the number of the input(s) changing.

Table 5. Comparison of the exhaustive and the simplified model in terms of number of parameters and characterization complexity. If n_c is the number of glitch collisions, the number of parameters is $3n_c$ and the number of transient analysis is stimated as $400n_c$. n_c is $n(n+1)$ for the exhaustive model and $2n$ for the simplified model.

n	no. of parameters		no. of tran analysis	
	exhaustive	simplified	exhaustive	simplified
1	6	6	800	800
2	18	12	2400	1600
3	36	18	4800	2400
4	60	24	8000	3200
5	90	30	12000	4000

number of parameters and the characterization complexity (measured as the number of transient analysis) for both models, applied to gates with up to five inputs. The benefits of the simplified model are clear, specially when increasing the number of inputs.

6   Conclusions

A way to extend the degradation delay model to the gate level has been presented Those input collisions that may cause degradation effect (glitch collisions) have bee analyzed and classified. Two models are presented: an exhaustive one which assigns set of degradation parameters to each glitch collision, and a simplified one which asso ciates a set of parameters to each input, instead to each collision. The simplifies mod has similar accuracy but reduces both the number of parameters and the complexity c the characterization process. This model allows the accurate simulation of the degrada tion effect at the gate level. An experimental simulator which implements this model currently under development.

References

1. L. Bisdounis, S. Nikolaidis, O. Koufopavlou. "Analytical Transient Response ar Propagation Delay Evaluation of the CMOS Inverter for Short-Channel Devices". IEEE J. Solid-State Circ. pp. 302-306. Vol. 33, no. 2, Feb. 1998.
2. J.M. Daga, D. Auvergne. "A Comprehensive Delay Macro Modeling for Submicromet CMOS Logics". IEEE J. of Solid State Circuits. Vol. 34, No. 1, Jan. 1999.
3. A.I. Kayssi, K.A. Sakallah, T.N. Mudge. "The Impact of Signal Transition Time on Pa Delay Computation". IEEE Trans. on Circuits and Systems-II: Analog and Digital Sign Processing, Vol. 40, No. 5, pp. 302-309, May 1993.
4. D. Auvergne, N. Azemard, D. Deschacht, M. Robert. "Input Waveform Slope Effects CMOS Delays". IEEE J. of Solid-State Circ., Vol. 25, No. 6, pp. 1588-1590. Dec. 1990
5. E. Melcher, W. Röthig, M. Dana. "Multiple Input Transitions in CMOS Gates Microprocessing and Microprogramming 35 (1992) pp. 683-690. North Holland.
6. C. Metra, M. Favalli, B. Riccò. "Glitch power dissipation model". In Proc. PATMOS'95. p 175-189
7. M. Eisele, J. Berthold. "Dynamic Gate Delay Modeling for Accurate Estimation of Glit Powe at Logic Level". In Proc. PATMOS'95. pp. 190-201.
8. S. H. Unger. "The essence of logic circuits". Ed. Prentice-Hall International, Inc. 1989
9. L.R. Marino. "General Theory of Metastable Operation". IEEE Trans. on Computers, C-. n.2, pp. 107-115, Feb. 1981.
10. L. Kleeman, A. Cantoni. "Metastable Behavior in Digital Systems", IEEE Design and Test Computers, vol. 4. Dec. 1987
11. L.M. Reyneri, L.M. del Corso, B. Sacco. "Oscillatory Metastability in Homogeneous ar Inhomogeneous Flip-flops". IEEE J. of Solid-State Circ. Vol.25. n.1. Feb. 1990.
12. J. Calvo, M. Valencia, J.L. Huertas. "Metastable Operation in RS Flip-flops". Int. Electronics, Vol. 70 n.6. 1991.
13. D. Rabe, B. Fiuczynski, L. Kruse, A. Welslau, W. Nebel. "Comparison of Different Ga Level Glitch Models". In Proc. PATMOS'96. pp. 167-176.
14. J. Juan-Chico, M.J. Bellido, A.J. Acosta, A. Barriga, M. Valencia. "Delay degradation effe in submicronic CMOS inverters". In Proc. PATMOS'97. pp. 215-224. Louvain-la-Neuv Belgium, 1997.
15. M.J. Bellido, J. Juan-Chico, A.J. Acosta, M. Valencia, J.L. Huertas. "Logical Modelling Delay Degradation Effect in Static CMOS Gates". IEE Porc. Circuits, Devices and System In Press.

Second Generation Delay Model
for Submicron CMOS Process

M. Rezzoug, P. Maurine, and D. Auvergne

LIRMM, UMR CNRS/Université de Montpellier II, (C5506),
161 rue Ada, 34392 Montpellier, France

Abstract. The performance characterization and optimization of logic circuits under rapid process migration is one of the big challenges of nowadays submicron CMOS technologies. This characterization must be robust on a wide design space in predicting the performance evolution of designs. In this paper we present a second generation of analytical modeling of delay performance, considering speed carrier desaturation induced non linear variation of delay, I/O coupling, load and input ramp effects. A first model is deduced for inverters and then extended to logic gates through a reduction protocol of the serial transistor array. Validations are given, on a 0.18μm process, by comparing values of simulated (HSPICE) and calculated delay for different configurations of inverters and gates.

1 Introduction

The design complexity afford by actual submicron processes implies to increase the level of circuit abstraction to manage this complexity. But the need of accuracy imposes to get available, at the highest level of abstraction, accurate physical level information on the performance of the structures used in the design.

Accurate timing circuit characterizations must be available at all the abstraction levels. Considering the external operating conditions they also may be able to predict the circuit performance evolution during process migration, voltage scaling or any alternative used for design optimization.

Speeding up the design time implies using logic cells or macro cells with well characterized performances. Standard look up tables with linear interpolation are too time consuming and no more sufficient to model the delay performances of today designs implemented in submicron processes.

An accurate modeling of this performance necessitates reliable data on the structure switching time together with their transition time. An accurate prediction of these data must be obtained when varying the structure or its operating conditions such as the load, the controlling input slew or the supply voltage.

Different methods have been proposed to model the delay at gate level. In the empirical method [1] the delay is represented as a polynomial expression with parameters calibrated from electrical simulations. This results in an empirical representation without any design information allowing design performance prediction or optimization. A complex modeling of the output waveform can also be used to obtain a good evaluation of the delay variation [2].

D. Soudris, P. Pirsch, and E. Barke (Eds.): PATMOS 2000, LNCS 1918, pp. 159-167, 2000.

The look up table method constitutes a discrete approach of the delay performance representation in which switching delay and output slope are listed versus the load and the slew of the input control. The final value is obtained from interpolation between the tabulated one. These tables feed from electrical simulations suffer of the uncertainty in defining the scale extent of the axis of the table to be characterized. Moreover they are of no help to the designer in defining optimization criteria as well as clearly showing reasonable limits to be considered for load and slew.

In the third method which can be considered, a design oriented modeling of the switching delays and transition times is developed from a careful study of the switching process and currents of the logic structures. This method constitutes a new alternative for design and timing tools developers because it may be accurate, fast and gives opportunities for performance optimization. An accurate modeling at inverter level is usually given and then generalized to gates by reduction of the serial array of transistors to an equivalent one [3], [4], [5], [6], [7], [8].

The goal of this paper is to extend a former global modeling of delay [9], to the general characterization of delay performances of deep submicron processes in which preceding approximations are removed to consider the successive reduction of transistor channel length. We propose a second generation of delay model for library cells validated on 0.18μm CMOS process.

In section 2 we present this new model developed for inverters. The extension to gates is given in section 3. In each case validations are given through Spice simulations of different configurations of inverters and gates implemented in a 0.18μm CMOS process. Finally a conclusion on this work and the future extensions is given in section 4.

2   Second Generation Delay Model for Inverters

It is generally observed that short channel effect related high electric fields and carrier speed desaturation effects during the switching process induce non linear variation of the delay with the external loading and controlling parameters. As a result the delay of CMOS structures depends not only on the structure but on the size of the gates, the load, the controlling input slope and the rank of the switching gate input [10].In Fig. 1 we represent the variation of the switching delay of inverters (gates) with different configuration ratio values versus the input ramp duration τ_{IN}/T_{HLS} normalized with respect to the structure step response used as a metric for performance, that we will defined later.

As shown, depending on the strength of the switching transistor, the delay variation appears highly non linear with the input ramp duration. This slow input ramp effect is difficult to be considered using look up tables and is responsible of large discrepancies between prediction and measurements.

As a result the need for the definition of a timing performance model including an accurate representation of switching delays and signal transition times with clear evidence of the structure, the supply voltage value, the size of the transistors the output load and the input signal transition time.

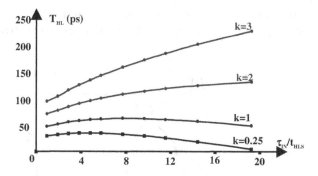

Fig. 1. Illustration of the non-linear variation of switching delay of an inverter for different control and load conditions.

We develop here the model for inverters focusing mainly on the falling edge, the rising one being deduced easily by inspection.

2.1   Metric Definition

As previously mentioned in Fig. 1. We defined metrics for delay in order to get easy calibration and design space representation. The first one τ_{ST}, is characteristic of the speed performance of the process and is independent of the transistor width [9]. The second one, t_{HLS} represents the step response of an inverter with a real load

$$t_{HLS} = \tau_{ST} \cdot \frac{C_L}{2C_N}$$

$$t_{LHS} = \tau_{ST} R_\mu \cdot \frac{C_L}{2C_P} \tag{1}$$

where C_L is the total output loading capacitance, C_N, C_P represent the gate input capacitance of the switching transistor and R_μ is the dissymetry factor between N and P transistors.

These parameters are used in a process characterization phase in which the transistor threshold voltage, conduction factor and I/O coupling parameters are calibrated following a well-defined protocol.

2.2   Second Generation Model

As illustrated in Fig. 2, using the inverter step response as a reference and considering linear variation of the output wave form, the switching time corresponding to an output falling edge is defined by:

$$T_{HL} = t_{SP} + t_{HLS}^* - \overline{\tau IN} \tag{2}$$

where T_{SP} is the time of occurrence of the maximum P transistor short circuit current [11], τ_{in} the input ramp duration time and t_{HLS}^* the cell step response corrected for slow ramp effects.

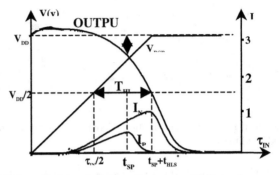

Fig. 2. Inverter delay definition for an output falling edge.

Considering that the P transistor short circuit current reaches its maximum value in linear operating mode at the edge of saturation, the time T_{SP} can be defined easily from the derivative of the current expression with respect to time. This gives:

$$t_{SP} = \frac{\tau_{IN}}{2} \cdot \left(1 - v_{TPL} + \frac{tov}{\tau_{IN}}\right) \tag{3}$$

where v_{TPL} is the threshold voltage value defined with respect to the supply voltage and T_{OV} the overshoot duration time [12] defined by:

$$t_{ov} = \tau_{in} \cdot v_{TNL} + \frac{C_M \cdot \tau_{ST} \cdot (1 - v_{TNL})}{C_n} \cdot (1 + t_{ovcor}) \tag{4}$$

with:

$$t_{ovcor} = \sqrt{1 + \frac{2 \cdot C_n \cdot v_{TNL} \cdot \tau_{in}}{C_M \cdot \tau_{ST} \cdot (1 - v_{TNL})}} \tag{5}$$

where C_M represents the I/O coupling capacitance.

In the same way using the preceding metric slow ramp effects on delay can be reproduced from [9]:

$$T_{HLS}^* = T_{HLS} \cdot \left(1 - 2 \cdot \frac{V_{DSp}}{V_{--}}\right) \tag{6}$$

where V_{DSp} represents the drain source voltage value of the P short circuiting transistor. Developing the three terms of eq.2 results in a complete design oriented delay modeling.

As shown this model exhibits explicit delay dependency on the technological parameters (τ_{ST}), the cell physical parameters (C_N, C_P, C_M) and the load and control conditions (C_L, τ_{IN}).

Despite its accuracy, the development of eq.2 is still too complicated to be used for performance evaluation and optimization. We propose here a simplified expression in which, conserving the design oriented delay representation we introduce pseudo empirical non linear correcting terms such as:

$$
\begin{aligned}
T_{HL} = {} & \delta_n \cdot v_{TN} \cdot \tau_{IN} \\
& + \frac{C_M \cdot \tau_{ST} \cdot (1 - v_{TN})}{CN} \\
& + t_{HLS}\left(1 - \left(\alpha_n k^2 + \beta_n k + \varepsilon_n\right)\left(\frac{\tau_{IN}}{t_{HLS}}\right)^{\gamma_n}\right)
\end{aligned}
\tag{7}
$$

where the output slope is still obtained from [9] as:

$$
\tau_{out} = 2t_{HLS,LHS} \cdot \frac{(1 - v_{TN,TP})}{\left(0.5 - v_{TN,TP} + \dfrac{T_{HL,LH}}{\tau_{in}}\right)}
\tag{8}
$$

The three terms of eq.7 represent respectively the input slope effect, the I/O coupling responsible of the overshoot and the loading effect through the step response t_{HLS} with the correcting factor for slow input ramp induced non linear effects.

These expressions have been validated on different processes (0.35, 0.25 and 0.18μm) resulting in an accurate modeling of the switching delays and output transition times over a large range of design space (less than 10% of discrepancy with respect to HSPICE simulations performed using the foundry model and simulation level)

Illustration of this comparison between calculated and simulated delay and transition time values for inverters with different configuration ratios is given in Fig. 3 and 4.

3   Extension to Gates

We present in this section an extension of the preceding model to simple Nand, Nor gates. The idea is to treat the gates as "equivalent" inverters through the evaluation of reduction factors reflecting the effect of the serial array of transistors on the current possibilities of the gate.

Let us for example consider a two input Nand gate with equal width W_{GEO} for the N transistors of the serial array. For identical load and input controlling ramp the switching delays are different depending on the switching transistor in the array. Due to their biasing or controlling conditions the two transistors of the serial array have different current possibilities.

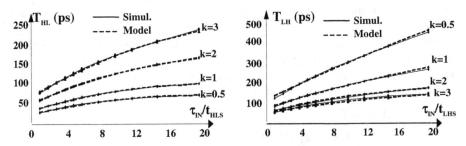

Fig. 3. Comparison of simulated and calculated falling and rising delay values for inverters with configuration ratio ranging from 0.25 to 3, implemented in a 0.18μm CMOS process.

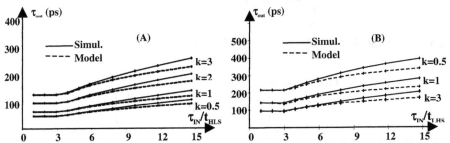

Fig. 4. Comparison of simulated and calculated falling (A) and rising (B) output transition time values for inverter with configuration ratio ranging from 0.25 to 3, implemented in a 0.18μm CMOS process.

The gate control voltage of the transistor close to the output (the top transistor of the array) is lower than the output voltage of the controlling gate due to the ohmic voltage drop of the conducting bottom transistor (close to the ground).

In the same way this bottom transistor suffers from a power supply reduction due to the threshold voltage reduction through the top transistor working as a transmission gate.

In this condition each input must be considered separately in order to develop an equivalent inverter representation from the evaluation of the switching current possibilities of the gate.

Let us consider that the gate controlling input is the top input. The current supplied by the top transistor of width W_{GEO} is smaller than that of an inverter of same size. This is due to the reduction of the applied gate source voltage due to the ohmic drop occurring in the bottom transistor. This current reduction can be easily evaluated from eq.1 and found to be equivalent to a reduced width transistor.

In this way the inverter to the Nand gate controlled on the top input that is with the same current possibility can be defined with an N transistor of width:

$$W_{Eq} = \frac{W_{G\acute{e}o}}{Red_{Sat}} \tag{9}$$

where Red_{sat} is the current reduction factor previously discussed.

Using the delay model developed for inverters this equivalent inverter will exhibit the same delay performance than the Nand gate controlled on the top input.

Let us consider the situation where the bottom transistor is switching. Due to the threshold voltage level degradation introduced by the top transistor the equivalent inverter can be easily deduced replacing the serial array by a transistor of width W_{GEO} but supplied through a reduced voltage equal to $(V_{DD}-V_{TN})$. Note here that the top transistor and the output capacitance as explained in [9] will load the equivalent inverter.

In fact as observed, the working mode of this bottom transistor varies depending on the value of the input slew. For fast input ramps the intermediate node is discharged faster than the output one, in this case the current in the array is limited by the top transistor. For slow input ramp value the current is controlled by the bottom one. This is illustrated in the Table1 where we can verify that for fast input ramps the step responses are identical for bottom or top controls. For slow input ramp conditions the reduction in supply voltage results in faster speed desaturation effects of the carriers in the bottom transistor, resulting in a τ_{STBot} smaller than that of the top one.

Table 1. Comparison of top, middle and bottom step response values of a 3 input Nand.

3 inputs Nand Falling Step Responses				
Load		CL=5Cin	CL=10Cin	CL=15Cin
Top Input	Simulation	76	146	216
	Model	*73*	*145*	*215*
Middle Input	Simulation	78	150	221
	Model	*77*	*153*	*219*
Bottom Input	Simulation	78	150	220
	Model	*77*	*152*	*228*

The last case to consider is a transition controlled by the middle input. For fast input ramps, as shown in Table1, the top transistor fixes the middle step response too. For slow input ramp, the transition is treated in two times as a combination of a top and a bottom commutation, as shown in Fig. 5.

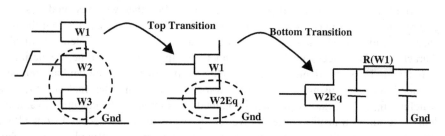

Fig. 5. Two steps middle input transition modeling.

The validation of this reduction protocols for the delay and the output transition time has been obtained by comparison with respect to SPICE simulations for 2 to 4 input Nand and Nor gates. As illustrated in Fig. 5 and 6 good agreement has been observed between simulated and calculated values over a large design range (less than 10% discrepancy for the delay and the slopes).

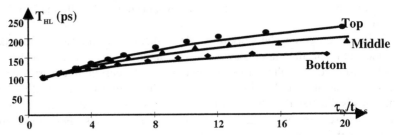

Fig. 6. Comparison of simulated and calculated falling delay values for 3 input Nand gate for Top, Middle and Bottom input implemented in a 0.18μm CMOS process.

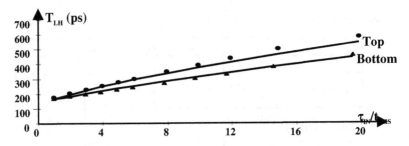

Fig. 7. Comparison of simulated and calculated rising delay values for 2 input Nor gate for Top and Bottom input implemented in a 0.18μm CMOS process.

4   Conclusion

Severe challenge in deep submicron design is to accurately predict timing performance of designs at all level of synthesis. For that we presented a second generation of delay performance modeling of CMOS structures considering the non-linear dependency of delay on controlling parameters. Based on a metric defined to characterize the process performance and the output transition time we defined a complete and design oriented model of delay for inverters. This model has then been extended to gates using a reduction protocol considering the rank of the gate switching input. Validations has been obtained by comparing the calculated values of delay and output transition time of inverter and gates to values deduced from SPICE simulations using the foundry card and simulation model defined for a 0.18μm CMOS process.

Extension to complex gates and to the management of timing closure during place and route is under development.

References

1. J. T. Kong and D. Overhauser, "Methods to Improve Digital MOS Macromodel Accuracy", IEEE Trans. on CAD of ICs and systems, vol. 14, no. 7, pp.868-881 July 1995.
2. L. Bisdounis, S. Nikolaidis, O. Koufopavlou " Propagation Delay and Short-Circuit Power Dissipation Modelling of the C.M.O.S. Inverter " IEEE Transactions on Circuits and Systems –I :Fundamental Theory and Applications, Vol. 45, n°3, March 1998.
3. T. Sakurai and A. Richard Newton, "Delay Analysis of Series-Connected MOSFET Circuits", IEEE Journal of Solid-State Circuits, Vol. 26, No. 2, Feb. 1991, pp 122-131.
4. T. Sakurai and A. R. Newton, "Alpha-power model, and its application to CMOS inverter delay and other formulas", IEEE JSSC vol. 25, pp 584-594, April 1990.
5. A. Nabavi_Lishi, N. C. Rumin "Inverter models of CMOS gates for supply current and delay evaluation", IEEE trans. On CAD of Integrated Circuits and systems, vol.13, n°10, pp.1271-1279, 1994.
6. K.O. Jeppson, "Modeling the influence of the transistor gain ratio and the input to output coupling capacitance on the CMOS inverter delay", IEEE J.Solid State Circuits, vol. 29, n°6, pp. 646-654, June 1994.
7. A. Hirata, H. Onodera, K. Tamaru "Estimation of Propagation Delay Considering Short Circuit Current for Static CMOS Gates". IEEE Transactions. On Circuits and Systems -I-, vol.45, n°11, pp.1194-1198, November 1998.
8. A. Chatzigeorgiou, S. Nikolaidis "Collapsing the Transistor Chain to an Effective Single Equivalent Transistor" DATE'98, pp. 2-6, Paris. March 1998.
9. J. M. Daga, D. Auvergne "A comprehensive delay macromodeling for submicron CMOS logics" IEEE J. of Solid State Circuits Vol. 34 n°1, Jan 1999, pp 42-56.
10. A.I. Kayssi, K.A. Sakallah and T. Mudge, "The impact of signal transition time on path delay computation" IEEE Trans. on Circuits and Systems II: analog and digital proce, ssing, vol.40, n°5, pp.302-309, May 1993.
11. Ph. Maurine, M. Rezzoug, D. Auvergne "Design Oriented Modeling of Short Circuit Power Dissipation for Submicronic CMOS" pp645-650. DCIS'99. November 1999.
12. S. Turgis, D. Auvergne "A novel macromodel for power estimation for CMOS structures" IEEE Trans. On CAD of integrated circuits and systems vol.17, n°11, pp1090-1098, nov.98.

Semi-modular Latch Chains
for Asynchronous Circuit Design

N. Starodoubtsev*, A. Bystrov, and A. Yakovlev

Department of Computing Science, University of Newcastle upon Tyne,
NE1 7RU, U.K.

Abstract. A structural discipline for constructing speed-independent
(hazard-free) circuits based on canonical chains of set-dominant and
reset-dominant latches is proposed. The method can be applied to decompose complex asymmetric C-gate generated by logic synthesis from
Signal Transition Graphs, and to map them into a restricted gate array ASIC library, such as IBM SA-12E that consists of logic gates with
maximum four inputs and includes AO12, AOI12, OA12 and OAI12.
The method is illustrated by new implementations of practically useful asynchronous circuits: a toggle element and an edge-triggered latch
controller.

1 Introduction

Asynchronous circuits offer promising advantages for circuit design in deep-
submicron technology, amongst which the most attractive are low power, EMC,
modularity and operational robustness. As systems-on-chip become a reality,
design of asynchronous control circuits that can tolerate variations in timing parameters of components is particularly important. Examples of such circuits are
interface controllers [1]. A class of asynchronous circuits that are insensitive to
gate delay variations is Muller's speed-independent (SI) circuits [2]. An extensive research has been in methods and algorithms for synthesis of SI circuits in
the last decade [3]. A software tool, called Petrify [4], can synthesise a SI circuit
from its Signal Transition Graph (STG) specification [5] if the latter satisfies the
basic implementability conditions [3]. The result of synthesis is a circuit in which
each non-input signal is a *generalised or asymmetric C-gate* [6] (see Section 2).

The property of *acknowledgement* is characteristic to SI circuits compared
to their less conservative counterparts, such as Burst-Mode circuits [7] or Timed
circuits [8,9]. According to this property, every transition of each gate output is
acknowledged by another signal, which allows the circuit to operate correctly for
unbounded gate delays. Guaranteeing this property, however, is a difficult task,
particularly if the circuit realisation is restricted by a given gate library. Petrify
can perform logic decomposition using a gate and latch library, in which components can be restricted to a given number of input literals. In order to preserve

---

* On leave from: Institute for Analytical Instrumentation, Russian Academy of Science, St. Petersburg, Russia; work in Newcastle supported by EPSRC GR/M94359.

D. Soudris, P. Pirsch, and E. Barke (Eds.): PATMOS 2000, LNCS 1918, pp. 168–177, 2000.

SI property after logic decomposition, Petrify seeks for the newly emerging gate outputs to be acknowledged by other signals, or in the case of complements of signals assumes the delay of input inverters ("bubbles") to be equal to zero. This is a limitation that is not present in our canonical decomposition of generalised C-gates.

This paper addresses the problem of the gate level realisation of SI circuits for CMOS ASIC libraries in which cells may have a limited number of inputs, e.g. three. A regular method for constructing a class of speed-independent circuits composed of 3-input gates AO12, AOI12, OA12 and OAI12 is presented. These gates implement monotonic Boolean functions $d = a+be$, $d = \overline{a + be}$, $d = a(b+e)$ and $d = \overline{a(d+e)}$, respectively. Most CMOS gate libraries include these elements. For example, the IBM SA-12E gate array library [10] offers such gates with high speed and low power consumption, and compared to other 3-input (simple) gates, such as AND3 and OR3, their functional capabilities are greater – one can construct latches out of them. E.g., a simple state-holding element $d = a + bd$ (*set-dominant latch*) can be built out of just one AO12 if output d connected to its input e. We present examples of the application of our construction method, by showing two new implementations of practically useful circuits, one is a toggle element and the other is a pipeline stage (latch) controller. Both circuits are built as chains of the above mentioned positive and negative gates. They are totally speed-independent, they do not have zero delay inverters, and thus compare favourably to the existing solutions.

Negative gate circuits attracted attention about three decades ago since it was noticed that basic CMOS gates have inherent output inverters and thus implement decreasing, or negative monotonic, Boolean function [11,12,13]. Later, interest to negative asynchronous circuits arose when it became clear that they consume less power than their non-negative counterparts [14,15,16].

The rest of the paper is organised as follows. Basic latches are introduced in Section 2. Positive latch chains for the implementation of asymmetric C-gates are described in Section 3. Negative chains and circuit reduction methods are presented in Section 4. Section 5 illustrates applications, a toggle element and edge-triggered latch control circuits. Analysis of behavioural correctness of our circuits is discussed in Section 6. Section 7 contains the conclusion.

2   Basic Latches and Notations

A latch built of a single AO12 element, known as *a set-dominant latch*, is shown in Fig. 1(a). Its behaviour is described by the STG depicted in Fig. 1(b), where "+" denotes the rising signal edge and "-" the falling edge. Signals a and b are inputs, signal d is an output. The solid arcs depict casualty relations within the circuit, whereas the dotted arcs describe the environment behaviour.

Following the STG in Fig. 1(b), transition $a+$ causes transition $d+$ while transition $d-$ is caused by the firing of $a-$ and $b-$. The signalling discipline between the latch and the environment assumes that transitions at a and b inputs may only occur after signal d becomes stable. This latch can be considered as a

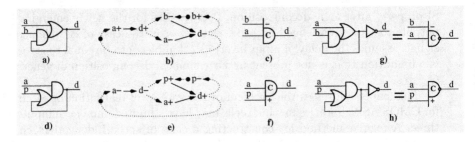

Fig. 1. Basic latches: b-gate(a), its behaviour(b) and notation(c); p-gate(d), its behaviour(e) and notation(f); g, h - negative $\bar{b}$- and $\bar{p}$-gates and their notations

simple case of an asymmetric C-gate [6] as shown in Fig. 1(c). Input a in this drawing, being connected to the main body of the symbol, controls both $d+$ and $d-$, while input b, being connected to the extension marked "-", controls only $d-$.

A dual circuit (*reset-dominant latch*) can be built of OA12 gate. Its schematic, STG and symbolic notation are shown in Fig. 1(d,e,f). In this case, input a controls both edges of d while p controls only the rising edge of d. The shapes of symbols in Fig. 1(c,f) look similar to Latin characters "b" and "p", which we will use to denote the asymmetric C-gates as b-*gate* and p-*gate* respectively. The latches with inverted outputs, shown in Fig. 1(g,h), will be denoted as $\bar{b}$-gate and $\bar{p}$-gate respectively. In the following sections the latches of b, p, $\bar{b}$ and $\bar{p}$ types are used as building blocks to construct more complex components of SI circuits.

3   Generalised Asymmetric C-gates

3.1   Homogeneous Positive Latch Chains: Generalised Latches

A homogeneous chain comprising b-gates only is shown in Fig. 2(a). Such a circuit, denoted as b^n, where n is the number of stages, implements a generalised C-gate with single input a controlling both edges of signal d and n signals $b_1, \ldots b_n$ controlling $d-$ only (set-dominant latch). A dual circuit, denoted as p^m, where m is the number of stages, is shown in Fig. 2(b). Note, that b^n and p^m chains are transitive, so any pair of gates within a chain can be swapped places without affecting the external specification of the chain.

Similar chains of more complex latches can be constructed. An example of a three input p-gate (a, p_1, p_2) is shown in Fig. 2(c). Its transistor-level implementation could be simple, being just one transistor pair larger than a b-gate. Such an element is not present in most gate array libraries and, therefore, will not be considered. However, it can be implemented as a p^2-chain.

3.2   Heterogeneous Positive Latch Chains: C-gates

Any asymmetric C-gate can be constructed as a composition of two generalised b and p-latches (see Fig. 3(a)), which results in a heterogeneous latch chain. Two

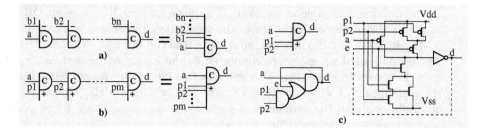

Fig. 2. Homogeneous b^n-chain (a); p^m-chain (b); single gate realisation of p^2-chain (c)

examples of a 3-input asymmetric C-gate, based on two simple b and p-latches, is shown in Fig. 3(b).

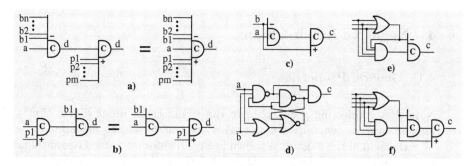

Fig. 3. Heterogenous latch chains: $b^n p^m$-chain for generalised asymmetric C-gate implementation (a); pb and bp-chains (b); 2-input symmetrical bp-chain C-element (c); Mayevsky C-element; 3-input (e) and 4-input (f) C-element

Both chains in Fig. 3(b) are equivalent in their functionality, though having different signal delays from input $b1$ (or $p1$) to output d. The chain function is preserved under any transposition of b and p-gates. Hence, any heterogeneous chain consisting of n b-gates and m p-gates is functionally equivalent to the $b^n p^m$-chain. Both bp and pb-chains can be used to implement a two-input symmetric (Muller) C-element [18] as shown in Fig. 3(c). This realisation favorably compares to the known Mayevsky [19] C-element shown in Fig. 3(d). The following list contains pairs of parameters for comparison of the pb-chain against Mayevsky C-element in the CMOS AMS-3.11 0.6μ realisation: 2/5 gates, 16/25 transistors, 1.51/1.85(ns) cycle time, 11.9/21.5 pJ energy per cycle and 699/1748 μ^2 area.

3.3   Chain Length Reduction

Serial connection of elements in a $b^n p^m$-chain may cause a significant delay. In many cases the chain length can be reduced by using a simpler generalised C-gate (with less inputs) and an expansion circuit comprising AND/OR gates.

A traditional expansion solution [17], shown in Fig. 3(f), uses an OR gate to detect the all-zeroes input state (the condition of switching to zero) and an AND gate to detect the all-ones input state (the condition of switching to one). The outputs of these gates are connected to the inputs of the symmetrical C-gate. It is easy to check that all signals in this circuit are acknowledged under the assumption of wires having no delay (Muller's SI model). This method is applicable only to the symmetrical C-gate inputs, i.e. to those which control both events of output switching to 1 and to 0.

A new compact solution to the expansion problem is shown in Fig. 3(e). It uses a single b-latch instead of the symmetric C-gate. This improvement is achieved at the expense of an additional circuit (connecting the output of the OR-gate to an additional input of the AND gate) providing the acknowledgement of 1 at the output of the OR-gate. A disadvantage of this solution is the number of possible inputs reduced by 1 in comparison with Fig. 3(f).

4   Negative Latch Chains

4.1   General Properties

Note that connecting p or b-gate to the symmetrical input a of a $b^n p^m$-chain implementing a generalised C-gate is equivalent to adding an input to "+" or "-" extension of the C-gate, as shown in Fig. 4(a) for a b-gate. The rule of adding inputs to the extensions for negative latch chains is more complicated.

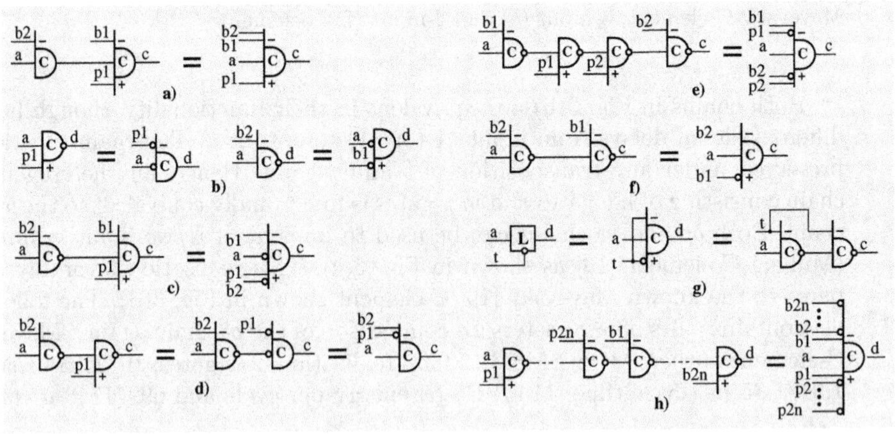

Fig. 4. Negative latch chains transformations: connecting b-gate to C-gate input (a); duality (De Morgan's) rule for $\bar{b}$- and $\bar{p}$-gates (b); connecting $\bar{b}$-gate to C-gate (c), conversion of $\bar{b}\bar{p}$-chain to an asymmetric C-gate, $\bar{b}\bar{p}^2\bar{b}$-chain (e), $\bar{b}^2$-chain (f); transparent latch implementation (g), example of complex negative chain (h)

The functionality of a $\overline{p}$-gate is equivalent to that of a b-gate with all inputs inverted. Output d of a $\overline{p}$-gate (see Fig. 4(b)) gets 1 as soon as $a = 0$ and it gets 0 as soon as $a = p1 = 1$. The same for a b-gate with inverted inputs: if $\overline{a} = 1$, then output d gets 1 and if $\overline{a} = \overline{p1} = 0$, then output d gets 0. This corresponds to DeMorgan's laws.

Connecting $\overline{p}$ or $\overline{b}$-gate to the input of a generalised C-gate is equivalent to adding an inverted input to "-" or "+" extensions of the C-gate, respectively, and inverting its input a. The example in Fig. 4(c) illustrates this for the $\overline{b}$-gate.

Let us consider chains consisting only of $\overline{p}$- and $\overline{b}$-gates, starting with the simplest case of a heterogeneous negative $\overline{b}\overline{p}$-chain shown in Fig. 4(d). Using the above transformations one can see that such a chain results in an asymmetric C-gate with two inputs connected to "-" extension. The symmetric chain $\overline{b}\overline{p}^2\overline{b}$ leads to a more symmetric generalised C-gate depicted in Fig. 4(e). In general, each $\overline{p}$ and $\overline{b}$-gate contributes to either "+" or "-" extensions, respectively, without input inverters if the signal path leading from the input of this gate to the chain output includes odd number of inverters. If such a path includes an even number of inverters (bubbles), then $\overline{p}$-gate ($\overline{b}$-gate) contributes an inverted input to the "-" ("+") extension.

The immediate consequence of this claim is that any transposition of odd gates in a negative chain preserves its functionality. The same takes place for even gates. Hence, each negative chain is functionally equivalent to $\overline{p}^n(\overline{p}\overline{b})^m(\overline{b}\overline{p})^k\overline{b}^l$, where $m, k = 0, 1, 2, \ldots$; $n, l = 0, 2, 4, \ldots$

A special case of such a chain, namely $\overline{b}^2$-chain (see Fig. 4(f)), is useful for transparent latch realisations. The transparent latch with "enable" input t, data input a and output d, which is transparent when $t = 0$ and opaque for $t = 1$, can be implemented as a generalised C-gate shown in Fig. 4(g) with t connected to both extensions.

Finally, as a more complex example, a heterogeneous $(\overline{p}\overline{b})^n(\overline{b}\overline{p})^n$-chain for a generalised C-gate with n inputs in both "+" and "-" extensions, with half of them being inverted, is given in Fig. 4(h).

4.2   Reduction of Negative Chains

Negative chains comprising $\overline{b}$ and $\overline{p}$-gates look more complex than those comprising b and p-gates. However, the structure of $\overline{b}$ and $\overline{p}$-gates, if implemented by CMOS circuits, includes two inverters: the first is the inherent inverter of b or p-gate and the second is output inverters depicted in Fig. 1(g,f). These inverters can be removed without changing the chain function in semi-modular applications. Such a circuit is simpler and faster than its positive counterpart.

A new realisation of inverting transparent latch can be derived from Fig. 4(g). The circuit in Fig. 5(a) is obtained by refining the $\overline{b}$ and b-gates. Note, that signal transitions propagate from left to right in that negative gate chain in such a way that, in any cycle which starts after signal d assumes a new value, signal g accepts the value of signal f with some delay. Therefore, the feedback from f to the input of the left-most $\overline{b}$-gate can be replaced by the feedback from output g. Further,

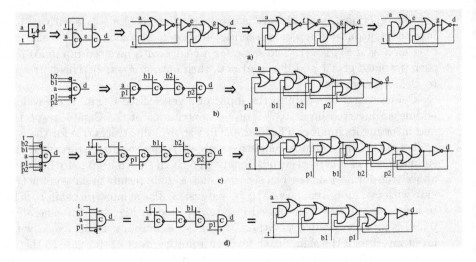

Fig. 5. Reduction of negative chains for: transparent latch application (a), C-gate with a few inputs inverted (b), the same with built-in transparent latch (c), 2-input C-element with built-in transparent latch

the transitions at f and e do not affect any other signal in the circuit. Hence, the inverters at f and e can be safely removed without affecting the circuit function.

This approach can be applied to any negative chain, as shown in examples in Fig. 5(b,c,d). A three-input C-element with two inverted inputs (see Fig. 5(b)) can be realised as $\bar{p}\bar{b}^2\bar{p}$-chain. A similar C-gate with an additional built-in transparent latch can be obtained from a reduced $\bar{b}^2\bar{p}\bar{b}^2p$-chain as shown in Fig. 5(c). A mixed negative-positive $\bar{b}^2bp$-chain, consisting of negative and positive latches, realises a two-input symmetric C-element with a built-in transparent latch (see Fig. 5(d)). This solution can be seen as a Muller C-element enhanced with the enabling/blocking input t.

5   Examples Based on Reduced Negative Chains

We have, so far, considered only applications mapped easily on latch chains. We will now consider other applications, which are compositions of two chains. These implementations present new solutions to the known practical circuit designs. They illustrate the power of the reduction approach applied to a chain that is a backbone of the application.

Toggle. A toggle is one of the key elements in constructing self-timed micropipeline controllers [20], with two-phase signalling discipline. The STG of a toggle element is shown in Fig. 6(a). It responds to each even (odd) transition on input x with a transition on output $y1$ ($y2$).

A known solution for toggle circuit [21] based on two transparent latches with different polarity of the control signal x is shown in Fig. 6(b). We propose

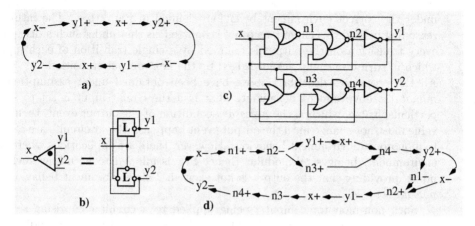

Fig. 6. A toggle circuit: STG (*a*), transparent latch-based realisation (*b*), reduced negative circuit(*c*), refined STG (*d*)

the implementation shown in Fig. 6(c), which is based on the transparent latch shown in Fig. 5(a). Its STG is given in Fig. 6(d).

Being implemented in IBM SA-12E gate array library (0.25μ, 2.5V), this circuit has the following delays from input x to outputs y_1 and y_2: $d(y_1+) = 0.29ns$, $d(y_2+) = 0.19ns$, $d(y_1-) = 0.30ns$, $d(y_2-) = 0.24ns$.

Edge-triggered latch control circuit. The edge-triggered latch control circuit described in [6] has the STG shown in Fig. 7(a). We refine the implementation based upon asymmetric C-gates [6] using our basic negative chains. The circuit in Fig. 7(b) is obtained by further reduction and simplification.

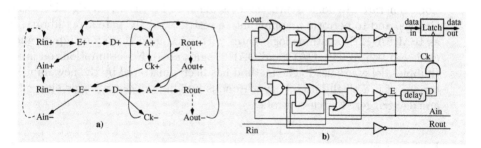

Fig. 7. Edge-triggered latch control: STG (*a*), circuit (*b*)

6   Behavioural Correctness

Semi-modularity [2] of two above examples was checked by Versify tool. All other our proposed solutions are also semi-modular, i.e. no hazards are possible under the correct environment behaviour defined in Fig. 1(b,e). The intuitive reasoning behind this claim (the proof is omitted) is that under such a discipline every transition on the output is followed by a single transition of each input, which in turn eventually acknowledged by the next output transition.

All the properties described above have been obtained under assumption of *monotonic* environment behaviour. That is if the circuit input is set to some particular value, which is the *necessary* condition of the output event, then this value must not change until the output event happens. All our circuits are robust to monotonic environment behaviour. However, there are applications where the environment, being semi-modular (hence SI), is allowed to withdraw such an input, providing that the output is not excited. This environment behaviour is *non-monotonic*.

Such non-monotonic inputs, being applied to a circuit comprising several stages may cause switching of the internal signals. Under the above condition of the output being not excited the events on internal signals are not acknowledged at the circuit output, which may result in hazards.

Latches and chains shown in Fig. 1-5 may produce hazards in a non-monotonic environment. The robustness analysis of the proposed circuits in non-monotonic environments is the subject of the future work.

7   Conclusion

A method of speed-independent asynchronous controllers design, using a limited fan-in gate library, has been developed. It is based on chains of set-dominant and reset-dominant latches. Several regular structures comprising positive and negative chains are studied and a reduction technique is used at the latch level. Our method can be applied to decompose complex asymmetric gate implementations generated by logic synthesis tools (such as Petrify) from Signal Transition Graphs, and to perform mapping into a restricted ASIC gate array library, such as IBM SA-12E (contains logic gates with maximum three-four inputs and includes AO12, AOI12, OA12 and OAI12 logic gates). No assumptions on inverter (bubble) delay are used. The method has been illustrated by the new implementations of practically useful asynchronous building blocks: a toggle element and an edge-triggered latch controller.

References

1. M. Kishinevsky, J. Cortadella, A. Kondratyev and L. Lavagno. Asynchronous interface specification, analysis and synthesis, Proc. DAC'98, pp. 2–7.
2. D. E. Muller and W. S. Bartky. A theory of asynchronous circuits. In Proceedings of an International Symposium on the Theory of Switching, pp. 204-243. Harvard University Press, April 1959.
3. A. Kondratyev, J. Cortadella, M. Kishinevsky, L. Lavagno, and A. Yakovlev. Logic Decomposition of Speed-Independent Circuits, In Proceedings of the IEEE/, Vol.87, No.2, February 1999, pp. 347-362.

4. J. Cortadella, M. Kishinevsky, A. Kondratyev, L. Lavagno and A. Yakovlev. Petrify: a tool for manipulating concurrent specifications and synthesis of asynchronous controllers, IEICE Trans. Inf. and Syst., Vol. E80-D, No.3, March 1997, pp. 315-325.

5. L.Y.Rosenblum, and A.V.Yakovlev. Signal graphs: From self-timed to timed ones, in Proc. Int. Workshop Timed Petri Nets, Torino, Italy, 1985, pp. 199–207.

6. S.B. Furber and J. Liu. Dynamic logic in four-phase micropipelines. Proc. of the Second Int. Symp. on Advanced Research in Asynchronous Circuits and Systems (ASYNC'96) March, 1996 Aizu-Wakamatsu, Japan, pp.11–16.

7. Steven M. Nowick. Automatic Synthesis of Burst-Mode Asynchronous Controllers. PhD thesis, Stanford University, Department of Computer Science, 1993.

8. Chris J. Myers. Computer-Aided Synthesis and Verification of Gate-Level Timed Circuits. PhD thesis, Dept. of Elec. Eng., Stanford University, October 1995.

9. Ken Stevens, Ran Ginosar, and Shai Rotem. Relative timing. In Proc. International Symposium on Advanced Research in Asynchronous Circuits and Systems (ASYNC'99), Barcelona, Spain, pages 208-218, April 1999.

10. http://www.chips.ibm.com/techlib/products/asics/databooks.html

11. T.Ibaraki, S.Muroga. "Synthesis network with a minimal number of negative gates", IEEE Trans. on Computers, Vol. C-20. No. 1, January 1971

12. G.Mago, "Monotone function in sequential circuits", IEEE Trans. on Computers, Vol. C-22. No. 10, October 1973, pp.928 - 933.

13. N.A.Starodoubtsev. Asynchronous processes and antitonic control circuits, In: Soviet Journal of Computer and System Science (USA), English translation of Izvestiya Akademii Nauk SSSR. Technicheskaya Kibernetika (USSR), 1885, vol.23, No.2, pp.112-119 (Part I. Description Language), No.6, pp.81-87 (Part II. Basic properties), 1986, Vol.24, No.2, pp.44-51 (part III. Realisation).

14. C.Piguet. Logic synthesis of race-free asynchronous sequential circuits. IEEE JSSC, vol.26, No 3, March 1991. pp. 371-380.

15. C.Piguet. Synthesis of Asynchronous CMOS Circuits with Negative Gates. Journal of Solid State Devices and Circuits, vol.5, No.2, July 1997.

16. C.Piguet, J.Zahnd. Design of Speed-Independent CMOS Cells from Signal Transition Graphs. PATMOS'98. Oct.1998, Copenhagen, pp.357-366.

17. V. Varshavsky (Ed.), Aperiodic Automata, Nauka, Moscow, 1976 (in Russian).

18. D. E. Muller. Asynchronous logic and application to information processing, Switching Theory in Space Technology, H. Aiken and W. F. Main, Eds. Stanford, CA; Stanford Univ. Press, 1963, pp. 289-297.

19. J. A. Brzozowski and K. Raahemifar. Testing C-elements is not elementary. In Asynchronous Design Methodologies, pp. 150-159. IEEE Computer Society Press, May 1995.

20. Ivan E. Sutherland. Micropipelines. In: Communications of the ACM. June 1989, vol.32, N6, pp.720-738.

21. M. Josephs. Speed-independent design of a Toggle. Handouts of ACiD-WG/EXACT Workshop on Asynchronous Controllers and Interfacing. IMEC, Leuven, Belgium, September 1992.

Asynchronous First-in First-out Queues

Francesco Pessolano[1] and Joep WL Kessels[2]

[1] South Bank University, 103 Borough Road, London, UK,
francesco.pessolano@sbu.ac.uk
[2] Philips Research Laboratories, Prof. Holstlaan 4, 5656AA Eindhoven, Netherlands,
joep.kessels@philips.com

Abstract. Improving processor performance pushes designers to look for every possible design alternative. Moreover, the need for embedded processing cores exhibiting low power consumption and reduced EM noise is leading to changes in system design. This trend has suggested the adoption of self-timed systems, whose energy and noise characteristics depend upon the processed data and the processing rate. In this paper, we explore the design space of first-in first-output queues, which are fundamental component in most recent proposals for asynchronous processors. Different strategies have been refined and evaluated using the handshake circuits methodology.

1   Introduction

The adoption of sub-micron technologies is raising questions about the processor design methodology to be adopted [1]. Wires stretching from one side of the chip to another are starting to behave as if transmission lines with delays (quantized in clock cycles) varying with both layout and process technology [2]. Processor architectures and implementations offering critical paths dominated by gates instead of interconnections will be thus preferred to actual implementations, which focus on reducing gate delay on the critical path. Within this scenario, the asynchronous circuit design discipline may prove helpful, thanks to its reliance on resource and communication locality [3].

In order to exploit asynchronous system design, recent proposals have been evaluated which aim at the development of an asynchronous-friendly architectural template [4-6]. Such asynchronous-friendly architecture should exploit features like decentralized control, de-coupling and data-dependent computation. The ideal situation would be to have an architectural template with all resources divided among fully de-coupled clusters, which only interact for data communication. Communication is generally implemented by means of first-in first-out queues (FIFO's) in order to improve elasticity and de-coupling among resource clusters. Such FIFO's can be either transparent to the compiler (i.e. dynamically allocated) or treated as simple registers. In the latter case, they also prove valuable in optimizing streaming computation such as in DSP applications [4,5].

Therefore, the choice of FIFO architecture is an important design parameter, which may affect processor performance in its normal working load. In this paper, we explore the design space of asynchronous FIFO queues. Different solutions, among

D. Soudris, P. Pirsch, and E. Barke (Eds.): PATMOS 2000, LNCS 1918, pp. 178-186, 2000.

which two interesting schemes, are analyzed and compared on the basis of their latency, throughput, area and power consumption. In this way, a complete picture of asynchronous FIFO design can be drawn and used in selecting what approach is more suitable for a given design.

This paper is organized as follows. The design methodology is described in Section 2. The FIFO architectures under analysis are described thorough Section 3, 4 and 5. Two novel FIFO architectures, which refine a more conventional one, are also discussed in Section 5. Results and comparison are given in Section 6, whilst some conclusions are drawn in Section 7.

2   Introducing Handshake Circuits

The methodology adopted for the designs described in this paper is based on handshake circuits [7]. A handshake circuit is a connected graph that consists of so-called handshake components, which communicate and synchronize each other through handshake channels.

In general, a handshake circuit may features control components, which only have communication channels not carrying information, and data components, which carry information. In between there are the so-called interface components; these can perform handshaking with or without carrying information at the same time.

The behavior of a handshake channel in each component can be classified into active or passive channel ports. An active channel port raises the request signal and waits for acknowledgment, whilst a passive one waits for the request and raises the acknowledgment. Components can features a different mix of channels leading to complex behavior such as pull components (passive input and active output ports), push components (active input and passive output ports), passive components (passive input and output ports), and so on. This allows to build a minimum set of components featuring the required basic operation upon which build more complex asynchronous systems.

Such basic components can be thus implemented using different asynchronous design styles or even a synchronous approach as recently discussed [7]. In this paper, we have considered only single-rail four-phase handshake circuits, because they previously proved a more efficient choice in terms of power budget and performance [8].

3   Standard FIFO Architecture

A standard asynchronous first-in first-out queue is based on cascaded buffering stages, which behaves like a pull component (Figure 1). Each stage waits for a new input data on its unique passive input channel and, then, outputs it through its active output channel. The functionality of each stage is thus minimal and requires very low design complexity – it is equivalent to two logic gates and a register.

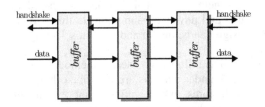

Fig. 1. Standard ripple FIFO architecture

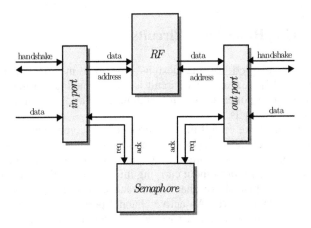

Fig. 2. Circular FIFO architecture with mutual exclusion arbitration

The standard –sometimes called ripple – FIFO queue presents the advantage of being extremely modular, since it can be expanded by simply cascading additional stages. Nevertheless, its main fault is evident: pushed data has to go through – viz. ripples – all stages in the FIFO queue before being popped. Therefore, minimum latency and power consumption are determined by the number of stages in the queue, whilst its throughput is determined by the number of full stages (i.e. tokens) and empty stages (i.e. bubbles). If the number of tokens is constantly equal to the number of bubbles, throughput is at its peak and latency is at its minimum. If the number of tokens is higher, both throughput and latency worsen. Otherwise, only throughput worsens. Power consumption per token is constant, while overall power budget depends on both throughput and number of stages.

4   Arbitrated Circular FIFO Architecture

A different approach is to implements asynchronous FIFO queues adopting a scheme based on a memory-like scheme (Figure 2). Push and pop operations are executed through an input and an output port, which respectively store data in and read data from an internal memory implementing the queue slots. In this case, each port has knowledge of the last read and written slots – i.e. memory addresses – in form of tail and head pointers. These pointers are updated depending on the current operation generally leading to performance reduction.

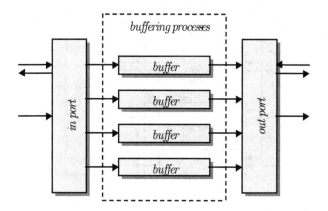

Fig. 3. De-coupled circular FIFO using buffering processes and centralized control.

This approach may improve performance: in fact, each token goes through only each port and a single slot. Therefore, latency and power demand should improve whilst throughput should worsen. Unfortunately, such a picture is easily drawn in synchronous systems, where pointers can be transparently updated once per clock cycle. In the case of an asynchronous design, both ports must be explicitly synchronized in order to ensure that tail and head pointers are consistent in the current operation for any possible sequence of operations. This implies that these pointers are a shared resource that should be accessed in mutual exclusion. Therefore, arbitration – through a semaphore as in Figure 2 – is required in order to grant access to this shared resource to only one port at a time. If multiple concurrent requests were raised, arbitration could incur in additional speed penalty because of internal metastability [3].

5   The De-coupled Circular FIFO Architecture

Both the ripple and the arbitrated FIFO schemes present a single distinctive feature, which is mostly responsible for their different performance. In the ripple FIFO, each slot behaves like a buffering process that does not accept new data before the last one has been transmitted. In the arbitrated scheme, the slot itself is just a memory location and correct sequence of events is ensured by means of memory pointers (tail and head) which are handled by coupled I/O ports. The former features lead to a scheme without arbitration, whilst the latter one to smaller latency and power budget thanks to absence of rippling.

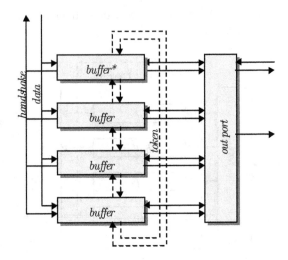

Fig. 4. Circular FIFO using buffering processes and distributed input control (token lines). The marked slot is the first active one after reset.

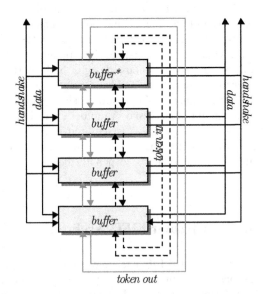

Fig. 5. Circular FIFO using buffering processes and fully distributed control (token in/out lines). The marked slot is the first active one after reset.

An evident improvement on both schemes could be achieved when the process-like slot – ripple scheme – is combined with the notion of memory pointer – arbitrated scheme – as in Figure 3. In such a FIFO queue, both input and output ports are decoupled, since they do not interact by means of a semaphore. The correct sequence of operation is ensured by the fact that each slot will automatically lock itself until its latest data has not being transmitted. Such a scheme avoid the penalty of a semaphore by adopting a more complex FIFO slot, which should result in improved performance.

This scheme can be further refined by moving the functionality of each port into the process-like slots. Each port could be replaced by a communication ring by which a slot is initially enabled as next active slot for a given action. In this way, we may obtain a scheme with either only one (Figure 4) or no I/O ports (Figure 5). The advantage of such schemes is obvious: no pointer is required thus improving latency and reducing power consumption. Moreover, the additional handshaking required to enable the following slot along the rings is easily hidden in the normal operation cycle. Nevertheless, they require broadcast of the input and output channels as well as an increase in slot complexity that could penalize the throughput.

6   Analysis and Comparison

We have evaluated the different schemes implementing a ripple FIFO (*SF*), an arbitrated circular FIFO (*AF*), a de-coupled FIFO with a single output port (*iRF*) and one with no ports (*ioRF*). All designs are 32-bit wide with variable depth and running in self-oscillation mode. Results are reported in Figure 6 for latency, Figure 7 for throughput, Figure 8 for power demand and Figure 9 for area. All design are based on a slow 0.8um@5v technology, which makes absolute values of no interest whilst useful for comparison purposes.

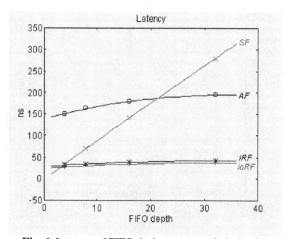

Fig. 6. Latency of FIFO designs respect their depth

Simulating FIFO queues in self-oscillating mode has implications on the obtained simulation results that have to be taken into consideration when analyzing the obtained results. A self-oscillating ripple FIFO will reach a stable equilibrium corresponding to maximum throughput and minimum latency – viz. equal number of tokens and bubbles. Therefore, in a realistic execution mode its performance would be worse than here considered with lower power budget eventually. A self-oscillating arbitrated FIFO instead reaches a stable equilibrium, which will never cause metastability in the semaphore. Therefore, the values in Figure 6, 7 and 8 represent the peak throughput, minimum latency and minimum power budget.

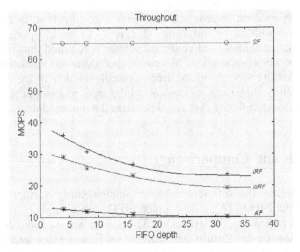

Fig. 7. Throughtput of FIFO designs respect their depth

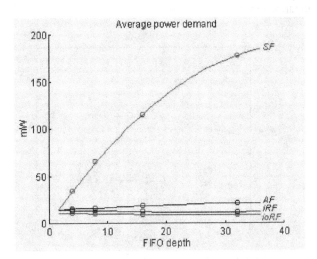

Fig. 8. Power of FIFO designs respect their depth

Self-oscillating de-coupled FIFO's will generally reach a stable equilibrium with the output always starving for new data. In this case, throughput will be the minimum one and distributing the output port will not result in a sensible improvement for the *ioRF* design. Therefore, in a different execution situation we expect the throughput of the de-coupled schemes to gain on the other two schemes – especially for the fully distributed *ioRF* scheme over the *iRF* one. Latency and power budget are not sensibly affected

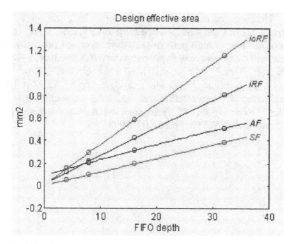

Fig. 9. Area of FIFO designs respect their depth

The de-coupled scheme proves more efficient in terms of latency and power budget, whilst the ripple one leads to higher (peak) throughput and smaller area. However, we expect the de-coupled scheme to prevail in generic execution that can sensibly differ from the self-oscillating mode. The arbitrated scheme is generally worse and its performance is expected to worsen once metastability penalty is considered.

Conclusions

In this paper, we have explored the design space of asynchronous FIFO queue. Different solutions, among which two novel de-coupled schemes, have been analyzed and compared on the basis of their latency, throughput, power consumption and area. In this way, a complete picture of asynchronous FIFO design has been drawn. The novel de-coupled schemes prove a good choice when latency and power budget are at stake, whilst sensible area penalty is introduced. Throughput is smaller respect to the peak one of a standard ripple scheme: however, when the fill and empty rates of the queue differ, the throughput gap is expected to lower sensibly.

References

1. M.J. Flynn, P. Hung, K.W. Rudd, "Deep-submicron Microprocessor Design Issues", IEEE Micro, pp.2-13, July-Aug.1999.
2. D Matzke, "Will physical scalability sabotage performance gains?", *IEEE Computer*, vol.30, No.9, pp.37-39, 1997.
3. MB Josephs, SM Nowick, K van Berkel, "Modelling and design of asynchronous circuits", *IEEE Proceedings*, Vol.87, No.2, 1999.
4. F. Pessolano, L. Augisteijn, H. van Gageldonk, J.W.L. Kessels, J. Leijten, A. Peeters, K. Moerman, "MDSP: Multi-process DSP architecture - Proposal and first evaluation", *Philips Research Technical Note TN355/99*, Nov. 1999.

5. F Pessolano, "Heterogeneous Clustered Processors: Organization and Design", *Springer-Verlag Lecture Notes in Computer Science Euro-Par99*, LNCS vol. 1685 , Sept. 1999.
6. R Kol, R Ginosar, "Kin: a high performance asynchronous processor architecture", *Proc. of the International Conference on Supercomputing ICS'99*, July, 1998.
7. A Peeters, K van Berkel, "Single-rail handshake circuits", *Proc. of the 2nd Working Conference on Asynchronous Design Methodologies*, pp.53-62, May 1995.
8. H van Gageldonk, K van Berkel, et al, "An asynchronous low-power 80C51 microcontroller", *Proc. of the 4th Symposium on Advanced Research in Asynchronous Circuits and Systems ASYNC'98*, pp.96-107, 1998.

Comparative Study on Self-Checking Carry-Propagate Adders in Terms of Area, Power and Performance[1]

A. P. Kakaroudas, K. Papadomanolakis, V. Kokkinos, and C. E. Goutis

VLSI Design Laboratory, Department of Electrical Engineering & Computers
University of Patras, 26500 Patras, Greece
{kakaruda, papk, kokkinos, goutis}@ee.upatras.gr

Abstract. In this paper, several self-checking carry-propagate adders are examined and compared in terms of area integration, power dissipation and performance. Real-time detection of any single fault, permanent or transient, is ensured for all the presented circuits while the characteristics of each adder are illustrated. The results indicate that the characteristics of the adders change when safety mechanisms are applied. The constraints, also, of the required system design dictate the appropriate adder.

1   Introduction

Low-power dissipation has become a critical issue, in the VLSI design area, due to the wide spread of portable and wireless applications and their need for extended battery life. Especially, in the increasing 8-bit market, low-voltage and low-power microcontrollers have made their appearance, challenging even dominant 8-bit conventional architectures such as 68HCxx, 8051/8031.

Portable systems, targeting the medical applications market, require highly safe operation (fail-safe systems), apart from the low-power dissipation. Erroneous functionality of the system, due to system failures, is not acceptable and on-line detection and indication of the error is desirable. The real-time system constraints must also be satisfied.

The design of highly reliable and safety systems leads to the use of additional hardware and/or software overhead (safety mechanisms). The required safety levels of the targeted application, which are derived from the international safety standards, affect notably the needed overhead [2]. In general, the conventional approach of such requirements employs either double-channeled (or multi-channeled) architectures (e.g. two microcontrollers in parallel), which continuously compare their data, or the use of safety mechanisms. These mechanisms detect faulty operation, for each functional unit of a specific architecture.

The first approach leads to a significant increase of the hardware requirements and the power dissipation of the system, thus it is not recommended when low-power dissipation is of great importance. The other approach leads to low-power dissipation of the system when several low-power techniques are applied.

---

[1] This work was supported by COSAFE project, ESPRIT 28593, ESD Design Cluster

D. Soudris, P. Pirsch, and E. Barke (Eds.): PATMOS 2000, LNCS 1918, pp. 187-194, 2000.

This new approach, which allows on-line error detection, is based on the use of coding scheme techniques to obtain redundant information able to detect transient and permanent faults in a significantly lower cost than that of the previous approach. Thus, the class of self-checking circuits and systems has been created. A self-checking circuit consists of a functional unit, which produces encoded output vectors and a TSC checker, which checks the vectors to determine if an error has occurred. The TSC checker has the ability to give an error indication even when a fault occurs in the checker itself.

Although a lot of safety mechanisms, that fulfill almost all the possible functional units have been presented [1], [2], [3], [4], [5], [6], [7], no concern is given for their power dissipation and optimized implementations depending on the specifications. Though less hardware is required for such systems, nothing guarantees that its power dissipation is minimized. In [2], the hardware and power requirements of self-checking architectures for common data path and data storage circuits were examined. The circuits examined were implemented for different coding schemes and use standard cell technology. The detection of any single fault, permanent or transient was ensured for all the proposed circuits while the effectiveness of each coding scheme in the detection of double and triple faults is also determined.

In [8], various implementations of adders are examined in terms of area, power dissipation and performance. In [8], also, the effect of each of these three major terms is illustrated. Although, the effect of these terms is well known, no concern was given to create safety mechanisms for these units, nor to explore for the existing ones, their characteristics. In this paper, a study on several, fault-secure carry-propagate adder implementations, in terms of area, power dissipation and performance, takes place. The examined adders are the one proposed in [7], a slightly modified of the latter implementation and the Carry-Complete full adder, which is adapted to the safety requirements. The rest of this paper is organized as follows: in section 2, basic background is presented. In section 3, the examined implementations are described and in section 4, results from the comparison of these implementations, in terms of area, power dissipation and performance is illustrated. Finally, in section 5 several conclusions are offered.

2   Basic Properties of Self-Checking Circuits

Self-checking circuits are used to ensure concurrent error detection for on-line testing by means of hardware redundancy. All these circuits aim at the so-called totally self-checking goal; i.e. the first erroneous output of the functional block provokes an error indication on the checker outputs. To achieve this goal, checkers have been defined to be Totally Self-Checking (TSC) and they have to be combined with TSC or Strongly Fault Secure (SFS) functional circuits. The terminology of the safety properties of a circuit is following, in order to provide the ability to understand terms as the ones above. Typical architectures to achieve the fault-secure property are also provided in the second subsection.

The design of a secure circuit has many aspects regarding the needs in safety. A circuit G provides the characterization of safety with respect to a fault set F.

If, for circuit G, with respect to F, for each $f \in F$, there is an input vector applied during at least one of the circuit operation modes, that detect F, then this circuit is called *self-testing (ST)*. This property of safety is characterizing the BIST (Built-In-Self-Testing) technique. According to this technique, a set of input vectors is capable to detect all the single errors presented to a circuit. The disadvantage of this technique is that it cannot be applied in safety critical applications because testing is not realized in real-time.

A circuit G is called *fault-secure (FS)* with respect to a fault set F if, for each $f \in F$, when the circuit never produces an incorrect output codeword for any input codeword. The above two properties, the self-testing and the fault-secure, when combined in a circuit, characterize it as *totally self-checking (TSC)*. The meaning of the codeword is strongly related with the code-disjoint property, which will be defined later on.

A circuit G is *strongly fault-secure (SFS)* if, with respect to F, for every fault in F, either:

1) G is self-testing and fault-secure or,
2) G is fault-secure, and if another fault from F occurs in G, then, for the obtained multiple faults, Case 1 or 2 is true.

When a coding algorithm is utilized for the inputs/outputs of a circuit, then, if the circuit always maps code words inputs into code words outputs and non-codeword inputs into non-codeword outputs, the circuit is *code disjoint (CD)*. A circuit that is both totally self-testing and code disjoint is a TSC checker.

A circuit G is *strongly code-disjoint (SCD)* with respect to a fault set F if:
-    before the occurrence of any fault, G is code-disjoint.
-    for every fault f in F, either:
1) G is self-testing or
2) G under fault f always maps non-codeword inputs to non-codeword outputs and if a new fault in F occurs, for the obtained multiple faults, Case 1 or 2 is true.

The above definitions are the properties, in terms of safe operation, of a circuit. Also, for the fault-secure circuits, a hypothesis is made for multiple errors, all along this paper. When an error is present in a circuit, a second one is possible to appear, after enough time, so the first one has been detected. This hypothesis may seem convenient but it is fully realistic. It is very hard that two errors appear simultaneously. When an error appears, the circuit should detect it, in order to be self-testing.

3   Self-Checking Adders

An adder, in order to be characterized as self-checking, must be constructed of two basic blocks, the fault-secure functional block and the TSC checker block. In [7], a study on parity prediction arithmetic operators is presented. The basic idea to design self-checking adders based on parity prediction is illustrated by the authors that propose a self-checking ripple-carry adder. This adder, fig.1, is one of the three implementations that are explored in this paper. The main characteristics of this adder are the performance, which is proportional to the bitwidth (n) of the input, and the low area integration.

A simple modification to the adder of fig. 1, to the parity prediction logic that calculates PCp(i), slightly reduces the glitch effect. The replacement of the cascading

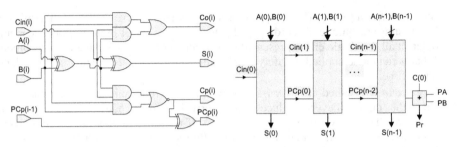

Fig. 1. Self-checking ripple-carry full adder

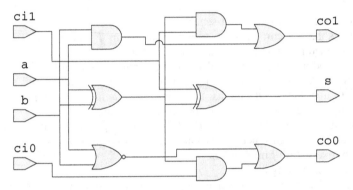

Fig. 2. Carry-complete full adder cell

XOR gates with a balanced tree of XOR gates, which is a basic low-power method, to minimize the "extra" transitions and power in a design, balances all signals and reduces the logic depth. Thus, spurious transitions due to finite propagation delays are minimized.

The third adder is the carry-complete, which appears to be similar to the ripple-carry adder of [7]. The carry-complete full adder cell is illustrated in fig. 2. The main characteristics of this adder are the performance, which is in average proportional to the log2 of the bitwidth (n) of the input, and the low power dissipation due to the balanced logic stages. Additional information and in depth analysis of the carry-complete adder can be found in [9]. Below, follows the proof of the self-checking property for the carry-complete adder.

Following the design methodology of the fault-secure ripple-carry adder found in [7], the prediction of the parity is used for the carry-complete full adder as well. To predict the parity of an adder, the well-known relationship $PS = PA \oplus PB \oplus PC \oplus C_0$ is utilized. This relationship provides the predicted parity of an addition, which must coincide with the parity of the produced sum. Any single error to inputs A, B or C_0 inverts this signal, generating an error indication from a TSC checker. Errors of the carry propagated along the stages must, also, generate an error indication.

The carry-complete full adder is based on the ripple-carry adder modified in such a way to include the propagation-complete detection logic. The two carry-in signals are given by the equations:

$$C_i^1 = A_i B_i + (A_i \oplus B_i) C_{i-1}^1 \qquad (1)$$
$$C_i^0 = \overline{A_i}\ \overline{B_i} + (A_i \oplus B_i) C_{i-1}^0$$

The above equations, if analyzed, prove that C_i^0 is complementary to C_i^1. Note that when both operands are equal to 0 or 1, the "carry/no-carry" decision can be made without waiting for the incoming carry. This property of the carry-complete adder makes the completion of the carry propagation faster, depending on the inputs.

Lemma 1. If any carry-out pair is guided to a TSC checker, any single error on the carry-out signals produces an error indication.

Proof. The property of the TSC checker determines that any non-codeword input produces non-codeword output.

Lemma 2. If a single error occurs to the output of $(A_i \oplus B_i)$, or AB, or $\overline{A}\ \overline{B}$ then the output is safe.

Proof. If the output of the XOR changes then either an erroneous sum is calculated and the correct carry is propagated, so a single error occurs in PS, or the two carry-outs are assigned the same value, which from lemma 1, produces an error indication.

The above statement proves that, the carry-complete full adder, when combined with an n-variable TSC checker, to check the carry-out signals, and the parity prediction mechanism, is self-checking.

4   Experimental Results

Implementations for the ripple-carry adder and for the double-channelled architecture are also considered in this paper, as a reference to the magnitude of area, power and performance of the non-safe and the commonly used fault-secure adder respectively. The rest of the implementations are the fault-secure ripple-carry full adder presented in [7], a modified version of this adder to achieve reduction of glitch effect and the proposed fault-secure carry-complete full adder. All adders are implemented for operand bitwidth of 4,8,16,32 and 64.

The first characteristic, of the adders, that is examined is the required area. Measures for several technologies are taken using Mentor Graphics DVE tool and then their average values are normalized to the non-safe ripple-carry adder. In fig.3 the results are illustrated.

In fig.3 the implementation of the modified adder of [7] is not presented, due to the same area requirements as the ripple-carry adder of [7].

The higher area requirements of the carry-complete full adder was expected, due to the significant area occupied by the n-variable TSC checker. The double-channelled architecture also present a factor of 2,9 compared to the non-safe ripple-carry adder due to the contiguous size of the adder and the n-variable TSC checker.

In [10], a methodology to estimate the power dissipation of circuits is presented, using logic simulators and synthesizers. The experimental results concerning the power dissipation of the examined implementations are illustrated in fig.4. The power dissipation of the double-channelled architecture is greater than 2 due to the power dissipated on the n-variable TSC checker. The experimental results of the double-channelled architecture concerning area requirements and power dissipation are not

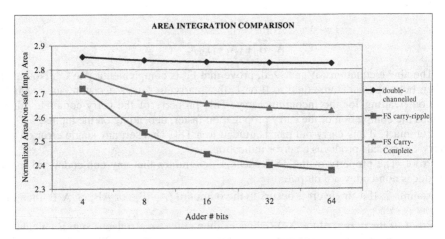

Fig.3. Area requirements of the adders

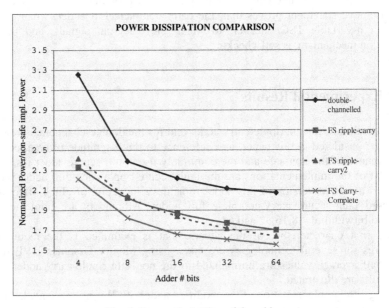

Fig. 4. Power dissipation of the adders

expedient for real applications. The multi-channelled technique is applicable to the outputs of the whole system, thus a systems' power and area are increased by a factor of 2. The last experimental results concern the performance of the implemented adders. It would be useful to mention the average delays of the ripple-carry adder and the carry complete. The ripple carry full adder, present a delay of $(2n-1)\tau$. In contrast, the carry-complete full adder in the worst-case operation is still proportional to n, but the best and average cases are improved considerably, the former being constant and the latter being proportional to $\log_2 n$.

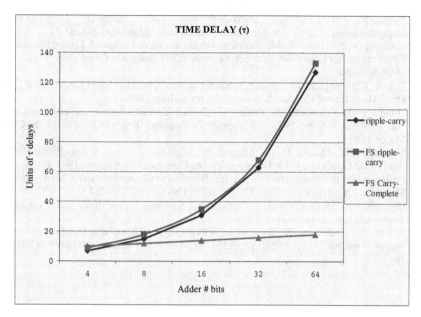

Fig.5. Performance of the adders

The adders are examined in term of performance and the results are illustrated in fig. 5. Note that the results imply that an extra delay is added due to the TSC mechanisms at the end of the adders' stages. Although the influence factor of these mechanisms is not of great importance, it must be considered when designing fault-secure systems targeting to performance critical applications.

5   Conclusions

In this paper, the experimental results, in terms of area, power and performance, of the implementations of several self-checking adders, have been illustrated. A fault-secure carry-complete full adder was proposed and compared to the other implementations. It was proved that for power and performance critical applications the carry-complete full adder is advisable, but not when the area overhead must remain low.

References

1.   Mine, H., Koga, Y.: Basic Properties and a Construction Method for Fail-Safe Logical Systems, IEEE Trans. Electronic Computers, vol. 16, June 1967, 282-289
2.   Kakaroudas, A. P., Papadomanolakis, K. S., Karaolis, E., Nikolaidis, S., Goutis, C. E.: Hardware/Power Requirements versus Fault Detection Effectiveness in Self-Checking Circuits, Proc. of Patmos'99, 387-396

3. Nicolaidis, M.: Fail-Safe Interfaces for VLSI: Theoritical Foundations and Implementation, IEEE Trans. On Comp., vol. 47, no. 1, January 1998, 62-77
4. Nikolos, D., Paschalis, A. M., Philokyprou, G.: Efficient Design of Totally Self-Checking Checkers for all Low-Cost Arithmetic Codes, IEEE Trans. On Comp., vol. 37, no. 7, July 1998, 807-814
5. Jha, N. K., Wang, S.: Design and Synthesis of Self-Checking VLSI Circuits, IEEE Trans. On CAD of Integr. Cir. And Syst., vol. 12, no. 6, June 1993
6. Duarte, R.O., Nicolaidis, M., Bederr, H., Zorian, Y.: Fault-Secure Shifter Design: Results and Implementations, 1997 – European Design and Test Conference ED&TC, Paris, March 1997
7. Nicolaidis M., Duarte, R. O., Manich, S., Figueras, J.: Fault-Secure Parity Prediction Arithmetic Operators, IEEE Design & Test of Computers, April-June 1997, 60-71
8. Callaway, T. K., Swartzlander E.E., Jr.: The Power Consumption of CMOS Adders and Multipliers, Low-Power CMOS Design, IEEE Press, 1998, 218-224
9. Omondi, A. R., Computer Arithmetic Systems: algorithms, architecture, and implementation, Prentice Hall, 1994
10. Zervas, N., Theoharis, S., Soudris, D., Goutis, C.E., Thanailakis, A.: Generalized Low Power Design Flow, ESPRIT 25256 Deliverable Report LPGD/WP1/UP/D1.3R1, Jan.1999

VLSI Implementation of a Low-Power High-Speed Self-Timed Adder

Pasquale Corsonello [1], Stefania Perri [2], and Giuseppe Cocorullo [2,3]

[1] Department of Electronic Engineering and Applied Mathematics,
University of Reggio Calabria, Loc. Vito de Feo, 88100 Reggio Calabria, ITALY
Pascor@deis.unical.it
[2] Department of Electronics, Computer Science and Systems
University of Calabria, Arcavacata di Rende - 87036 - Rende (CS), ITALY
Perri@deis.unical.it
g.cocorullo@unical.it
[3] IRECE-National Council of Research
Via Diocleziano 328, 80125 – Napoli, ITALY

Abstract. Usually, self-timed modules for asynchronous system design are realized by means of dynamic logic circuits. Moreover, in order to easily detect the end-completion, dual-rail encoding is preferred. Therefore, dynamic differential logic circuits (such as Differential Cascode Voltage Switch Logic (DCVSL)) are widely used because they intrinsically produce both true and inverted values of the output. However, the use of dynamic logic circuits presents two main difficulties: i) design and testing is more complex, ii) often it is not possible to use standard design methodology. This paper presents a new static logic VLSI implementation of a high-speed self-timed adder based on the statistical carry look-ahead addition technique. A 56-bit adder designed in this way has been realized using 0.6μm AMS Standard Cells. It requires about 0.6mm^2 silicon area, has an average addition of about 4 ns, and dissipates only 20.5 mW in the worst case.

1 Introduction

Self-timed systems are often attractive as they can compute in mean time, reduce power consumption, and avoid long clock connections [1,2]. They use variable time computational elements by running just when a request and data word arrive. However, designing a self-timed system is not a straightforward task. This is due to the fact that events must be logically ordered avoiding races and hazards by means of an appropriate handshaking protocol. A handshaking circuit is used to guarantee that the computational elements will have stable data inputs during the evaluation phase and to allow overlapping of alternating initialization-evaluation phases in adjacent computational elements.

In many applications, the computational elements are self-timed adders. Efficient variable-time adders have been widely studied [3,4,5]. Typically, they are realized

D. Soudris, P. Pirsch, and E. Barke (Eds.): PATMOS 2000, LNCS 1918, pp. 195-204, 2000.
© Springer-Verlag Berlin Heidelberg 2000

using CMOS dynamic logic circuitry (e.g. Domino, DCVS). The latter are faster and also occupy an area smaller than that required by traditional static circuits. However, they are very sensitive to noise, circuit and layout topology. Moreover, they suffer from charge leakage, charge sharing and cross talk. Therefore, their usage in an asynchronous self-timed system greatly increases the effort required to verify the functionality and the performance of the whole system.

In this paper, we demonstrate that it is possible to realize a high-speed self-timed adder using conventional VLSI design methodologies and static logic cells. In this way, all the above problems are removed and the designer can redirect his effort toward system-level verification of the asynchronous design.

The proposed adder is based on the statistical carry look-ahead addition (SCLA) technique that was recently introduced as a new technique to carry out efficient N-bit self-timed adders whose average delay is much lower than $\log_2(N)$ [6]. One of the peculiarities of this technique is that it does not require dual-rail signaling in order to detect operation completion. Such an implemented 56-bit adder allows an average addition time of about 4 ns to be achieved consuming less than 50% of the power dissipated by conventional dynamic logic designs [3].

2   Brief Background on Statistical Carry-Look-Ahead Addition

The statistical carry look-ahead addition technique allows a self-timed addition between two N-bit operands A and B to be performed using end-completion sensing radix-b full adders as basic elements. Being $b = 2^M$, the adder consists of $n = \lceil N/M \rceil$ M-bit end-completion sensing radix-b full adders. Let's suppose the latter compute propagate terms p_i for each bit position i such that $p_i = A_i \oplus B_i$ (i=0...M-1). Each radix-b full adder can perform its sum operation either waiting for the valid carry-in or without waiting for it. These events can be identified computing the term $\overline{p_{NW}} = p_0 \cdot p_1 \cdot p_2 \cdot \cdots \cdot p_{M-1}$, which is high if the radix-b full adder can proceed without waiting for an incoming carry-in, otherwise it is low.

It is easy to understand that, supposing a uniform distribution of the operands, for a non-least significant radix-b full adder, the probability of having $\overline{p_{NW}} = 1$ is $pr = (b-1)/b$.

Note that the carry-out of the least significant radix-b full-adder is always known after the presentation of the operands and carry-in. Due to this, the probability of computing its carry-out bit without waiting for the valid carry-in is equal to 0. Therefore, in a N-bit adder the $\overline{p_{NW}}$ signals can be computed for all non-least significant radix-b full adders and their composition can be used to represent a (n-1)-bit binary number j.

The probability of a given configuration of j is

$$p_j = pr^{u(j)} \cdot (1 - pr)^{n-u(j)-1} \tag{1}$$

where u(j) is the number of 1s figuring in j.

As it has been demonstrated in [6], the average number of cascaded radix-b full adders waiting for a carry can be obtained by (2)

$$AVG_{FA} = 1 + \sum_{j=0}^{2^{(n-1)}} z(j) \cdot p_j \tag{2}$$

Where z(j) is the length of the longest string of 0s figuring in j, and the 1 at the beginning is due to the fact that at least one radix-b full adder (theoretically the least significant one) is waiting for the carry-in. From numerical calculation of (2), it can be concluded that an adder implemented using the above principle will show an average delay much lower than $\log_2(N)$ [6].

The average time needed for the adder to compute the carry out of all radix-b full-adders (τ_{CARRY}) is obtained by (3), where τ_{LSFA} and τ_{MSFA} are the average delays of the least significant and of a non-least significant radix-b full-adders, respectively. As shown in Section 4, $\tau_{LSFA} < \tau_{MSFA}$ then (3) has to be modified as (4) indicating that at least one of the more significant radix-b full-adders contributes to the average delay.

$$\tau_{CARRY} = \tau_{LSFA} + \left(\sum_{j=0}^{2^{(n-1)}-1} z(j) \cdot p_j \right) * \tau_{MSFA} = \tau_{LSFA} + 0.614123 * \tau_{MSFA} \tag{3}$$

$$\tau_{CARRY} = \tau_{MSFA} + 0,614123 * \tau_{MSFA} \tag{4}$$

It is worth pointing out that τ_{CARRY} does not take into account the average time required to compute the sum bits of a radix-b full-adder when an incoming carry ripples into it for some bit positions after the computation of its carry-out.

In the previous dynamic implementations of adders based on the SCLA technique [6,7], this amount of time has been considered as a constant [6]. Therefore, a further fixed delay is added to (3). However, in the implementations shown in [6,7], the amount of time needed to generate end-completion signal is large enough to suppose that during this time all sum bits have been calculated.

In this paper, for the first time, the additional delay computing sum bits will be fully taken into account also considering its variability. In fact, additional delay is not constant, because it depends on how many bit positions an incoming carry ripples through.

Let's suppose that M=4, that is each M-bit block represents a radix-16 full adder. Let's also suppose that the i-th radix-16 full adder generates the carry-out completion earlier having $p_3=0$. If the (i-1)-th radix-16 full adder has a high carry propagation, this carry will ripple through three bit positions in the i-th radix-16 full adder. Thus, the sum bits change after the i-th full adder has flagged the validity of the carry-out. It can be easily verified that the probability of the above event occurring is 16/256.

Moreover, similar events can occur for the cases in which the i-th radix-16 full adder has $p_3=1$ and $p_2=0$, or $p_3=p_2=1$ and $p_1=0$, or $p_3=p_2=p_1=1$ and $p_0=0$.

A software routine has been built up to compute the probability of all the above cases occurrences. Then, taking into account that a rippling through 1-bit position at least will occur, the actual average addition time is

$$\tau_{AVG}= \tau_{CARRY}+\tau_{rip1}+ 24/256*(\tau_{rip2}-\tau_{rip1}) + 16/256*(\tau_{rip3}-\tau_{rip1}) + \tau_{GEND} \tag{5}$$

Where τ_{ripK} is the time required to a rippling through K-bit positions (i.e. $\tau_{rip0}=0ns$) and τ_{GEND} is a fixed time needed to obtain the end-completion signal.

3   The Proposed Implementation

We have investigated the possibility of efficiently implementing a 56-bit self-timed adder based on SCLA technique using AMS 0.6μm Standard Cells [8]. A completely new appropriate architecture has been designed. In accordance with [6,7], M=4 has been chosen.

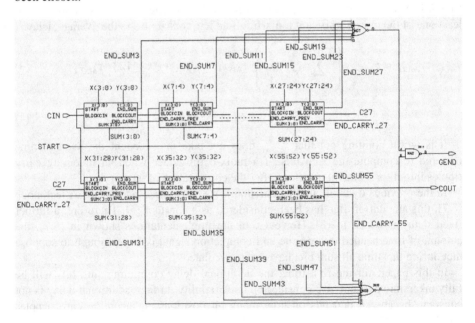

Fig. 1. Top-level architecture of the implemented 56-bit adder

Lowering the START signal starts adder activity. In the following we will suppose that the handshaking modules, which are not detailed here, assure that this event happens as soon as operands appear on the input lines. (Note that lowering START simultaneously at the operands arrival is the worst condition).

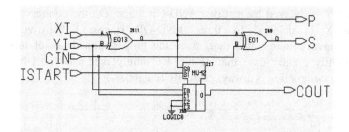

Fig. 2. 1-bit full adder circuit. Its carry-out signal is low initialized (COUT)

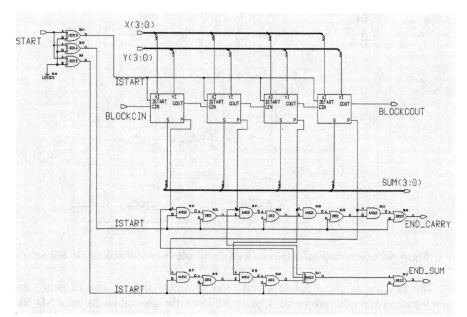

Fig. 3. Schematic diagram of the least significant variable-time radix-16 full adder

In Fig.2, the 1-bit full adder circuit with low initialization phase is reported. Note that the 1-bit full-adder is in its initialization phase (i.e. its carry-out is low independently of operands) when the signal ISTART is high. As soon as the ISTART signal is lowered, the 1-bit full-adder is able to compute its carry-out and sum bits. The above scheme is used in both the least significant and non-least significant radix-16 full adders shown in Fig.3 and Fig.4, respectively. It can be seen that to accommodate loads, some logic gates are either duplicated or strengthened.

Two appropriate rippling chains are used to form END_CARRY and END_SUM signals, which flag the validity of the carry-out and sum bits, respectively. Both chains are high initialized and are realized by means of a proper number of AND-OR stages. Referring to the END_CARRY chain, its output is lowered (after ISTART becomes low) with a delay dependent on which p_i signal is low. Thus, if only $p_3=0$

END_CARRY is delayed by just one AND-OR stage. On the contrary, if only $p_1=0$ END_CARRY is delayed by three AND-OR stages. Exhaustive post-layout simulations have demonstrated that, since the propagation delay of this AND-OR stage is slightly greater than that of the 4:1 multiplexer used in 1-bit full adder circuits, carry completion is always correctly flagged.

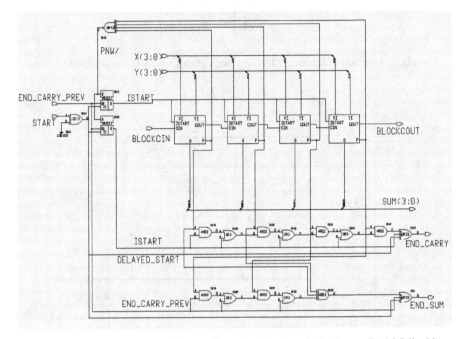

Fig. 4. Schematic diagram of the non-least significant variable-time radix-16 full adder

The running of the rippling chain used to signal the validity of sum bits is analogous. There, the propagate signals influence the generation of the END_SUM signal in an opposite manner. The END_SUM signals (together with the END_CARRY of the most significant radix-16 full adder) are used to determine the whole operation completion. Thus, the production of the END_SUM signals is anticipated to partially compensate for the delay introduced by NOR-NAND logic gates shown in Fig.1. This has been done considering p_0 and p_1 having the same weight (i.e. reducing the maximum rippling path of the END_SUM signal from 4 to 3 AND-OR stages).

The delay introduced by NOR-NAND logic gates shown in Fig.1 can be actually considered as constant and corresponds to the above mentioned τ_{GEND}.

To analyze the running of the adder, let's suppose that at time t_0 valid operands appear and the START signal is lowered. After the delay introduced by a XOR gate (τ_{XOR}) all p_i signals are valid and after a further delay due to a 4-input NAND (τ_{NAND4}) all $\overline{p_{NW}}$ signal are determined.

Observing Fig.3, it can be seen that the ISTART signal is delayed with respect to START by means of a XOR gate. Therefore, ISTART falls when all the p_i signals are valid. Then, the 1-bit full adders start to compute the sum and carry bits. The chain computing the validity of the carry-out (END_CARRY) and of the sum bits (END_SUM) start rippling. END_CARRY signal is lowered when the carry-out bit is valid.

In the meantime, each radix-16 full-adder in more significant position is able to know whether its carry-out can be computed independently of carry-in or not. In the former case, the i-th radix-16 full-adder starts the evaluation of its carries at the time $t_0 + \tau_{XOR} + \tau_{MUX2}$, where τ_{MUX2} is the delay of the 2:1 multiplexer depicted in Fig.4. In the opposite case, the i-th radix-16 full-adder is left in its initialization phase until a valid carry-in arrives. The (i-1)-th radix-16 full-adder will flag the validity of the carry-out bit lowering the END_CARRY$_{i-1}$ signal. Thus, the latter signal will be selected by the above mentioned 2:1 multiplexer to start the evaluation of the i-th radix-16 full-adder. It is worth pointing out that, since $\tau_{XOR} + \tau_{MUX2}$ is slightly greater than $\tau_{XOR} + \tau_{NAND4}$, glitches are avoided on the 2:1 multiplexer output.

When all END_SUM$_i$ and END_CARRY$_{55}$ become high GEND rises signaling operation completion. Then, the circuit is re-initialized by a rising edge of the START signal.

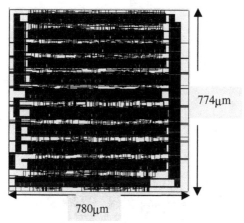

774μm

780μm

Fig. 5. Layout of the 56-bit proposed adder

The completion of the initialization phase is signaled by the subsequent falling edge of the GEND signal. As shown in Fig.1, a NAND gate instead of a typical Muller-C element is used to generate the GEND signal. This choice allows partial overlapping between the initialization phase of the adder and handshake signaling.

4   Results

The circuit described above has been realized using the Austrian Mikro System p-sub, 2-metal, 1-poly, 5 V, 0.6μm CMOS process [8]. The layout of the 56-bit adder for testing purposes is organized on 11 Standard Cells rows and it is reported in Fig.5. It requires about 780μmx780μm silicon area.

Digital and transistor level (using BSIM 3v3 device models) simulations have been performed. In order to calculate average addition times of least significant and non-least significant radix-16 full adders, their exhaustive simulations have been carried out using worst delay models. From above simulations τ_{LSFA}=1.54ns, τ_{MSFA}=2.04ns, τ_{rip3}=2ns, τ_{rip2}=1.46ns, τ_{rip1}=0.8ns and τ_{GEND}=0.12ns were measured. Thus, using (5) an average addition time of about 4.3ns is obtained. In order to confirm the theoretical results, the 56-bit adder was also simulated with a large number of random operands.

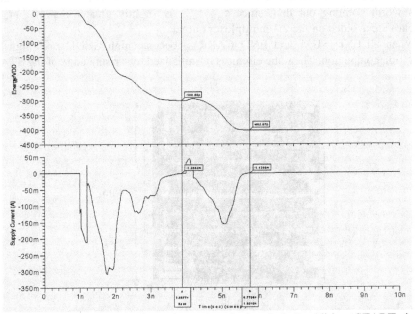

Fig. 6. Energy dissipation and supply current during 0+0+0 addition. START signal falls at 1ns and rises after operation completion re-initializing the circuit

In accord with [3], power dissipation measurements have been performed in two specific cases: a) without carry propagation (minimum value), b) with the longest carry propagation path (maximum value).

At 1ns valid operands and START falling edge (falling time 200ps) are contemporaneously imposed on input lines. After operation completion the START signal rises re-initializing the circuit. This action corresponds to the precharge phase of a dynamic circuitry and for the 56-bit adder it requires about 1.7ns. Re-initializing the circuit, power dissipation of about 5mW and 9.5mW were measured, after operations of the cases a) and b) have been performed at 10 MHz, respectively.

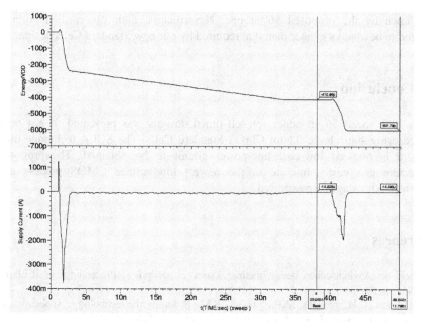

Fig. 7. Energy dissipation and supply current during 0+FFFFFFFFFFFFFF+1 addition. START signal falls at 1ns and rises after operation completion re-initializing the circuit

In Table 1, performance comparison between the proposed adder and recent efficient adders realized using dynamic logic gates is summarized. All data reported in Table1 is referred to laid out designs. Thus, interconnection parasistics were taken into account. Since the adders described in [3] have been realized using 1.0μm CMOS process, scaled values of the average addition time were added to Table 1

Table 1. Performance comparison between the new adder and previously proposed adders. *Data reported in [3] is related to 32-bit adders, their average addition time for the 0.6μm process was estimated by means of $delay \square length^{1.5}$ relationship. 5V supply voltage was used for all designs

Type of adder	Area [μm²]	Power [mW] min/max @10MHz	Avg addition time [ns] Process 0.6μm	Avg addition time [ns] Process 1.0μm
RC in [3]*	274x2430	39.9/41.8	4.6	10
CLA in [3]*	304x2567	45.5/49.3	5.8	12.5
BCL in [3]*	1020x2265	74.1/79.3	4.2	9
New 56-bit adder	780x780	15.0/20.5	4.3	9.2

From these results it can be concluded that the new adder allows very high speed with very low power dissipation. The scaled average addition times, which can be used just as a rough indicator, allow us to claim that if the adders described in [3] are designed using the 0.6μm CMOS process their speed will be, however, lower than that shown by the proposed 56-bit one. Nevertheless, their power dissipation is expected to be always greater than that required by our new Standard Cells design.

5   Conclusion

A new high-speed 56-bit adder for self-timed designs was presented. It has been realized using static logic 0.6μm CMOS Standard Cells. The SCLA technique used allows a high-speed low-cost low-power circuit to be obtained. The proposed architecture can easily migrate to the newest low-voltage CMOS process and augmented advantages are expected.

References

1. Hauck, S.: Asynchronous design methodologies: an overview. Proceedings of IEEE **83** (1995) pp. 69-93
2. Van Berkel, C.H., Josephs, M.B., Nowick, S.M.: Scanning the technology: Application of asynchronous system. Proceedings of IEEE **87** (1999) pp. 223-233
3. Ruiz, G.A.: Evaluation of three 2-bit CMOS adders in DCVS logic for self-timed circuits. IEEE J. Solid State Circuits **33** (1998) pp. 604-613
4. Kinniment, D.J.: An evaluation of asynchronous addition. IEEE Trans. on VLSI **4** (1996) pp. 137-140
5. Kinniment, D.J.: A comparison of power consumption in some CMOS adder circuits. Proc. of PATMOS Conf. (1995)
6. De Gloria, A., Olivieri M.: Statistical carry look-ahead adders. IEEE Trans. on Comp. **45** (1996) pp.340-347
7. Corsonello, P., Perri, S., Cocorullo, G.: A new high performance circuit for statistical carry-look-ahead addition. Int. J. of Electronics **86**, (1999) pp. 713-722
8. Austrian Mikro system, Support Information Center, http://asic.vertical-global.com/

Low Power Design Techniques
for Contactless Chipcards

Holger Sedlak

Infineon Technologies AG, Business Division Security and Chip Card ICs
P.O. Box 80 09 49, D-81 609 Munich, Germany
holger.sedlak@infineon.com

The history of chipcards begun in the eighties of the last century. The first chips consist of a non-volatile memory (NVM), a serial I/O-channel and a finite state machine, offering the necessary security to enable only secure access to the stored data. The memory size was in the range of some 10 to 100 bytes, the clock rate around 200 KHz and the power consumption more than 100 mA @ 5V.

Today, the chipcard market is grown up to a billion Euro business. Powerful dedicated 32-bit security controllers like the 88-Core of Infineon Technologies are ready to revolutionize chipcard based solutions. The 88-Core is a straitforward RISC architecture with caches for data and instructions. In addition, the first time in chipcard world it offers virtual memory with an efficient translation look-aside buffer which enables optimally organisational security. To realize also a quantum leap of physical security, several independent mechanism are implemented, for example hard encryption of memories.

Chipcards based on this architecure, like the 88-family, support the 88-Core with ROM and NVM of up to 256 Kbytes each, as well as up to 16 Kbytes of RAM, a variety of powerful coprocessors like a DES accelerator and the Advanced Crypto Engine (ACE), and of course a set of peripherals. The internal clock rate is up to 66 MHz, but nevertheless these chipcards are able to operate in a proximity contactless environment specified by ISO 14443, i.e. distance is less than 10 cm, but no battery is available and the transmitted power is much less than 10 mW. How is this possible ?

The solution is not one great invention but the smart combination of a few techniques. The base is a leading egde quarter micron technology. Unfortunately, for cost reasons it have to be a standard process, but it is adjusted at the best tradeoff between performance and *lowest* power consumption. Next, the chips are developed with an unconventional design methodology. It enables the flexible integration of hard macros in a VHDL design. Of course, the hard macros are described in a dedicated high level language, too. Design parts having a certain regularity and a relatively high switching frequency are selected to become a hard macro. But what is the advantage of these hard macros ? They are designed in a switching current free design style called dual rail logic with precharge. This design style reduce power consumption dramatically and, if well designed, does not increase transistor count.

D. Soudris, P. Pirsch, and E. Barke (Eds.): PATMOS 2000, LNCS 1918, pp. 205–206, 2000.

Last but not least, a revolutionary power balancing technique is introduced. The internal voltage regulator does no longer use a shunt transistor to balance the voltage but the power consuming circuit itself. The internal clock rate is the control value. Thus, the chip consumes only that power which is transmitted while wasting nothing. Using all that techniques Infineon Technologies is able to fulfill the hardest requirements for contactless applications.

Dynamic Memory Design
for Low Data-Retention Power

Joohee Kim and Marios C. Papaefthymiou

Advanced Computer Architecture Laboratory
Department of Electrical Engineering and Computer Science
University of Michigan, Ann Arbor, MI 48109
{jooheek, marios}@eecs.umich.edu

Abstract. The emergence of data-intensive applications in mobile environments has resulted in portable electronic systems with increasingly large dynamic memories. The typical operating pattern exhibited by these applications is a relatively short burst of operations followed by longer periods of standby. Due to their periodic refresh requirements, dynamic memories consume substantial power even during standby and thus have a significant impact on battery lifetime.

In this paper we investigate a methodology for designing dynamic memory with low data-retention power. Our approach relies on the fact that the refresh period of a memory array is dictated by only a few, worst-case leaky cells. In our scheme, multiple refresh periods are used to reduce energy dissipation by selectively refreshing only the cells that are about to lose their stored values. Additional energy savings are achieved by using error-correction to restore corrupted cell values and thus allow for extended refresh periods. We describe an exact $O(n^{k-1})$-time algorithm that, given a memory array with n refresh blocks and two positive integers k and l, computes k refresh periods that maximize the average refresh period of a memory array when refreshing occurs in blocks of l cells. In simulations with 16Mb memory arrays and a (72,64) modified Hamming single-error correction code, our scheme results in an average refresh period of up to 11 times longer than the original refresh period.

1 Introduction

Mobility imposes severe constraints on the design of portable electronic systems, particularly with respect to their power dissipation [1]. A popular approach to minimizing power consumption in portable devices is to employ a standby mode in which almost all modules are powered down. Large-density dynamic random access memory (DRAM) dissipates energy even during standby, however, due to its periodic refresh requirement. Such dissipation is of particular concern in the case of data-intensive applications, due to their large dynamic memory requirements.

The charge stored in dynamic memory cells must be periodically refreshed to counter the corrupting effects of leakage currents. Due to local process perturbations, each cell has different leakage currents, resulting in a distribution of

D. Soudris, P. Pirsch, and E. Barke (Eds.): PATMOS 2000, LNCS 1918, pp. 207–216, 2000.

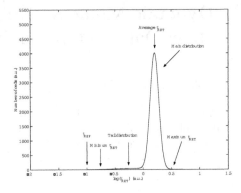

Fig. 1. Distribution of data-retention time for DRAM cells. Data adapted from [5].

data-retention times t_{RET} similar to the one shown in Figure 1. Conventional DRAMs use a single periodic refresh signal to restore the charge level in each cell capacitor to its original value. To prevent errors, refreshing must be done at the minimum refresh period t_{REF}. This simple approach inevitably dissipates more power than necessary. First, since t_{REF} is set with respect to the few "bad" cells, most memory cells are refreshed too early, thus dissipating unnecessary power. Second, due to its strong dependency with the leakage current, t_{REF} is determined at the highest operating temperature, resulting in unnecessary dissipation at lower operating temperatures.

In this paper, we investigate the use of multiple refresh periods to eliminate the power associated with refreshing good cells too often. We also explore the use of error correcting codes (ECC) to further extend the average refresh period t_{REF}. We give an exact $O(n^{k-1})$-time algorithm for computing an optimal set of refresh periods for a memory array with n refresh blocks. Specifically, given positive integers k and l, our algorithm computes k refresh periods that maximize the average refresh period of the memory array, when memory is refreshed in blocks of l cells. The addition of ECC enables to further increase the average refresh period by correcting the errors occurring during the extended refresh. In simulations of a 16Mb memory array with a Single Error Correcting Code (SEC), our proposed multirate refresh scheme results in 11-fold increase of the average refresh period with respect to a conventional single-period refresh scheme without ECC.

The remainder of this paper has six sections. Section 2 gives an overview of leakage-current induced errors and refreshing in DRAMs. Error correcting codes are briefly introduced in Section 3. The proposed multirate ECC-enhanced refresh scheme is described in Section 4. Our algorithm for the optimal selection of k refresh periods is described in Section 5. Section 6 presents simulation results from the application of our methodology to a 16Mb DRAM array. We conclude our paper in Section 7 with a brief discussion of future work.

2   DRAM Refresh

Conventional single-transistor DRAM cells are composed of one transistor and a capacitor. Due to its simplicity, this structure can be used to fabricate high-density memories. Unlike static random access memory (SRAM), however, the stored charge is not retained by a continuous feedback mechanism, and leakage current reduces the stored voltage level over time. There are many known leakage paths in a DRAM cell. The junction leakage current from the storage node, which increases exponentially with the operation temperature, is known to be the major leakage mechanism [5].

Leakage current can be expressed using the simple empirical formula

$$I = A \, exp \left(\frac{E_a}{kT} \right) , \tag{1}$$

where E_a is the activation energy, k is the Boltzmann constant, T is the operating temperature, and A is constant factor [5]. From this equation it follows that the leakage current is a function of the activation energy, which depends on fabrication processes such as ion implantation [6]. Due to local process fluctuations, activation energies vary among cells [7,8]. A study has showed that the $log(t_{RET})$ of the cells follows a bimodal distribution. The large main distribution is composed of good cells, and a small tail distribution is composed of bad cells [5]. To restore their intended voltage levels, DRAM cells need to be periodically refreshed at a period not exceeding their minimum data-retention time t_{RET}.

3   ECC for DRAM

Error correcting codes are traditionally used in communications to battle the corruption of transmitted data by channel noise. Extra information is added to the original data to enable the reconstruction of the original data transmitted. The encoded data, or codewords, are sent through the channel and decoded at the receiving end. During decoding the errors are detected and corrected if the amount of error is within the allowed, correctable, range. This range depends on the extra information, parity bits, added during encoding.

In DRAMs, saving data in memory corresponds to sending it down a noisy channel. Figure 2 shows the usage of ECC to correct errors in a memory system. Traditionally, ECC has been used to correct hard errors introduced during fabrication, thus increase yield. It has also been used to correct soft errors caused by α-ray during operation [2,3]. Due to the random distribution of the error in DRAMs, HV parity code and Hamming code were most commonly used.

With improvements in modern process technologies, the number of hard errors has decreased. The remaining few errors are usually dealt with by bypassing the row or column containing the hard error and using redundant rows or columns. Moreover, as the junction area in the device deceases due to scaling, the occurrence of soft errors has decreased [4]. Hence ECC is seldom used for general purpose DRAM in recent years.

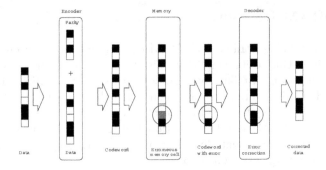

Fig. 2. Data flow in ECC added memory

4   ECC-Enhanced Multirate Refresh Scheme

Power consumption in DRAM memories is given by the expression

$$P = P_{Array} + P_{Aux} , \qquad (2)$$

where P_{Array} is the power dissipated to read/write data and retain data, and P_{Aux} is the power consumption of auxiliary modules such as internal voltage generator. P_{Array} is mainly due to the switching activity in the cell capacitors, bit lines, sense amplifiers and decoders and is hence frequency dependent. On the other hand, P_{Aux} is less frequency dependent.

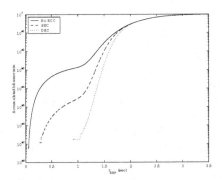

Fig. 3. Bit error rate versus t_{REF} in the presence or absence of error correction.

Data-retention power can be decreased by extending t_{REF}. ECC technology can be used to correct the errors caused by not refreshing within the required time. Figure 3 shows simulated bit error rates (BER), defined as the number of errors over the total number of cells, for a 16Mb memory, with respect to t_{REF}. The simulation was based on a leakage current distribution reported in [5]. The three graphs show results when the memory is operated with no ECC, with a row-based single error-correcting code (denoted by SEC), and a row-based double error-correcting code (denoted by DEC). Since ECCs reduce the number of generated errors, a longer t_{REF} is possible at any given error rate. The overall extent to which t_{REF} can be prolonged depends on the tolerable error levels of the application for which the memory is used.

In addition to the modified dissipation from the conventional sources, our ECC-enhanced approach incurs the dissipation of the ECC circuitry and the additional parity bits:

$$P' = P'_{Array} + P_{Aux} + P_{ECC} + P_{Parity}. \qquad (3)$$

The power consumption due to the ECC, P_{ECC}, and to the parity bits, P_{Parity}, are also frequency dependent and will thus offset the decrease in the array power P'_{Array}. The size of the ECC circuitry and associated parity bits depends on the choice of an ECC.

The introduction of ECC is not guaranteed to extend t_{REF} when a single refresh period is used. In the case of single error correction, for example, if two worst-case bits appear in a single codeword, they will still determine the extended t_{REF} for the entire memory. Since the geometric location of the bad cells cannot be controlled, the resulting extended t_{REF} can not be controlled either.

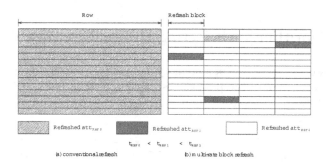

Fig. 4. Multirate block refresh scheme.

Using a collection of discrete refresh periods t_{REF} to selectively refresh memory blocks can increase the average refresh period and reduce power dissipation. The minimum t_{REF} within the set can be set to the t_{REF} without ECC. Memory blocks comprising "bad" cells will still be refreshed at this rate. The longer refresh periods can be used to refresh the blocks with good cells. Once the refresh periods are selected, the variability is in the number of the memory blocks refreshed at a particular t_{REF}.

Figure 4 shows the application of our multirate scheme on a memory array. In this figure, our approach is applied at a fine granularity level by segmenting the refresh block, which conventionally is a row, into smaller blocks. In this case, the total dissipation is given by the equation

$$P'' = P''_{Array} + P_{Aux} + P'_{ECC} + P'_{Parity} + P_{PA}, \qquad (4)$$

where P_{PA} denotes the energy dissipated for the partial activation of a row. In this approach, the additional energy required for refreshing smaller refresh blocks, partially activating a row for each different t_{REF}, is traded off to increase the average refresh period. Therefore, total savings depend on the size of the refresh block and the associated overhead.

The implementation of the proposed ECC-enhanced multirate refresh scheme requires to store the refresh period t_{REF} of each block in a refresh controller. The implementation of two refresh periods t_{REF} for a memory without ECC has been reported in [12]. Additional circuitry is required for partial row activation if the refresh block is smaller than a row. The implementation of memory arrays with partial row activation to reduce word line capacitance has been reported in [13]. The information about the required t_{REF} can be obtained after manufacturing and can be stored in many forms. For example, it can be hard-wired using electrical fuses. Alternatively, it can be stored in re-writable memory elements if post-fabrication modification is desired. During memory operation, the refresh controller uses the stored information to refresh blocks at their required t_{REF}. If the multiple refresh periods are multiples of the minimum refresh period ($t_{REF\ MIN}$), than refreshing can be achieved by simple consecutive refreshes at $t_{REF\ MIN}$, activating only the refresh blocks that need to be refreshed and skipping the ones that do not.

5   Algorithm for Selecting Optimal Refresh Periods

The power consumption of a memory array under multirate refreshing is proportional to the sum of the power consumption of each block at its refresh period. Hence, total power consumption is given by the expression

$$P = A \sum_{i=1}^{n} N_i \cdot \frac{1}{t_{REFi}} , \qquad (5)$$

where N_i is the number of blocks that are refreshed at a refresh period t_{REFi}, and A is a proportionality factor. It should be noted that power consumption depends on the size of the refresh block, the number of refresh periods, and the refresh periods themselves.

Figure 5 demonstrates the basic idea behind the computation of an optimal set of refresh periods for a memory array. This graph shows the number of blocks that have a given retention time. Each vertical line corresponds to a refresh period. Between any two consecutive vertical lines, the total area under the curve gives the total number of blocks refreshed at the shorter of the two periods. The refresh periods must be chosen so that the sum of the individual area/period ratios is minimized.

Figure 6 gives the pseudocode of our algorithm that computes an optimal set of refresh periods for a memory array with M rows of N bits, given the required refresh period for each refresh block of l cells (DB). For simplicity, our procedure is described for $k = 4$ refresh periods. The minimum refresh period is set to

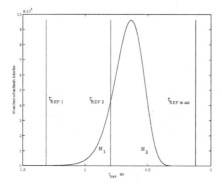

Fig. 5. Optimal selection of multiple refresh periods

1: STREF_OPT (DB, k=4)
2: STREF = ∅
3: $temp = \infty$
4: **for** $p = t_{REF\ MIN}$ to $t_{REF\ MAX}$ **do**
5:    $N[p] = A_p$ ▷ number of refresh blocks refreshed at t_{REFp}
6:    **for** $q = t_{REFp}$ to $t_{REF\ MAX}$ **do**
7:      $N[q] = A_q$ ▷ number of refresh blocks refreshed at t_{REFq}
8:      **for** $r = t_{REFq}$ to $t_{REF\ MAX}$ **do**
9:        $N[r] = A_r$ ▷ number of refresh blocks refreshed at t_{REFr}
10:       $N[MIN] = N_{TOT} - (N[p] + N[q] + N[r])$
          ▷ number of refresh blocks refreshed at $t_{REF\ MIN}$
11:       $P = \frac{N[MIN]}{t_{REFMIN}} + \frac{N[p]}{t_{REFp}} + \frac{N[q]}{t_{REFq}} + \frac{N[r]}{t_{REFr}}$
12:       **if** $temp < $ P **then**
13:         $temp = $ P
14:         $STREF = \{t_{REFMIN}, t_{REFp}, t_{REFq}, t_{REFr}\}$
15:       **end if**
16:     **end for**
17:   **end for**
18: **end for**
19: return STREF

Fig. 6. Algorithm for finding optimal set STREF of block refresh periods.

the single-period refresh period t_{REFMIN}. The nested loop structure iteratively assigns possible values to the three remaining refresh periods, computing the corresponding power of each assignment using Equation 5. For arbitrary k, there are $k - 1$ nested loops and the complexity of this scheme is $O(n^{k-1})$, where n is the number of refresh blocks in the memory array.

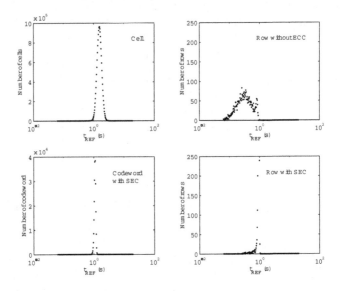

Fig. 7. Distribution of required t_{REF} for different size refresh blocks.

6   Simulation Results

We evaluated the effectiveness of our ECC-enhanced multirate refresh scheme using a (72,64) modified Hamming SEC [9] and a 16Mb DRAM whose t_{REF} distribution and electrical characteristics are reported in [5] and [10], respectively.

Figure 7 shows the impact of refresh block granularity and ECC on the number of refresh blocks with short t_{REF}. The two graphs on top give the number of blocks at each minimum refresh period for cell-based and row-based refresh, respectively, with no error correction. Row-based SEC greatly reduces the number of rows that require short t_{REF}. The refresh periods can be extended even further by reducing the size of a refresh block from a 4608-bit row (4096 data bits + 512 ECC bits) to a 72-bit codeword (64 data bits + 8 ECC bits).

Figure 8 shows the trend of power consumption with the introduction of a second refresh period t_{REF}. As the second t_{REF} increases toward the maximum refresh period shown in the distribution of Figure 7, power consumption decreases below that of the single-refresh scheme at t_{REFMIN}. Moreover, power dissipation decreases with the application of ECC and increase in the refresh granularity, since the fraction of blocks requiring short t_{REF} decreases.

Figure 9 shows the positive effect of multiple refresh periods on power dissipation. The dissipation of row-refresh with SEC is close to the ideal minimum of cell-refresh. The use of ECC results in significant power reductions with fewer periods than without ECC. When two refresh periods are used, setting the second period at a multiple of the original t_{REF} of 64ms [10] reduces the complexity of refresh control. Since variations of power dissipation are more gradual at short periods, selecting a refresh period of $64 \times 11 = 704ms$, which is slightly

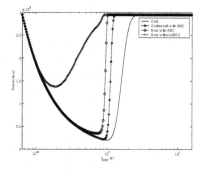

Fig. 8. Power consumption versus period of second refresh cycle.

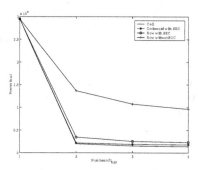

Fig. 9. Effect of block size and number of refresh periods on power.

smaller than the optimal $735ms$, will increase the average refresh period (and thus decrease dissipation) by approximately 11 times.

7 Conclusion

This paper describes an ECC-enhanced multirate refresh scheme for low data-retention power in dynamic memories and presents an algorithm for selecting an optimal set of refresh periods. Simulation results with a 16Mb DRAM show that simple Hamming SEC can extend the average refresh period by up to 11 times over conventional single-cycle refresh. We are currently evaluating the energy efficiency of our scheme including the control and ECC overhead. We are also investigating efficient algorithms for computing optimal refresh periods.

Acknowledgments

This research was supported in part by the US Army Research Office under Grant No. DAAD19-99-1-0304.

References

1. K. Itoh, K. Sasaki and Y. Nakagome. Trends in Low-Power RAM Circuit Technologies. In *Proceedings of the IEEE*, 83(4):524–543, April 1995.
2. H. Kotani, T. Yamada, J. Matsushima and M. Inoue. 4Mbit DRAM Design Including 16-bit Concurrent ECC. In *1987 Symposium on VLSI Circuits. Digest of technical Papers. Bus. center for Acad. Soc. Japan, Tokyo, Japan*, pages 87–88, 1987.
3. H. L. Kalter et al. A 50-ns 16-Mb DRAM with a 10-ns Data Rate and On-Chip ECC. *IEEE J. Solid-State Circuits*, 25(5):1118–1128, October 1990.
4. K. Itoh, Y. Nakagome, S. Kimura and T. Watanabe. Limitation and Challenges of multigigabit DRAM Chip design. *IEEE J. Solid-State Circuits*, 32(5):624–634, May 1997.
5. T. Hamamoto, S. Sugiura and S. Sawada. On the retention Time Distribution of Dynamic Random Access Memory (DRAM) *IEEE Transactions on Electron devices*, 45(6):1300–1309, June 1998.
6. M. Ogasawara, Y. Ito, M. Muranaka, Y. Yanagisawa, Y. Tadaki, N. Natsuaki, T. Nagata and Y. Miyai. Physical Model of Bit-to-bit Variation in Data retention time of DRAMs. In *1995 53rd Annual Device research Conference Digest.IEEE,New York,NY,USA*, pages 164–165, 1995
7. P. J. Restle, J. W.Park and B. F. Lloyd. DRAM Variable retention Time. *International Electron Devices Meeting 1992. Technical Digest.IEEE, New York, NY, USA*, Pages 807–810,1992.
8. E. Adler et al. The evolution of IBM CMOS DRAM technology *IBM J. Develop.*, 39(1/2):167–188, March 1995.
9. M. Y. Hsiao. A Class of Optimal Minimum Odd-weight-column SEC-DED Codes. *IBM J. Develop.*, 14(14):395–401, July 1970.
10. Toshiba. *16,777,216-word X 1-bit DYNAMIC RAM Data sheet.*
11. Y. Katayama et al. Fault-Tolerant Refresh Power Reduction of DRAMs for Quasi-Nonvolatile Data Retention. *International Symposium on Defect and Fault Tolerance in VLSI Systems*, 311:318, 1999.
12. S. Takase and N. Kushiyama. A 1.6GB/s DRAM with Flexible Mapping Redundancy Technique and Additional Refresh Scheme. *International Solid-State Circuits Conference*, 410:411, 1999.
13. T. Murotani et al. Hierchical Word-Line Architecture for Large Capacity DRAMs. *IEICE TRANS. ELECTRON.*, E80-C(4), 550:556, 1997.

Double-Latch Clocking Scheme
for Low-Power I.P. Cores

Claude Arm , Jean-Marc Masgonty, and Christian Piguet

CSEM Centre Suisse d'Electronique et de Microtechnique SA
Jaquet-Droz 1, 2007 Neuchâtel, Switzerland
claude.arm@csem.ch
www.csem.ch

Abstract. This paper describes the design of VHDL-based I.P. (Intellectual Property) cores using a Double-Latch clocking scheme instead of single-phase clock and D-Flip-Flops. This Double-Latch clocking scheme with two non-overlapping clocks provides several advantages in deep submicron technologies, i.e. a much larger clock skew tolerance, clock trees easy to generate, efficient clock gating and in some examples, such as an 8-bit CoolRISC microcontroller, a reduced power consumption.

1 Introduction

More and more I.P. cores (Intellectual Property) are available on the market. They are more and more "soft" cores written in VHDL or Verilog languages and synthesizable using Synopsys. One can find 32-bit RISC cores, DSP cores and 8-bit microcontroller cores, for instance, many 8051 cores. The main issue in such cores implemented in deep submicron technologies is the reliability. As they are synthesized using Synopsys, the soft core has to work for any customer with any constraint. It is therefore more difficult to guarantee that there is no timing violation in the synthesized "soft" core than with hard cores (layout provided to the customer). Furthermore, enhanced reliability generally increases the power consumption. It is therefore a major issue to increase reliability as well as to decrease power consumption.

2 I.P. Cores

As mentioned in the introduction, the main issue in the design of "soft" cores [1] is reliability. In deep submicron technologies, gate delays are smaller and smaller compared to wire delays. Complex clock trees have to be designed with clock tree generation tools linked with routers to satisfy to the required timing after the place and route step, mainly the smallest possible clock skew, and to avoid any timing violation.

Furthermore, "soft" cores have to present a low power consumption to be attractive to the possible licensees. If the clock tree is a major issue to achieve the required

D. Soudris, P. Pirsch, and E. Barke (Eds.): PATMOS 2000, LNCS 1918, pp. 217-224, 2000.

clock skew, requiring strong buffering, its power consumption could be larger than desired. The clocking scheme of I.P. cores is therefore a major issue, both for its functionality and for its power consumption.

Today, most I.P. cores are based on a single-phase clock and are based on D-Flip-Flops. Another approach than the conventional single-phase clock with D-Flip-Flops (DFF) is presented in this paper. It is based on a Double-Latch approach with two non-overlapping clocks. This clocking scheme has been used for the 8-bit CoolRISC microcontroller I.P. core [2] as well as for other cores, such a DSP core and other execution units [3]. The advantages as well as the disadvantages will be presented.

3   CoolRISC Microcontroller

The CoolRISC is a 3-stage pipelined core (Fig. 1). The branch instruction is executed in only one clock [2], [4], [5]. In that way, no load or branch delay can occur in the CoolRISC core, resulting in a strictly CPI=1 (Clock Per Instruction). It is not the case of other 8-bit pipelined microprocessors (PIC, AVR , Scenix, MCS-251, Flip8051). It is known that the reduction of CPI is the key to high performances.

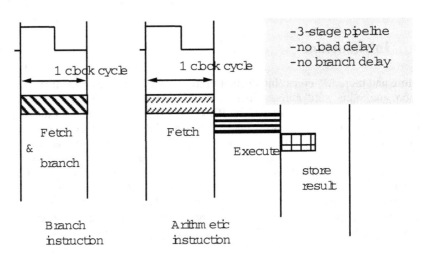

Fig. 1. CoolRISC Pipeline

For each instruction, the first half clock is used to precharge the ROM program memory. The instruction is read and decoded in the second half of the first clock (Fig. 1). A branch instruction is also executed during the second half of this first clock, which is long enough to perform all the necessary transfers. For a load/store instruction, only the first half of the second clock is used to store data in the RAM memory. For an arithmetic instruction, the first half of the second clock is used to read an operand in the RAM memory or in the register set, the second half of this second clock to perform the arithmetic operation and the first half of the third clock to store the result in the register set.

Another very important issue in the design of 8-bit microcontrollers is the power consumption. The gated clock technique [2], [4], [5] has been extensively used in the design of the CoolRISC cores (Fig. 2).

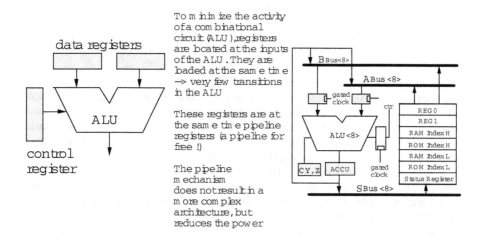

Fig. 2. Gated Clock ALU

The ALU, for instance, has been designed with input and control registers that are loaded only when an ALU operation has to be executed. During the execution of another instruction (branch, load/store), these registers are not clocked thus no transition occur in the ALU (Fig. 2). This reduces the power consumption. A similar mechanism is used for the instruction registers, thus in a branch, which is executed only in the first pipeline stage, no transitions occur in the second and third stages of the pipeline. It is interesting to see that gated clocks can be advantageously combined with the pipeline architecture; the input and control registers implemented to obtain a gated clocked ALU are naturally used as pipelined registers.

4   Latch-Based Design of I.P. Cores

Figure 3 shows the double-latch concept that has been chosen for such I.P. cores to be more robust to the clock skew, flip-flop failures and timing problems at very low voltage [6]. The clock skew between various Ø1 (respectively Ø2) pulses have to be shorter than half a period of CK. However, one requires two clock cycles of the master clock CK to execute a single instruction. It is why one needs, for instance in technology TSMC 0.25 μm, 120 MHz to generate 60 MIPS (CoolRISC with CPI=1), but the two Øi clocks and clock trees are at 60 MHz. Only a very small logic block is clocked at 120 MHz to generate two 60 MHz clocks.

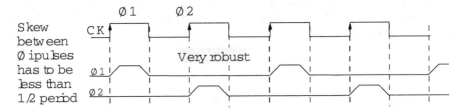

Fig. 3. Double-Latch Clocking Schemes

The design methodology using latches and two non-overlapping clocks has many advantages over the use of DFF methodology. Due to the non overlapping of the clocks and the additional time barrier caused by having two latches in a loop instead of one DFF, latch based designs support greater clock skew before failing than a similar DFF design (each targeting the same MIPS).

With latch-based designs, the clock skew becomes relevant only when its value is close to the non-overlapping of the clocks (so half the period of the master clock). When working at lower frequency and thus increasing the non-overlapping of clocks, the clock skew is never a problem. It can even be safely ignored when designing circuits at low frequency. However, a shift register made with DFF can have clock skew problems at any frequency.

This allows the synthesizer and router to use smaller clock buffers and to simplify the clock tree generation, which will reduce the power consumption of the clock tree.

Example: A DSP core synthesized with a low-power library in TSMC 0.25 μm. The test bench A contains only few multiplication operations, while the test bench B performs a large number of MAC operations. The circuit was synthesised then routed, Table 1 shows the power consumption results for two different values of clock skew constraint given to CTGen, the first was done for a clock skew max of 3 ns, for the second one, a 10 ns clock skew max was chosen. Results show that, if the power is sensitive to the application program, it is also quite sensitive to the required skew: 50% of power reduction from 3 ns to 10 ns skew. This shows that major power savings can be obtained with latch based circuits when the clock frequency allows to lighten the clock skew constaints.

Table 1. Power consumption of the same core with various test benches and skew

Skew	Test bench A	Test bench B
10 ns	0.44 mW/MHz	0.76 mW/MHz
3 ns	0.82 mW/MHz	1.15 mW/MHz

Futhermore, if the chip has clock skew problems at the targeted frequency after integration, you are able with a latch-based design to reduce the clock frequency. It results in the fact that the clock skew problem will disappear, allowing the designer to test the chip functionality and eventually to detect other bugs or to validate the design functionality. This can reduce the number of test integration needed to validate the chip. With a DFF design, when a clock skew problem appears, one has to reroute and integrate again. This point is very important for the design of a chip in a new process not completely or badly characterized by the foundry, which is the general case as a new process and new chips in this process are designed concurrently for reducing the time to market.

Using latches for pipeline structure can also reduce power consumption when using such a scheme in conjunction with clock gating. The latch design has additional time barriers, which stop the transitions and avoid unneeded propagation of signal and thus reduce glitch power consumption. The clock gating of each stage (latch register) of the pipeline with individual enable signals, can also reduce the number of transitions in the design compared to the equivalent DFF design, where each DFF is equal to two latches clocked and gated together.

Another advantage with a latch design is the time borrowing (Fig. 4). It allows a natural repartition of computation time when using pipeline structures. With DFF, each stage of logic of the pipeline should ideally use the same computation time, which is difficult to achieve, and in the end, the design will be limited by the slowest stage (plus a margin for the clock skew). With latches, the slowest pipeline stage can borrow time from either or both the previous and next pipeline stage. The clock skew only reduces the time that can be borrowed. An interesting paper [7] has presented time borrowing with DFF, but such a scheme needs a complete new automatic clock tree generator that does not minimize the clock skew but uses it to borrow time between pipeline stages.

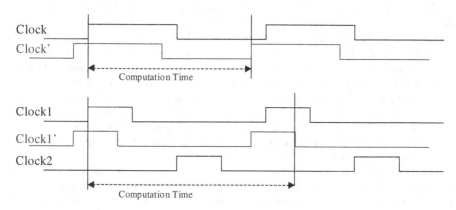

Fig. 4. Time Borrowing

Using latches can also reduce the number of MOS of a design. For example, a microcontroller has 16*32-bits registers, i.e. 512 DFF or 13'312 MOS (using DFF with 26 MOS). With latches, the master part of the registers can be common for all the registers, which gives 544 latches or 6'528 MOS (using latches with 12 MOS). In this example, the register area is reduced by a factor of 2.

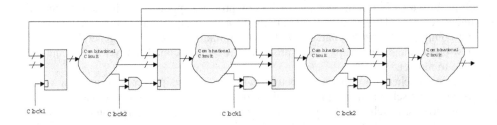

Fig. 5. Latch-based Clock Gating

5   Gated Clock with Latch-Based Designs

The latch-based design also allows a very natural and safe clock gating methodology. Figure 5 shows a simple and safe way of generating enable signals for clock gating. This method gives glitch free clock signals without the adding of memory elements, as it is needed with DFF clock gating.

Synopsys handles very nicely the proposed latch-based design methodology. It performs nicely the time borrowing and seems to analyze correctly the clocks for speed optimization. So it is possible to use this design methodology with Synopsys, although there are a few points of discussion linked with the clock gating.

This clock gating methodology cannot be inserted automatically by Synopsys. The designer has to write the description of the clock gating in his VHDL code. This statement can be generalized to all designs using the above latch-based design methodology. We believe Synopsys can do automatic clock gating for pure double latch design (in which there is no combinatorial logic between the master and slave latch), but such a design results in a loss of speed over similar DFF design.

The most critical problem is to prevent the synthesizer from optimizing the clock gating AND gate with the rest of the combinatorial logic. To ensure a glitch free clock, this AND gate has to be placed as shown in Figure 5. This can be easily done manually by the designer by placing these AND gates in a separate level of hierarchy of his design or placing a 'don't touch' attribute on them.

Forcing a 'don't touch' on these gates presents the drawback that this part of the clock tree will not be optimized for speed or clock buffering. Remark that the AND gate shown in Figure 5 represents a NAND gate followed by an inverting clock buffer. It would be interesting that the tool handles this gate in a special way to keep it in front of the latch clock input. Maybe by placing a specific attribute on it in such a way that it can recognize it as a clock gating gate, which forbid the optimizer to move logic between it and the latch, but still allows it to size the NAND and the clock buffer.

The second problem we encountered was the fact that the Design Compiler found timing loops going through the clock enables. Assume two registers A and B, each register having its clock gated by an enable signal (Fig. 6). The enable signal of register A depends on the value of register B and the enable of register B depends on the value of register A. This is seen as an open loop by the tool, although the clocks of register A and B are defined in such a way that they cannot be '1' at the same time.

The condition on the clocks ensures that there is no open loop. Design Compiler seems not to take the non-overlapping of the clock into account when analyzing this loop, and we found no way to declare it in such a way that it is taken into account. This loop has to be cut with the 'set_disable_timing' command. This work around is not good, because it disables the timing optimization on some paths of the design that should have been optimized. In the above example, there is an important timing path from the clock input of latch A to the enable input of the AND gate of the clock gating of latch A. There is a similar path for latch B, and those two path overlap. If you place a "set_disable_timing" somewhere in the loop, you cut at least one of those paths.

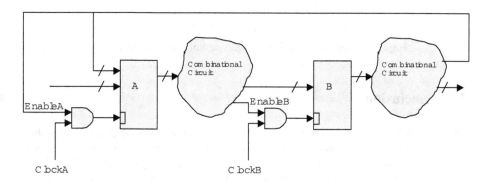

Fig. 6. Timing Loops

6   Results

A synthesizable by Synopsys CoolRISC–DL 816 core with 16 registers has been designed according to the proposed Double Latch (DL) scheme (clocks Ø1 and Ø2) and provides the estimated (by Synopsys) following performances (only the core, about 20'000 transistors) in TSMC 0.25 μm:

- 2.5 Volt, about 60 MIPS (but 120 MHz single clock). It is the case with the core only. If a program memory with 2 ns of access time is chosen, as the access time is included in the first pipeline stage, the achieved performance is reduced to 50 MIPS

- 1.05 Volt, about 10 μW/MIPS, about 100'000 MIPS/watt

The core "DFF+Scan" is a previous CoolRISC core designed with flip-flops [2, 4, 5]. The CoolRISC-DL "double latch" cores with or without special scan logic provide better performances.

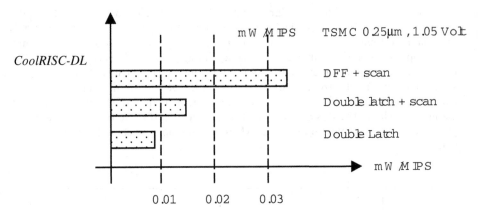

Fig. 7. Power consumption comparison of "soft" CoolRISC cores

7   Conclusion

The I.P. CoolRISC core has been licensed to one company. Furthermore, the Double-Latch clocking scheme has been used for other cores and execution units, such as in [3]. It was shown that it was more reliable and mandatory at very low voltage. Furthermore, it provides a power consumption reduction compared to a single-phase clock scheme with D-Flip-Flops.

References

1.   M. Keating, P. Bricaud, "Reuse Methodology Manual", Kluwer Academic Publishers, 1999.
2.   C. Piguet et al. "Low-Power Design of 8-bit Embedded CoolRISC Microcontroller Cores", IEEE JSSC, Vol. 32, No 7, July 1997, pp. 1067-1078.
3.   Ph. Mosch et al. "A 72µW, 50 MOPS, 1V DSP for a hearing aid chip set" ISSCC'00, San Francisco, February 7-9, Session 14, paper 5.
4.   J-M. Masgonty et al. "Low-Power Design of an Embedded Microprocessor", ESSCIRC'96, September 16-21, 1996, Neuchâtel, Switzerland
5.   www.csem.ch, www.xemics.ch, www.coolrisc.ch
6.   C. Piguet, "Low-Power Digital Design", invited talk, CCCD Workshop at Lund University, March 9-10, 2000, Lund, Sweden.
7.   J. G. Xi, D. Staepelaere, "Using Clock Skew as a Tool to Achieve Optimal Timing", Integrated System Magazine, April 1999, webmaster@isdmag.com

Architecture, Design, and Verification of an 18 Million Transistor Digital Television and Media Processor Chip

Santanu Dutta

Philips Semiconductors,
Sunnyvale, CA 94088.

Abstract. This paper describes the architecture, functionality, and design of **NX-2700** — a digital television (DTV) and media-processor chip from Philips Semiconductors. NX-2700 is the second generation of an architectural family of programmable multimedia processors that supports all eighteen United States Advanced Television Systems Committee (ATSC) [1] formats and is targeted at the high-end DTV market.
NX-2700 is a programmable processor with a very powerful, general-purpose Very Long Instruction Word (VLIW) Central Processing Unit (CPU) core that implements many non-trivial multimedia algorithms, coordinates all on-chip activities, and runs a small real-time operating system. The CPU core, aided by an array of autonomous multimedia co-processors and input-output units with Direct Memory Access (DMA) capability, facilitates concurrent processing of audio, video, graphics, and communication-data.

1 Architecture and Functionality of NX-2700

NX-2700 is a DTV processor chip targeted to be used in high or standard-definition television systems, digital set-top-boxes, and other DTV-based applications. A combination of hardware and software is used to implement the key DTV functionality. The chip features a very powerful general-purpose VLIW processor core (DSPCPU) and an array of DMA-driven multimedia and input/output functional units and co-processors that operate independently and in parallel with the DSPCPU, thereby making software media-processing of multimedia algorithms extremely efficient. As illustrated in the block-diagram in Figure 1, some key functional modules of the NX-2700 design are:

- **high-speed internal data-highway buses** used for memory-data transfers as well as Memory Mapped Input Output (MMIO) control register read/write transactions,
- a **Main Memory Interface (MMI) unit** that arbitrates accesses to the highway buses and manages the interface between the NX-2700 core plus its on-chip peripherals and the off-chip main memory (SDRAM),
- a **VLIW CPU core** that uses a general-purpose VLIW Instruction Set Architecture (ISA) enhanced by powerful multimedia-specific instructions,

D. Soudris, P. Pirsch, and E. Barke (Eds.): PATMOS 2000, LNCS 1918, pp. 225–232, 2000.

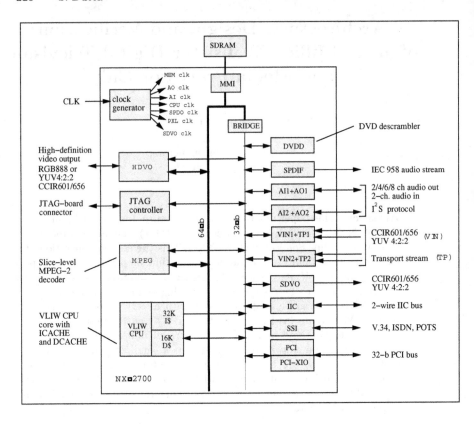

Fig. 1. Block diagram of NX-2700

- a **Transport-stream Processor (TP)** that can gluelessly connect to assorted demodulator/decoder chips and perform PID-based filtering of MPEG-2 transport packets as per the ISO/IEC 13818-1 standard,
- a **slice-level MPEG-2 decoder** that can decode the highest-resolution (*main profile at high level*) interlaced compressed video bitstream,[1]
- multiple **Audio In (AI)** and **Audio Out (AO) processors** that can capture audio data from external world, can produce upto 8 channels of audio output, can decode AC-3 and ProLogic audio, and can also connect to external audio amplifiers,
- a **Sony-Philips Digital Interface (SPDIF)** that not only supports one or more Dolby-Digital AC-3 6-channel data streams and/or MPEG-1 and MPEG-2 audio streams as per Project 1937, but also produces IEC958-compliant outputs,

---

[1] The MPEG pipeline consists of a Variable-Length Decoder (VLD), a Run-Length Decoder (RLD), an Inverse Scan (IS) unit, an Inverse Quantizer (IQ), an Inverse Discrete Cosine Transform (IDCT) block, and a Motion Compensation (MC) unit.

- a **micro-programmable High-Definition Video Out (HDVO) unit** that can mix multiple video and graphics planes and is capable of scaling vertically and horizontally pictures of the highest resolution (1920 × 1080) specified in the ATSC DTV standard,[2]
- a **DVD Descrambler (DVDD)** that supports both PC-based and stand-alone DVD players,
- a **standard-definition-video-in (VI) subsystem** that can capture a video stream directly from any CCIR656/601-compliant device,
- a **standard-definition-video-out (VO) subsystem** that can produce outputs in a PAL or NTSC format for driving monitors and a CCIR656-compliant format for recording in digital VCRs,
- a **two-wire Inter-Integrated Circuit (IIC) interface** for configuring and inspecting the status of various peripheral video devices such as digital multi-standard decoders, digital encoders, and digital cameras,
- a **Synchronous Serial Interface (SSI)** that is specially designed to connect to an off-chip modem-analog-front-end subsystem, a network terminator, an A/D, a D/A, or a Codec through a flexible bit-serial connection and perform full-duplex serialization/deserialization of a bit stream from any of these devices,
- a **Peripheral Component Interconnect (PCI) interface** that allows easy communication with high-speed peripherals,
- a **PCI External Input-Output (PCI-XIO) interface** that serves as a bridge between the PCI bus and XIO devices such as ROMs and flash EEPROMs, thereby allowing a PCI-like transaction to proceed between NX-2700 and an *inherently-non-PCI* device on the PCI bus,
- a **system-boot-logic block** that enables configuration of the various internal registers via *host-assisted* or *autonomous bootstrapping*,
- a **JTAG controller** that facilitates board-level testing by providing a bridge for asynchronous (to the NX-2700 system clock) data-transfer between the on-chip scannable registers and the external Test Access Port (TAP), and
- a **clock module** comprising Phase Locked Loop (PLL) filter circuits and Direct Digital Synthesizer (DDS) circuits for generating assorted clocks for the memory, the core, and the peripherals.

---

[2] The HDVO unit contains a set of pipelined filters and video processing units that communicate with a set of memory blocks via a Crossbar interconnection network and perform functions such as **horizontal scaling** (polyphase direct and transposed filtering), **horizontal filtering** (multi-tap FIR filtering), **panoramic zooming** (horizontal scaling using a continuously-varying zoom factor), **vertical filtering & scaling** (de-interlacing and median filtering), **129-level alpha blending** (to merge video and graphics planes), **chroma keying** (for computer-generated or modified overlays), **table lookup** (for color scaling and color modification, *e.g.*, RGB1/2/4/8/16 to RGB32 conversion), **color conversion** (for YUV to RGB and *vice versa*), and **horizontal chroma up/down-sampling**.

2   VLSI Implementation Highlights

Some characteristic features of the NX-2700 design, that deserve special mention, are as follows:

- **Multiple clock domains:** NX-2700 being a multi-clock design, specially-designed synchronizers, allowing both fast-to-slow and slow-to-fast clock-domain transitions, are used at almost all clock-domain crossings, except where the data and/or control are guaranteed to be stable by virtue of the design.
- **Clock routing:** Clock signals are routed all over the chip using a hierarchical clock-tree network where specially designed buffers, that equalize clock skews, feed the clocks to the storage elements (flip-flops, memory, *etc.*).
- **Power management:** Two different power-management schemes are followed in our design: *dynamic clock gating* and *software-controlled static powerdown*.
- **Silicon-debug aids:** In order to aid in the debugging of the final silicon, we have implemented, in our chip, a *SPY* mechanism that allows some important internal signals — the *SPY signals* — from each block to be observable at the top level at run-time.
- **GPIO functionality:** We have designed special on-chip circuitry to enable a large number of pins to operate as General Purpose Software Input Output (GPIO) pins and support functions such as infrared remote input, printer output, software-controllable switches in the system logic, software communication link, *etc.*
- **HDVO memory-system design:** The large number of HDVO memories have been organized into individual rows of multiple banks with two wide *Metal-4* (M4) wires for *vdd* and *ground* power distributions across the banks in each row. The rest of the memory (in each row) has been covered by a grounded M4 plate in order to minimize *crosstalk* by isolating and shielding the memory circuits from the signals routed in the next-higher *Metal-5* layer; the grounded metal plate acts as a large decoupling capacitor.
- **Package considerations:** The chip uses a Prolinx 352-pin Enhanced VBGA package that features two VDD (3.3V & 2.5V) and one GND ring. The thermal resistivity (θ_{ja}) of the package being $10 - 12°C/W$, a power dissipation of 6W at the room temperature ($25°C$) can potentially raise the junction temperature to $(25 + 6 \times 12) = 97°C$; therefore, to ensure correct operations at elevated temperatures, the timing and clock-speed analysis have been performed based on a worst-case operating temperature of $125°C$.

3   Design Tools

We have used state-of-the-art Computer-Aided-Design (CAD) tools for the bulk of the design process. From the suite of external design-automation tools that we have used, the most notable ones are:

- **Verilog** *(Cadence and OVI)*: for Register Transfer Level (RTL) designs,
- **Verilog-XL** and **NC-Verilog** *(Cadence)*: for Verilog simulations.
- **Design Framework II** *(Cadence)*: for design database and schematic entry,
- **Design Compiler** *(Synopsys)*: for logic synthesis,
- **PathMill, TimeMill, PowerMill** *(Synopsys/Epic)*: for transistor-level timing and power analysis,
- **Pearl** *(Cadence)*: for full-chip static timing analysis,
- **Fire & Ice** *(Simplex Solutions)*: for extraction of layout parasitics,
- **Chrysalis** *(Chrysalis Symbolic Design, Inc.)*: for formal verification,
- **VeriSure** *(TransEDA Ltd.)*: for determining code and branch coverage,
- **Cell3, Silicon Ensemble, Dracula** *(Cadence)*: for place-and-route and LVS/DRC (Layout Versus Schematic and Design Rule Check) tasks,
- **HSPICE** *(Meta Software)*: for transistor-level circuit simulation, and
- **Quickturn** *(Quickturn Design Systems, Inc.)*: for emulation.

4   Design Verification

Some key aspects of our verification methodology have been:

- using a combination of C, Verilog, C-shell scripts, and PERL routines to develop a hybrid testbed,
- writing self-checking test programs in assembly or C (or a combination thereof) that are compiled and loaded in the external SDRAM using in-house software tools,
- execution of the loaded binary on the Verilog model of the chip in order to program the block(s) under test in the desired mode via MMIO reads/writes,
- development and use of integrated Verilog-based checkers for capturing and comparing the run-time outputs from the blocks against expected outputs,
- automation of the regression runs,
- using the MPEG decoder from the MPEG Software Simulation Group at Berkeley to provide expected results for various public-domain MPEG-2 conformance streams and locally-generated synthetic stress streams,
- development of a co-simulation (based on Verilog and C) environment for testing the HDVO sub-blocks,
- development of a transaction-generator-based random-testing environment for block-level testing of the MMI,
- development of a random-transaction generator for PCI verification, and
- development of application tests for Quickturn-based emulation.

5   Design Summary

Table 1 presents some of the physical and electrical characteristics of the NX-2700 chip.

Table 1. Chip-level design parameters

Parameter	Value
Technology	0.25 μm CMOS
Metal layers used	5
Core supply voltage	2.5 volts
IO supply voltage	3.3 volts
System clock speed	130 MHz
Average power dissipation	8 watts
Design complexity	18 million devices
Package	Prolinx Enhanced VBGA
Package pins	352

6   DTV System Setup

An example reference design platform, based on NX-2700, is shown in Figure 2. The Network Interface Module (NIM) incorporates the VSB demodulator and Forward Error Correction (FEC) chips and performs all of the necessary demodulation and channel-decoding functions from tuning to Transport Stream (TS) generation. Once the TS is generated, it is processed by NX-2700, optionally, along with a separately-received and decoded standard-definition video and its corresponding audio. In a typical digital video application, NX-2700 performs the following key functions:

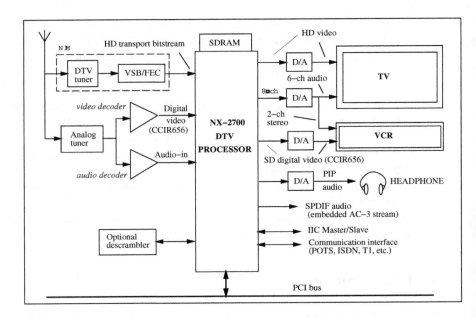

Fig. 2. NX-2700-based DTV receiver system

- transport stream capture, demultiplexing & PID filtering,
- bitstream buffer management,
- MPEG-2 video decoding,
- AC-3 audio decoding,
- clock recovery from the bitstream and video-audio synchronization,
- 2-D graphics for closed captioning, user interface, program guide, *etc.*,
- display-video format conversion including horizontal and vertical scaling, conversions between interlaced and non-interlaced formats, and blending of graphics and video surfaces, and
- processing of the CCIR656 video and its corresponding audio.

Outputs from NX-2700 drive a TV monitor, a VCR, an audio power-amplifier, and/or an audio headphone. The on-chip data and control flow, for an example DTV application, is shown in Figure 3.

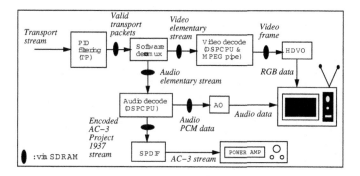

Fig. 3. Data & control flow for example DTV application

7   Conclusions

NX-2700 is the second generation of an architectural family of programmable multimedia processors from Philips Semiconductors. The DTV market is still evolving throughout the world and so there is a clear need for a programmable DTV processor that will allow manufacturers to not only quickly develop ATSC television sets, set-top boxes, and PC-TVs, but also add new features and support emerging services such as program guides, interactive advertising, and video telephony. NX-2700 provides all the above capabilities and can also act as an analog-, cable-, or ISDN-modem for use in fully-interactive services such as Web-browsing through the television set, video-on-demand, video teleconferencing, and interactive online gaming. The chip executes various digital-television applications and different media-processing tasks through a mixture of hardware support and software control. NX-2700 borrows the CPU core, the instruction and the data caches, and some peripheral units from the TM1100 [2]-[5] design; however, several new peripheral units have been added in order to provide the key functionality for DTV applications.

References

1. "Advanced Television Systems Committee," *http://www.atsc.org/*.
2. "TriMedia," *http://www.semiconductors.philips.com/trimedia/*
3. B. Case, "Philips Hopes to Displace DSPs with VLIW," *Microprocessor Report,* December 1994.
4. B. Case, "First Trimedia Chip Boards PCI Bus," *Microprocessor Report,* November 1995.
5. S. Rathnam and G. Slavenburg, "An Architectural Overview of the Programmable Multimedia Processor, TM1," *Compcon,* 1995.

Cost-Efficient C-Level Design
of an MPEG-4 Video Decoder

Kristof Denolf, Peter Vos, Jan Bormans, and Ivo Bolsens

IMEC, Kapeldreef 75, B-3001 Leuven, Belgium
kristof.denolf@imec.be

Abstract. Advanced multimedia systems intrinsically have a high memory cost, making the design of high performance, low power solutions a real challenge. Rather than spending most effort on implementation platform dependent optimization steps, we advocate a methodology and tool that involve C-level platform independent optimizations. This approach is applied to an MPEG-4 video decoder, leading to high performance, reusable C code. When mapped on (embedded) processors, this allows for lower clock rates, enabling low power realizations.

1   Introduction

Novel multimedia compression systems, like the object-based MPEG-4 standard [1], offer an interactive and user-friendly representation of information. However, the compact representation of audio, video and data comes with the cost of complex and data intensive algorithms. Increasingly, these new systems are also specified in software, next to the traditional paper specification. For MPEG-4, this reference code in C consists of several hundred thousands of lines of code spread over many files. Realizing a cost-efficient implementation from such a specification is a real design challenge.

Additional difficulties, like late specification modifications and ever-changing market requirements, can require changing the implementation target. Moreover, the design has to be completed within the right time-to-market. Typically, hardware/software partitioning is one of the first steps in the design process, followed by platform dependent optimizations. In contrast, we describe the application of a high level, platform independent methodology, with the support of the ATOMIUM tool [2]. This approach allows a late choice of the target platform and provides more flexibility to deal with the problems described above.

This paper first briefly summarizes the proposed optimization methodology and then explains the functionality of the ATOMIUM framework, which provides means to deal with the code complexity of modern multimedia systems and to support the optimizations. Subsequently, the design of an MPEG-4 natural visual (video) decoder illustrates the use of ATOMIUM and the impact of the platform independent optimizations on the memory complexity. Finally, we measure the performance increase of the optimized decoder on a PC platform and indicate the relation between the reduction of memory accesses and the resulting speed up factor.

D. Soudris, P. Pirsch, and E. Barke (Eds.): PATMOS 2000, LNCS 1918, pp. 233-242, 2000.

2   C-Level Design

Recent multimedia applications are almost by definition data dominated i.e. the amount of data transfer and storage operations are at least of the same order of magnitude as the amount of arithmetic operations [3]. This reflects itself in a dominant impact on the efficiency of the system realization: mainly the performance for software and the power and silicon estate for hardware realization.

2.1   DTSE Methodology

We have previously presented a Data Transfer and Storage Exploration (DTSE) methodology that provides a systematic way of reducing the memory cost [4]. It consists of a platform independent and a platform dependent part. The first part of the DTSE transformations is carried out at a platform independent level. These optimizations are hence not affected by possible changes in the implementation target and the resulting, optimized code (typically C code) is reusable. The target platform is chosen before the second, platform dependent part. This means that the outcome of the first design steps can be considered as reusable C-level IP (Intellectual Property). We show the results of the platform independent steps applied to the MPEG-4 video decoder.

2.2   ATOMIUM Tool

The huge C code complexity of multimedia systems makes the application of DTSE without additional help tedious and error-prone (see Section 0). To tackle this design bottleneck, the C-in-C-out ATOMIUM framework is being developed [2].

This framework consists of a scalable set of kernels and tools providing functionality for advanced data transfer and storage analysis, pruning and source-to-source code transformations. This paper focuses on the application of the first two items.

Using ATOMIUM in a design involves three steps: instrumenting the program, generation instrumentation data and postprocessing of this data.

Instrumentation. The input C files, together with ATOMIUM specific include files, are parsed and analyzed by ATOMIUM resulting in C++ output files. These files have the same input/output behavior as the original files, but also include additional instrumentation code. Compilation with a regular C++ compiler and linking with the ATOMIUM run time library creates an executable as shown in Fig 1.

Generation of Instrumentation Data. Running the previously generated executable with the (normal) input stimuli produces additional instrumentation data next to the normal output (Fig. 2).

Postprocessing. The instrumentation data is then used for memory analysis and code pruning (see next sections).

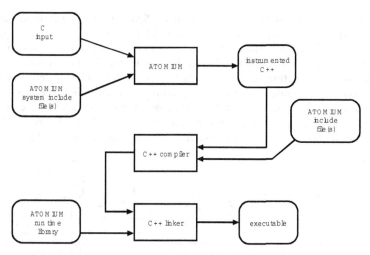

Fig 1. Instrumenting the code with ATOMIUM prepares memory analysis or code pruning

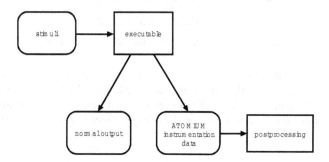

Fig. 2. Running instrumented code produces instrumentation data next to the normal output

3   MPEG-4 Natural Visual Decoder

MPEG-4 can be considered as the first true multimedia standard. It describes a scene as a composition of synthetic or natural audiovisual objects: audio, video and graphics. These objects are coded separately using the most efficient compression tool.

A specific device will only need a subset of the MPEG-4 tools to fulfill the need of the application. A profile in MPEG-4 is the definition of such a subset. A level restricts the performance criteria, like the computational complexity of the profile tool set [1], [5].

The MPEG-4 standard is divided in several parts: audio, systems, visual, etc. Next to the "classical" video objects, called natural visual objects, synthetic visual objects (such as facial animation) are distinguished. The MPEG-4 (natural visual) video decoder is a block-based algorithm exploiting temporal and spatial redundancy in subsequent frames. An MPEG-4 Visual Object Plane (VOP) is a time instance of a visual object (i.e. frame). A decompressed VOP is represented as a group of MacroBlocks (MBs). Each MB contains six blocks of 8 x 8 pixels: 4 luminance (Y), 1 chrominance red (Cr) and 1 chrominance blue (Cb) blocks. Fig. 3 shows a simple profile decoder, supporting rectangular I and P VOPs. An I VOP or intra coded VOP contains only independent texture information, decoded separately by inverse quantization and IDCT scheme. A P-VOP or predictive coded VOP is coded using motion compensated prediction from the previous P or I VOP. Reconstructing a P VOP implies adding a motion compensated VOP and a texture decoded error VOP.

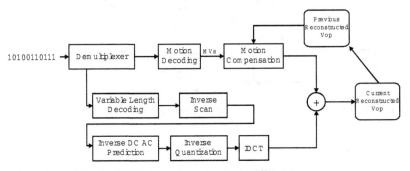

Fig. 3. MPEG-4 simple profile natural visual decoding

Next to a paper description, the MPEG-4 encoding or decoding functionality is also completely specified through Verification Models (VM), normative reference code implementing an MPEG-4 subpart (audio, visual, etc). This software, written in C, is the reference for the encoding and decoding tools of that part of the standard.

4   Pruning

The VM software used as input for this paper is the FDIS (Final Draft International Standard) natural visual part [6]. Having working code at the start of the design process can overrule the tedious task to implement a system from scratch. Unfortunately, the software specification contains many different coding styles and is often of varying quality.

Moreover, the VM has to contain all the functionality resulting in oversized C code distributed over many files. Table 1 lists the code size of the video decoder only: 93 files (.h and .c source code files) containing 52928 lines (without counting the comment lines).

Table 1. ATOMIUM pruning reduces the code size with a factor 2.5. This allows manual code rearrangement that further reduces the code complexity

Code version	Number of files	Number of lines	Reduction
FDIS	93	52928	-
Pruned	26	21340	2.5
Optimized	20	10221	5.2

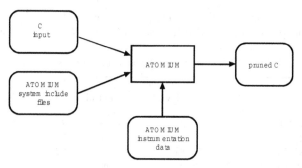

Fig. 4. ATOMIUM pruning extracts the required functionality from the source code based on the instrumentation data of the input stimuli

A necessary first step in the design is extracting the part of the reference code corresponding to the desired MPEG-4 functionality of the given profile and level. ATOMIUM pruning, shown in Fig. 4, is used for this error-prone and tedious task. The tool identifies functions that are never used in the code (static pruning) and functions that are never called according to the input stimuli and parameters used to produce the instrumentation data (dynamic pruning). Consequently, ATOMIUM removes these functions and their calls. This implies careful selection of the set of input stimuli, which has to exercise all the required functionality.

Applying automatic pruning with a testbench covering the MPEG-4 simple profile natural visual tools reduced the code to 40 % of its original size (2.5 x reduction, see Table 1). From this point, further manual code reorganization and rewriting is feasible and shrinks the number of lines to 19 % of the original (5.2 x reduction). This last reduction is obtained by flattening the hierarchical function structure and because the memory optimizations allow further simplification of the required functionality.

5   Memory Analysis

The C-level design approach requires an analysis of the data and transfer storage characteristics, initially for an early detection of possible implementation bottlenecks, subsequently to measure the effects of the optimizations. Traditionally, designers manually insert counter-based mechanisms. This is a valid, but time consuming error-prone approach. Profilers offer an alternative but use internally a flattened memory model and moreover, produce machine dependent results [7].

Postprocessing the instrumentation data with the ATOMIUM reporter generates an instrumentation database in a selectable output format. Using HTML as output offers an efficient and intuitive way of navigating through the memory access reports. The analysis results can be produced on array basis or on function basis. Table 2 lists the characteristics of the video bitstreams used as input stimuli for the creation of the instrumentation data. *Akiyo* is a typical head and shoulder sequence with little motion, *Foreman* is a medium motion sequence, whereas *Calendar and Mobile* is a highly complex sequence. When enabling rate control, the MPEG-4 encoder sometimes skips frames to obtain the specified bitrate. This explains the difference between the number of displayed VOPs and the number of coded VOPs (when the encoder skipped a frame, the decoder displays the previous one). The results listed are for the MPEG-4 simple profile for CIF (358 x 288) and QCIF (176 x 144) image sizes.

Table 2. Characteristics of the testbench video sequences

Test case	Number of VOPs	Rate Control	Number of coded VOPs	Bitrate (kbps)
1. Akiyo QCIF	81	yes	71	53
2. Foreman QCIF 1	81	none	81	95
3. Foreman QCIF 2	81	none	81	96
4. Foreman CIF 1	81	yes	62	104
5. Calendar and Mobile QCIF	81	none	81	1163
6. Foreman CIF 2	81	yes	58	104
7. Foreman QCIF 3	81	yes	81	51
8. Foreman CIF 3	101	none	101	274
9. Foreman CIF 4	101	none	101	465
10. Foreman CIF 5	101	none	101	764

Analysis of the access reports of the automatically pruned code allows early identification of bottlenecks. Table 3 lists the most memory intensive functions together with the relative execution time spent in this function for the Foreman CIF 3 test case. The timing results are obtained with Quantify [8] on a Pentium II 350 MHz PC (intentionally a low-end model since eventually embedded systems are targeted). As expected, memory bottlenecks popping up at this platform independent level also turn out to consume much time on the PC platform. The following list explains the behavior of the functions in Table 3:

- VopMotionCompensate: Picks the MB positioned by the motion vectors from the previous reconstructed VOP. In case of halfpell motion vectors, interpolation is required.
- BlockIDCT: Inverse Discrete Cosine Transform of an 8 x 8 block
- VopTextureUpdate: Add the motion compensated and texture VOP.
- CloneVop: Copies data of current to previous reconstructed VOP by duplicating it.
- VopPadding: Add a border to previous reconstructed VOP to allow motion vectors to point out of the VOP.
- WriteOutputImage: Write the previous reconstructed VOP (without border) to the output files.

Only the IDCT is a computationally intensive function, all the others mainly involve data transfer and storage. The motion compensation and block IDCT together cause more than 40 % of the total number of memory accesses, making them the main bottlenecks. Focusing on these functions during the memory optimizations (i.e. reduce the number of accesses) is hence logical.

The platform independent DTSE optimizations consist of global data flow, global loop and control flow transformations. These transformations reduce the number of memory accesses, improve the locality of the array accesses and decrease the amount of required memory [3], [4]. The listed results (Table 4) only include a part of the possible control and data flow and loop transformations. The reduction factor varies from 4.6 to 10.8 as the effect of some of the optimizations relies on the content of the bitstream.

Table 3. Motion compensation and the IDCT are the memory bottlenecks of the decoder. This analysis was done using the Foreman CIF 3 test case

Function name	# accesses/frame (10^6 accesses/frame)	relative # accesses	relative time (%)
VopMotionCompensate	3.9	25.4	14.6
BlockIDCT	2.8	18.0	17.7
VopTextureUpdate	1.7	10.7	5.4
CloneVop	1.2	7.5	5.0
VopPadding	1.1	7.0	6.4
WriteOutputImage	1.0	6.2	27.3
Subtotal	11.6	74.7	76.3
Total	15.5	100.0	100.0

Table 4. The memory optimization result varies from a factor 4.6 to 10.8

Test case	# accesses/frame pruned (10^6 accesses/frame)	# accesses/frame optimized (10^6 accesses/frame)	Reduction factor
1. Akiyo QCIF	2.8	0.3	10.8
2. Foreman QCIF 1	4.1	0.6	7.1
3. Foreman QCIF 2	3.9	0.6	6.5
4. Foreman CIF 1	11.4	1.4	8.1
5. Cal & Mob QCIF	4.8	1.0	4.6
6. Foreman CIF 2	10.8	1.3	8.0
7. Foreman QCIF 3	3.8	0.5	7.2
8. Foreman CIF 3	15.5	2.2	7.2
9. Foreman CIF 4	16.3	2.5	6.5
10. Foreman CIF 5	17.0	2.8	6.0

6   Evaluation of the Optimizations

The implemented memory optimizations have a positive effect on the platform dependent level, both for hardware and software. At the HW side the reduction in power consumption evaluates the gain, at the SW side the speed up of the code determines the effectiveness. This speed up can then be used to lower the clock speed hence reducing the power consumption.

The main part of the power consumption in data dominated applications is due to the memory [4]. The ATOMIUM instrumentation data together with the number of words and the width (in bits) of the used memory provides the necessary input to calculate a simple estimation of the power consumption 2:

$$P_{Tr} = E_{Tr} \times \frac{\#Transfers}{Second} \tag{1}$$

$$E_{Tr} = f(\#words, \#bits) \tag{2}$$

Doing this calculation for every memory block yields an estimate of the total power dissipation. Reducing the amount of necessary memory size allows the choice of memory blocks with a lower E_{Tr} energy per transfer. Combining this with a lower number of accesses (Table 4) leads to a lower overall power consumption of the optimized decoder. We have previously demonstrated this approach for HW by designing the OZONE, an ASIC for wavelet-based MPEG-4 visual texture compression [9].

Table 5. The speed up factor of the video decoder varies between 6.0 and 19.5

test case	pruned (fps)	optimized float IDCT (fps)	speed up	optimized integer IDCT(fps)	speed up
1. Akiyo QCIF	27.3	235.3	8.6	533.3	19.5
2. Foreman QCIF 1	16.5	95.2	5.8	187.9	11.4
3. Foreman QCIF 2	16.7	92.9	5.6	176.1	10.5
4. Foreman CIF 1	6.0	73.3	12.1	85.9	14.2
5. Cal & Mob QCIF	13.3	28.9	2.2	80.1	6.0
6. Foreman CIF 2	6.4	76.6	12.0	89.9	14.1
7. Foreman QCIF 3	18.0	147.0	8.2	213.2	11.8
8. Foreman CIF 3	4.3	30.4	7.1	52.1	12.1
9. Foreman CIF 4	4.0	21.3	5.3	43.1	10.8
10. Foreman CIF 5	3.8	15.7	4.1	36.5	9.7

Note that the VM saves the decoded video sequence to disk to allow for an assessment of the compression results. In real life applications, the decoded results are written to the video memory. To avoid this inconsistency, the speed up is measured here without writing to disk.

The speed improvement of the MPEG-4 video decoder due to the platform independent memory optimizations is listed in Table 5. The gain varies between 2.2 and 12.0. The number of cache hits is a crucial factor of the performance [10]. Lowering the amount of memory and the number of accesses and improving the data locality increases their probability. This gain comes in addition to the (well-known) gain achieved by replacing the floating point IDCT by a computationally more efficient integer version (resulting in an overall speed up factor between 6.0 and 19.5). This, together with the transformed control flow graph explains the speed increase of Table 5. Comparing these rates measured on a Pentium II, 350 MHz NT PC, with state-of-the-art results, like presented in [11] and [12] is not straightforward. The performance logically depends on the platform and the coding characteristics of the input sequences: the rate control method, the compressed bitrate, the quantization level etc. Unfortunately, insufficient details about the testbench in [11] and [12] are provided to make a detailed comparison, but globally our results achieve the same performance without the use of platform dependent optimizations.

The decrease of the total number of array accesses can also be used as an indication of the speed up, without the need to do the actual mapping on a platform (see also Fig. 5). Of course, this thesis only holds as long as the application remains data dominated. A more precise estimate can be obtained by combining the decrease of number of a certain function with a factor to indicate its data dominance level. *Calendar and Mobile* illustrates the effect of having the main reduction of accesses in the data dominated functions and only a small part of reduction of accesses is in the computation dominated IDCT functionality (i.e. the application is no longer data dominated). The speed up factor, using floating point IDCT is hence smaller than the access reduction factor. *Forman* CIF 1 and 2 illustrate the opposite case. Here, the main part of the reduction of accesses is due to the IDCT function and hence the speed up is higher then the reduction of accesses. Consequently, the replacement of the floating point IDCT by an integer one gives a proportionally larger speed improvement for *Calendar and Mobile* and a smaller one for *Foreman*.

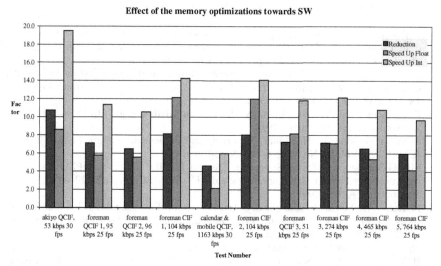

Fig. 5. The reduction of the number of accesses is an indication for the speed up factor

7   Conclusions

The MPEG-4 video decoder with a highly complex software specification has large data transfer and storage requirements. We have illustrated the use of the ATOMIUM tool for the automatic pruning and the advanced data transfer and storage analysis of the MPEG-4 video decoder specification. ATOMIUM gives designers the necessary support to deal with complex analysis and platform independent optimizations at the C-level. The effect of these optimizations is a reduction of the memory accesses with a factor 4.6 to 10.8. This optimized platform independent code results in a speed up between 6.0 and 19.5 when compiled on a PC platform. This performance increase creates the possibility to lower the clock frequency and hence to reduce the power consumption on (embedded) processors.

C-level, platform independent optimization is hence an approach that allows to re-use the optimization effort for different target platforms. A prediction of the resulting performance gain on a specific platform, taking into account the degree and distribution of the data dominance of the application, is possible without the effort of an actual implementation.

Acknowledgement

This work was partially funded by the Flemish IWT fund (HIPMUC project) and by the IMEC MPEG-4 IIAP program.

References

1. MPEG Requirements Subgroup, "Overview of the MPEG-4 Standard", ISO/IEC JTC1/SC29/WG11 N3156, Maui, December 1999.

2. J. Bormans, et al., "Integrating system-level low power methodologies into a real-life design flow", *Proc. IEEE Workshop on Power and Timing Modeling, Optimization and Simulation - PATMOS '99*, IEEE, pp. 19-28, Kos, Greece, 1999.
3. L. Nachtergaele, et al., "System-Level power optimisation of Video Codecs on Embedded Cores: a Systematic Approach", *Journal of VLSI Signal Processing*, Kluwer, Boston, Vol. 18, No. 2, pp. 89-111, February 1998.
4. F. Catthoor, et al. "Custom Memory Management Methodology", ISBN 0-7923-8288-9, Kluwer Academic Publishers, 1998.
5. J. Kneip, B. Schmale, H. Möller, "Applying and Implementing the MPEG-4 Standard", *IEEE Micro*, Vol. 19, No. 6, pp.64-74, Nov.-Dec. 1999.
6. MPEG Simulation Software Subgroup, "Text of ISO/IEC 14496-5 (MPEG 4 simulation software) final draft international standard", ISO/IEC JTC1/SC29/WG11 N2805, Vancouver, CA, July 1999.
7. P. Kuhn, W. Stechele, "Complexity Analysis of the Emerging MPEG-4 Standard as a Basis for VLSI Implementation", *Proc. International Society for Optical Engineering - SPIE*, Vol. 3309, pp. 498-509, San Jose, Jan. 1998.
8. http://www.rational.com/products/vis_quantify/index.jtmpl
9. B. Vanhoof, et al., "A Scalable Architexture for MPEG-4 Embedded Zero Tree Coding", *Custom Integrated Circuits Conference*, San Diego, US, May 1999.
10. R. Coelho, M. Hawash, "DirectX, RDX, RSX and MMX Technology, a Jumpstart Guide to High Performance APIs", Addison-Wesley Developers Press, 1998.
11. F. Casalino, G. Di Cagno, R. Luca, "MPEG-4 Video Decoder Optimisation", *Proc. IEEE International Conference on Multimedia Computing and Systems*, IEEE Comput. Soc, Vol. 1, pp. 363-368, Los Alamitos, US, 1999.
12. G. Hovden, N. Ling, "On Speed Optimisation of MPEG-4 Decoder for Real-Time Multimedia Applications", *Proc. Third International, Conference on Computational Intelligence and Multimedia Applications – ICCIMA '99*, IEEE Comput. Soc, Vol. 1, pp. 399-402, Los Alamitos, US, 1999.

Data-Reuse and Parallel Embedded Architectures for Low-Power, Real-Time Multimedia Applications

D. Soudris[1], N. D. Zervas[2], A. Argyriou[1], M. Dasygenis[1],
K. Tatas[1], C. E. Goutis[2], and A. Thanailakis[1]

[1] VLSI Design and Testing Center, Dept. of Electrical & Computer Eng.,
Democritus Univ. of Thrace, Xanthi 67100, Greece.
[2] VLSI Design Lab., Dept. of Electrical & Computer Eng.,
Univ. of Patras, Rio 26500, Greece.

Abstract. Exploitation of data re-use in combination with the use of custom memory hierarchy that exploits the temporal locality of data accesses may introduce significant power savings, especially for data-intensive applications. The effect of the data-reuse decisions on the power dissipation but also on area and performance of multimedia applications realized on multiple embedded cores is explored. The interaction between the data-reuse decisions and the selection of a certain data-memory architecture model is also studied. As demonstrator a widely-used video processing algorithmic kernel, namely the full search motion estimation kernel, is used. Experimental results prove that improvements in both power and performance can be acquired, when the right combination of data memory architecture model and data-reuse transformation is selected.

1 Introduction

The number of multimedia systems used for exchanging information is rapidly increasing nowadays. Portable multimedia applications, such as video phones, multimedia terminals and video cameras, are available. Portability as well as packaging, cooling and reliability issues have made power consumption an important design consideration [1]. For this reason there is great need for power optimization strategies, especially in higher design levels, where the most significant savings can be achieved.

Additionally, these applications also require increased processing power for manipulating large amounts of data in real time. To meet this demand, two general implementation approaches exist. The first is to use custom hardware dedicated processors. This solution leads to smaller area and power consumption. However, it lacks of flexibility since only a specific algorithm can be executed by the system. The second solution is to use a number of embedded instruction set processors. This solution requires increased area and power in comparison to the first solution. However, it offers increased flexibility and mainly meets easier the

D. Soudris, P. Pirsch, and E. Barke (Eds.): PATMOS 2000, LNCS 1918, pp. 243–254, 2000.
© Springer-Verlag Berlin Heidelberg 2000

time-to-market constraints. In both cases, to meet the real time requirements, the initial application description must be partitioned and assigned to a number of processing elements, which has to be done in a power efficient way. For multimedia applications realized in custom-processor platforms, the dominant factor in power consumption is the one related to data storage and transfer [2]. In programmable platforms though, the power consumed for instructions storage and transfers limits the dominant role of the power related to data storage and transfer [4].

The related work that combines partitioning of the algorithm and techniques for reducing the memory related power cost is relatively small [2][3][4][5]. More specifically, a systematic methodology for the reduction of memory power consumption is presented in [2][3]. According to this methodology, power optimizing transformations (such as data-reuse) are applied in the high level description of the application prior to partitioning step. These transformations mainly targets to reduction of the power due to data storage and transfer. Although, the efficiency of this methodology has been proved for custom hardware architectures [2] and for commercially available multimedia processors (e.g. Trimedia) [3], it does not tackle with the problem when an embedded multiprocessor architectures are used. The latter point has been stressed in [4] where the data-reuse exploration as proposed in [6] has been applied for uni-processor embedded architectures. The experimental results of [4] indicated that the reduction of the data memory-related power does not always come with a reduction of the total power budget for such architectures. Finally, a partitioning approach attempting to improve memory utilization is presented in [5]. However, this approach limited by the two-level memory hierarchy, does not explore the effect of the high-level power optimizing transformations, and its applicability is limited to a class of algorithms expressed in Weak Single Assignment Code (WSAC) form. Clearly, previous research work has not explored the effect on power, area, and performance of the high level transformations for the case of multiprocessor embedded architectures. In such architectures a decision that heavily affects power, area and performance is the one related to the data memory architecture-model (i.e. shared, distributed, share-distributed) to be followed.

The motivation of this work is to investigate the dependencies between the decision of adapting a certain data memory architecture-model and the high-level power optimizing transformations. The intuition is that these two high-level design steps, which heavily influence all design parameters are not orthogonal to each other. Consequently, in this paper we apply all possible data-reuse transformations [6] in a real-life application, assuming a LSGP partitioning scheme [11] and three different data memory architecture-models, namely Distributed, Shared, and Shared-Distributed. For all the data-memory architectures, the transformations' effect on performance, area and power consumption is evaluated. The experimental results prove that the same data-reuse transformations do not have similar effect on power and performance when applied for different data-memory architecture models. Thus, the claim that the application of these transformations in the first step can optimize power and/or performance, regardless the decisions related to data memory architecture that must follow is proved to be weak. Furthermore, the comparative study concerning power, performance and area of the three architectures and all the data reuse transformations indicate that an effective solution can be acquired from the right combination of data memory architecture model and data-reuse transformation. Finally, once more, the critical influence of the instruction power consumption on the total power budget is proved.

2   Target Architectures

We are working on multiple processor architectures each of which has its own single on-chip instruction memory. The size of the instruction-memory is strongly-depended on the code size executed by a processor. We name this scheme application specific instruction memory (ASIM). The instruction memory of different processors may have different size. Concerning the data-memory organization, application specific data memory hierarchy (ASDMH) is assumed. [2][7]. Since we focus on parallel processing architectures, we explore ASDMH in combination with three well-established data-memory architectures models: 1) *distributed data-memory architecture DMA*, 2) *shared data-memory architecture SMA*, and 3) *shared-distributed SDMA* data memory architecture. For all the data-memory architectures models a shared background (probably off-chip) memory module is assumed. Thus, in all cases special care must be taken during the scheduling of accesses to this memory, to avoid violating data-dependencies and to keep the number of memory ports as small as possible in order to keep the power per access cost as small as possible. With DMA, a separate data-memory hierarchy exists for each processor (Fig. 1). In this way all memories modules of the memory hierarchy are single ported, but also area overhead is possible in cases of large amount of common data to be processed by the N processors. The second data-memory architecture-model (i.e. SMA) implies a common hierarchy of memory levels for the N processors (Fig. 2). Since, in the data-dominated programmable parallel processing domain, it is very difficult and very performance inefficient to sequentially schedule all memory accesses, we assume that the number of ports for each memory block equals the maximum number of parallel accesses to it. Finally, SDMA is a combination of the above two models, where the common data to the N processors are placed in a shared memory hierarchy, while a separate data memory hierarchy also exist for the lowest levels of the hierarchy (Fig. 3). For experimental purposes, we have considered target models with $N=2$ without any restriction about memory hierarchy levels.

3   Data Reuse Transformations

The fact that in multimedia applications the power related to memory transfers is the dominant factor in total power cost, motivate us to find an efficient method to reduce them. This goal can be done by efficient manipulation techniques of memory data transfers. For that purpose, we performed an exhaustive data reuse

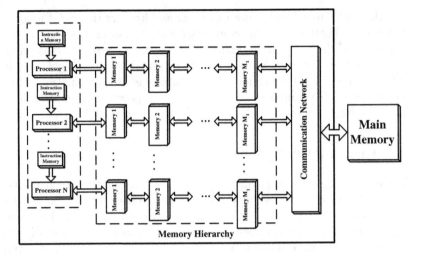

Fig. 1. The distributed memory data-memory architecture model

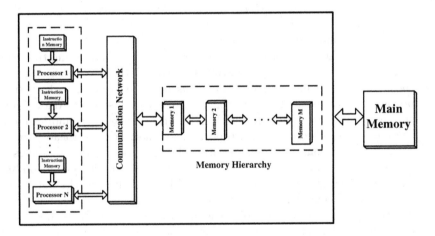

Fig. 2. The shared memory data-memory architecture model

exploration of the application's data. Employing data reuse transformations, we determine the certain data sets, which which are heavily re-used in a short period of time. The re-used data can be stored in smaller on-chip memories, which require less power per access. In this way, redundant accesses from large off-chip memories are transfered on chip, reducing power consumption related to data transfers. Of course, data reuse exploration has to decide which data sets are appropriate to be placed in separate memory. Otherwise, we will need a lot of different memories for each data set resulting into a significant area penalty.

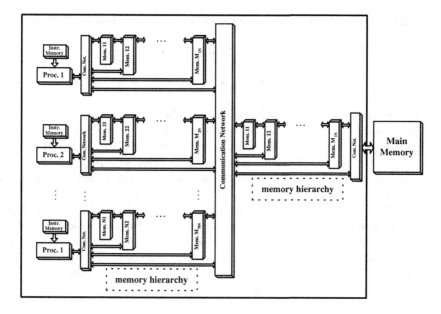

Fig. 3. The shared-distributed data-memory architecture model

Since our target architecture consists of programmable processors, we must take into consideration the power dissipation due to instruction fetching. Previous work [4] forms a sign that this power parameter is a significant part of total system's power, and thus, it should not be ignored. Also, it depends on both number of executed instructions and the size of the application code. Particularly, the number of executed instructions determines how many times the instruction memory is accessed, while the code size determines the memory size. The cost function used for our data reuse exploration on all target architectures is evaluated in terms of power, performance, and area, taking into account both data and instruction memories. The cost function for power is:

$$Power\_cost = \sum_{i=1}^{N} power\_cost_i \qquad (1)$$

where N is the number of processors and the i-th power estimate, $power\_cost_i$ is:

$$
\begin{aligned}
power\_cost_i = \sum_{c \epsilon CT} &[P_r(word\_length(c), \#words(c), f_{read}(c), \#ports(c)) \\
&+ P_w(word\_length(c), \#words(c), f_{write}(c), \#ports(c))] \\
&+ P_i(instr\_word\_length, code\_size, f) \qquad (2)
\end{aligned}
$$

where c is a member of the copy tree (CT) [6] , $P_r(\cdot)$, $P_w(\cdot)$, and $P_i(\cdot)$ is the power consumption estimate for read operation, write operation, and instruction fetch, respectively. For memory power consumption estimation we use the models reported in [2]and [8].

The total delay cost function is obtained by:

$$Delay\_cost = \max_{i=1,\dots,N} \{\#cycles\_processor_i\} \tag{3}$$

where $\#cycles\_processor_i$ denotes the number of the executed cycles of the i-th processor $(i = 1, 2, \cdots, N)$. Also, the maximum number of cycles is the performance of the system. In order to estimate the performance of a particular application, we use the number of executed cycles resulting from the considered processor core simulation environment. Here, for experimental reasons we will use the ARMulator [12].

High level estimation implies that a designer should decide, which possible solution of a certain problem is the most appropriate. For that purpose, we will use the measure of *power $\times$ delay product*. This measure can be considered as a generalization of the similar concept from circuit level design and allows the designer performing trade-offs among several possible implementations. That is, the power efficient architecture is:

$$Power\_eff\_arch = Power\_cost \times Delay\_cost \tag{4}$$

The corresponding area cost function is:

$$Area\_cost = \sum_{i=1}^{N} area\_cost_i \tag{5}$$

with

$$area\_cost_i = \sum_{c \epsilon CT} Area(word\_length(c), \#words(c), \#ports(c))$$
$$+ Area(instr\_word\_length, code\_size) \tag{6}$$

For the area occupied by the memories, Mulder's model is used [9]. The cost function of the entire system is given by:

$$Cost = a \cdot Power\_eff\_arch + b \cdot Area\_cost \tag{7}$$

where a and b are weighting factors for area/energy trade-offs.

4   Experimental Results-Comparative Study

In this section, we perform extensive comparative study of the relation between data-reuse transformations and data-memory models, assuming the application's partitioning. We begin with the description of our test vehicle and through its partitioning scheme, we will provide the experimental results after the application of the data-reuse transformations for all target architectures, in terms of power performance and area.

4.1   Demonstrator Application and Partitioning

Our demonstrator application was selected to be the full search motion estimation algorithm [10]. It was chosen this algorithm because it is used in a great number of video processing applications. Our experiments were carried out using the luminance components of QCIF frame (144x176) format. Reference window was selected to include 15x15 candidate blocks, while blocks of 16x16 pixels were considered. The algorithm structure is described in Figure 4(a) which has three double nested loops. A block of the current frame (outer loop) is compared to a number of candidate blocks (middle loop). In the inner loop, a distortion criterion is computed to perform the comparison.

Partitioning was done with the use of LSGP technique [11]. By applying this technique to a generalized for-loop structure, while assuming p partitions, the form of the partitioned algorithm becomes as shown in Fig.5.

```
for(x = 0; x < N/B; x++)
for(y = 0; y < M/B; y++)
   for(i = -p;i < p+1; i++)
   for(j = -p;j < p+1; j++)
      for(k = 0;k < B; k++)
      for(l = 0;l < B; l++)
         if((B*x+i+k) < 0 || (B*x+i+k) > N-1 ||(B*y+j+l) < 0 || (B*y+j+l)>M-1)
         \*conditional statement for the pixel of candidate block * \
```

Fig. 4. The full search motion estimation algorithm

```
Do in parallel:
Begin
for(x=0; x< ⌈N/pB⌉;x++) {sub-algorithm}
for(x=⌈N/pB⌉; x< ⌈2N/pB⌉;x++) {sub-algorithm}

⋮

for(x= ⌈(p-1)N/pB⌉; x< ⌈N/B⌉;x++) { sub-algorithm}
End
```

Fig. 5. The partitioned algorithm

The semantic *"Do in parallel"* imposes the parallel (concurrent) execution of p nested loops (i.e. sub-algorithm). From this above -code, it is apparent that the outermost loop is broken into p partitions, each of which is mapped to processor. The p processors execute the same algorithmic structure for different values of loop index x, i.e. different current blocks. Due to the inherent property of algorithm, a set of data should be used by two consecutive sub-algorithms. In other words, data from $(k\text{-}1)$-th processor should be used by k-th processor $(k = 1, 2, 3, \cdots, p)$. Our experiments were carried out assuming $p = 2$, meaning two partitions. Therefore, the loop index x has a range of nine. Due to QCIF format (144x176), the outermost index ranges from 0 to 8. The first and second processor execute the algorithm in parallel fashion, for loop index x ranging from 0 to 4 and from 5 to 8, respectively. We examined the impact of partitioning combined with 21 data reuse transformations on power, performance, and area. These transformations were applied after the partitioning process was finished in accordance with the previous section. They involved the insertion of memories for a line of current blocks (CB line), a current block (CB), a line of candidate blocks (PB line), a candidate block (PB), a line of reference windows (RW line) and a reference window (RW). These transformations were applied for all the three data-memory architecture models by taking into account each architecture's characteristics. In Fig. The copy tree [6]of the full search motion estimation algorithm is identical for processor 1 and 2, where the dashed lines show the memory levels. Each rectangle contains three labels, where the number determines the applied data reuse transformations associated to memory hierarchy level. The remaining two labels determine the size of an PB and CB line or block, RW line or reference window.

4.2   Experimental Results

Comparisons among the three target architectures, in terms of power, performance, and area are shown in Fig. 6,8, and 9.

Fig. 6 provide comparisons results of power consumption with respect to data-reuse transformations. The most power efficient design approach is the combination of SDMA and data-reuse transformations 4,5,15,19 and 20. In contrary, almost all data-reuse transformations increase the total power when DMA or SMA is assumed.

The effect of the data-reuse transformations on power consumption of data memory is shown in Fig. 7. As it can be seen, the largest effect is on the SMA, while the most efficient are the two other two data memory architecture models. Comparing Fig. 6 and 7, it is deducted that the power cost related to instruction memory have significant contribution on the total power budget, and in many cases overturns the power savings acquired in the data memory. Thus, the power component related to instruction-memory cannot be ignored during such high level power exploration. Fig. 8 shows that with DMA and SMA the data-reuse transformations barely affects performance, while with SDMA the transformations have a more significant impact on performance. The greater variation in performance when the SDMA is assumed results from the size of instruction code related to control operations, specifying which memories of the hierarchy should be accessed. However, it can be generally concluded that the transformations have similar effect on the performance for all data-memory architecture models (i.e. a certain transform positively/negativelly affects performance for all data-memory architecture models). Although this is true, the optimal transfor-

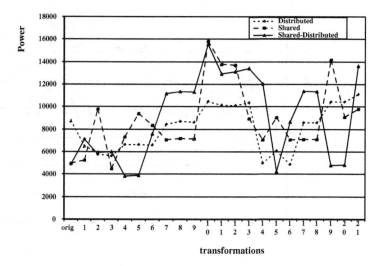

Fig. 6. Comparison results for total power.

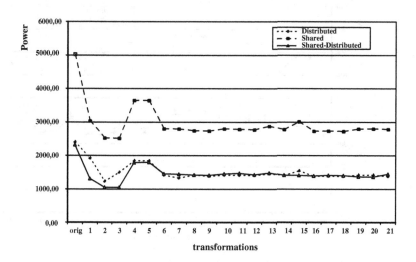

Fig. 7. The effect of data-reuse transf. on power of data memory.

mations in terms of performance are different for each different data-memory architecture model. Specifically 4,5,6,18,19 and 20 for SDMA, 6,7,8,9,13,16,17 and 18 for SMA and DMA are the near-optimal or optimal solutions in terms of performance.

In Fig.9 the effect of data-reuse transformations on area is illustrated. From that it can be inferred that each transformation influences area in almost iden-

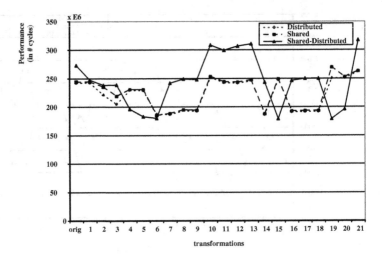

Fig. 8. Performance comparison results of the target architectures.

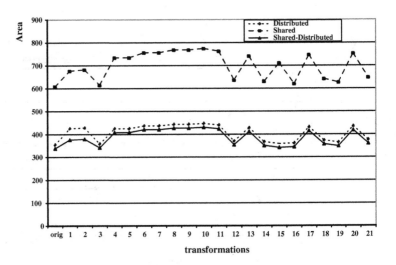

Fig. 9. Area comparison results of the target architectures.

tical manner for all data-memory architectural models. It is also clear that all transformations increase area, since they impose the addition of extra data memory hierarchy levels. Moreover, for both DMA and SDMA area cost is similar for each data-reuse transformation. With SMA the area occupation is larger in all cases. This due to the fact that several memory modules are dual ported, to be accessed in parallel by the processing elements. In contrary, most memory modules are single ported and thus, they occupy less area. As it can be seen,

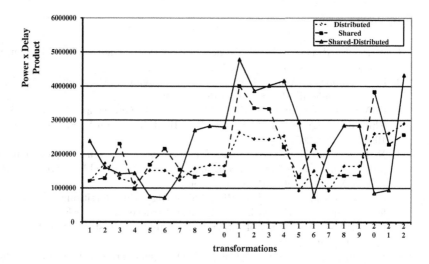

Fig. 10. Comparison results of the target architectures with respect to *power × delay product*.

the SDMA is the most area efficient, since with this data-memory architecture model there are no memories in the hierarchy with duplicate data.

In order to define which combination of data-memory architecture model and data-reuse transformation is the most efficient in terms of performance and power, we plot *power × delayproduct* (Fig.10). We infer that there exist enough possible solutions, which can be chosen by the designer. These solutions are: the transformation 3 with SMA, transformations 15 and 17 with DMA and transformations 4,5,15,19 and 20 with SDMA. If also the area dimension is taken into account, the effective solutions are transformations 15 and 17, and, 4,5,15,19 and 20 with DMA and SDMA, respectively.

5   Conclusions

Data-reuse exploration for the partitioned version of a real life application and for three alternative data-memory architecture models was performed. Application specific, data-memory hierarchy and instruction memory, as well as embedded programmable processing elements, were assumed. The comparison results prove that an effective solution either in terms of power or power and delay or power and delay and area, can be acquired from the right combination of data memory architecture model and data-reuse transformation. Thus, in the parallel processing domain for multimedia applications, the high-level design decision of adapting a certain data-memory architecture model and the application of high-level power optimizing transformations should be performed interactively and not in a sequential way (regardless the ordering) as prior research work proposed.

References

1. A. P. Chandrakasan, R. W. Brodersen, Low Power Digital CMOS Design, Kluwer Academic Publishers, Boston, 1998.
2. F. Catthoor, S. Wuytack et al., Custom Memory Management Methodology, Kluwer Academic Publishers, Boston, 1998.
3. K. Masselos, F. Catthoor, H. De Man, and C.E. Goutis, and "Strategy for Power Efficient Design of Parallel Systems", in IEEE Trans. on VLSI, vol. 7, No. 2, June 1999, pp. 258-265.
4. N. D. Zervas, K. Masselos, and C.E. Goutis, "Data-reuse exploration for low-power realization of multimedia applications on embedded cores", in Proc. of PATMOS'99, October 1999, pp. 71-80.
5. U. Eckhardt and R. Merker,"Hierarchical Algorithm Partitioning at System Level for an Improved Utilization of Memory Structures", in IEEE Transactions on Computer-Aided Design of Integrated Circuits and Systems, Vol. 18, No. 1, January 1999, pp. 14-23.
6. S. Wuytack, J.-P. Diguet, F. Catthoor, D. Moolenaar, and H. De Man "Formalized Methodology for Data Reuse Exploration for Low-Power Hierarchical Memory Mappings", in IEEE Trans. on VLSI Systems, Vol. 6, No. 4, Dec. 1998, pp. 529-537.
7. L. Nachtergaele, B. Vanhoof, F. Catthoor, D. Moolenaar, and H De Man,"System-level power optimazations of video codecs on embedded cores: a systematic approach", Journal of VLSI Signal Processing Systems, Kluwer Academic Publishers, Boston, 1998.
8. P. Landman, Low power architectural design methodologies, Doctoral Dissertation, U.C. Berkeley, Aug. 1994.
9. J.M. Mulder, N.T. Quach, and M.J. Flynn,"An Area Model for On-Chip Memories and its Application", IEEE Journal of Solid-State Circuits, Vol. SC26, No.1, Feb. 1991, pp.98-105.
10. V. Bhaskaran and K. Kostantinides, Image and Video Compression Standards, Kluwer Academic Publishers, Boston, 1998.
11. S. Y. Kung,"VLSI Array Processors", Prentice Hall, Eaglewood Cliffs, 1988.
12. ARM software development toolkit, v2.11, Copyright 1996-7, Advanced RISC Machines.

Design of Reversible Logic Circuits by Means of Control Gates

A. De Vos[1], B. Desoete[2],
A. Adamski[3], P. Pietrzak[3], M. Sibiński[3], and T. Widerski[3]

[1] Universiteit Gent and Imec v.z.w., B-9000 Gent, Belgium
[2] Universiteit Gent, B-9000 Gent, Belgium
[3] Politechnika Łódzka, PL-90-924 Łódź, Poland

Abstract. A design methodology for reversible logic circuits is presented. Any boolean function can be built using the three fundamental building blocks of Feynman. The implementation of these logic gates into electronic circuitry is based on c-MOS technology and pass-transistor design. We present a chip containing single Feynman gates, as well as an application: a chip containing a fully reversible four-bit adder. We propose a generalization of the Feynman gates: the reversible control gates.

1 Introduction

Classical computing machines using irreversible logic gates unavoidably generate heat. This is due to the fact that each loss of one bit of information is accompanied by an increase of the environment's entropy by an amount $k \log(2)$, where k is Boltzmann's constant. In turn this means that an amount of thermal energy equal to $kT \log(2)$ is transferred to the environment, having a temperature T. According to Landauer's principle [1] [2], it is possible to construct a computer that dissipates an arbitrarily small amount of heat. A necessary condition is that no information is thrown away. Therefore, logical reversibility is a necessary (although not sufficient) condition for physical reversibility.

It is widely known that an arbitrary boolean function can be implemented into logic using only NAND-gates. A NAND-gate has two binary inputs (say A and B) but only one binary output (say P), and therefore is logically irreversible. Fredkin and Toffoli [3] have shown that a basic building block which is logically reversible, should have three binary inputs (say A, B, and C) and three binary outputs (say P, Q, and R). Feynman [4] [5] has proposed the use of three fundamental gates:

- the NOT gate,
- the CONTROLLED NOT gate, and
- the CONTROLLED CONTROLLED NOT gate.

See Table 1. Together they form a set of three building blocks with which we can synthetize an arbitrary logic function. The NOT gate simply realizes $P =$ NOT A. The CONTROLLED NOT satisfies $P = A$, together with

$$\text{If } A = 0, \text{ then } Q = B, \text{ else } Q = \text{NOT } B \ . \tag{1}$$

D. Soudris, P. Pirsch, and E. Barke (Eds.): PATMOS 2000, LNCS 1918, pp. 255–264, 2000.

The CONTROLLED CONTROLLED NOT satisfies $P = A$, $Q = B$, together with

$$\text{If } A \text{ AND } B = 0, \text{ then } R = C, \text{ else } R = \text{NOT } C. \qquad (2)$$

The logic expressions of the CONTROLLED NOT are equivalent with

$$P = A$$
$$Q = A \text{ XOR } B,$$

where XOR is the abbreviation of the EXCLUSIVE OR function. The gate is thus the reversible form of the conventional (irreversible) XOR gate. The logic expressions of the CONTROLLED CONTROLLED NOT are equivalent with

$$P = A$$
$$Q = B$$
$$R = (A \text{ AND } B) \text{ XOR } C.$$

Table 1. Feynman's three basic truth tables: (a) NOT, (b) CONTROLLED NOT, (c) CONTROLLED CONTROLLED NOT

A	P
0	1
1	0

(a)

AB	PQ
0 0	0 0
0 1	0 1
1 0	1 1
1 1	1 0

(b)

ABC	PQR
0 0 0	0 0 0
0 0 1	0 0 1
0 1 0	0 1 0
0 1 1	0 1 1
1 0 0	1 0 0
1 0 1	1 0 1
1 1 0	1 1 1
1 1 1	1 1 0

(c)

The CONTROLLED CONTROLLED NOT is a universal primitive [6]. This means that any boolean function of any finite number of logic input variables can be implemented by combining a finite number of such building blocks.

In spite of the fact that the CONTROLLED CONTROLLED NOT is sufficient, we will use all of the three Feynman blocks for synthesis. The NOT block is trivial, as we make use of dual electronics. This means that any boolean variable A is represented by two electric signals: A and $\overline{A} = \text{NOT } A$. Therefore, a simple metal cross-over is sufficient to realize the NOT function: P being connected to $\overline{A}$, while $\overline{P}$ is connected to A. Function (1) leads to the implementation of

Figure 1a, whereas function (2) is realized as in Figure 1b. The latter circuit is deduced from the former. In the four sides of the square of Figure 1b, the single switches from Figure 1a are replaced either by a series or by a parallel connection. This extrapolation is inspired by conventional (restoring) digital electronics, where a similar extrapolation of the NOT gate (Figure 2a) leads to the NOR gate (Figure 2b) and to the NAND gate (Figure 2c).

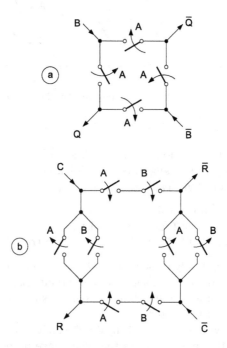

Fig. 1. Basic square circuits: (a) $Q = A$ XOR B, (b) $R = (A$ AND $B)$ XOR C. Here a switch is in the state indicated by the arrow if the logic variable next to it equals one

2   Electronic Implementation

Within the framework of the European multiproject-wafer service *Europractice*, silicon prototypes of some circuits have been fabricated, in the *Alcatel Microelec-tronics* n-well c-MOS 2.4 μm technology. The layout is designed with Cadence DesignFrameWork II 4.3.4 full-custom software. The n-MOS transistors have length L equal to 2.4 μm and width W equal to 2.4 μm, whereas the p-MOS transistors have $L = 2.4$ μm and $W = 7.2$ μm. The p-MOS transistors are chosen three times as wide as the n-MOS transistors in order to compensate for the fact that holes are three times less mobile in silicon than electrons.

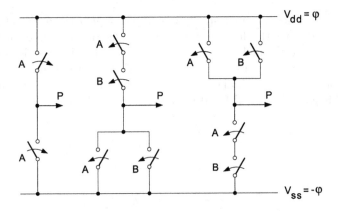

Fig. 2. Conventional c-MOS logic gates: (a) the NOT gate, (b) the NOR gate, (c) the NAND gate

For the implementation of the on-off switch, we use the transmission gate, consisting of two MOS transistors in parallel, i.e. one n-MOS transistor and one p-MOS transistor. This leads to the following number of transistors:

- the NOT gate: no transistors
- the CONTROLLED NOT gate: 8 transistors
- the CONTROLLED CONTROLLED NOT gate: 16 transistors.

We remark that not only these building blocks can be used to construct reversible circuits. Indeed, in the past some simple circuits have been implemented using another fundamental building block having hexagonal symmetry [7] [8] [9], but using, however, 24 transistors.

Figure 3 shows a prototype of the CONTROLLED NOT gate and the CONTROLLED CONTROLLED NOT gate. We stress that they have no power supply inputs. Thus there are neither V_{dd} nor V_{ss} nor ground busbars. Note also the complete absence of clock lines. Thus all signals (voltages and currents) and all energy provided at the outputs originate from the inputs. As a consequence, an inevitable signal loss occurs at the outputs. However, measurements indicate that the loss in our chip is always smaller than 10 mV for an input signal of 2, 3 or 4 volts [8].

This circuit is an example of dual-line pass-transistor logic, as opposed to conventional restoring logic. In conventional c-MOS, output pins are fuelled from a V_{dd} and a V_{ss} power line. See e.g. the conventional c-MOS gates in Figure 2.

3   Application

Higher levels of computation particularly need the implementation of the full adder. This can e.g. be realized with the help of two half adders. The latter circuit can easily be built from one CONTROLLED NOT block and one CONTROLLED CONTROLLED NOT block, as shown in Figure 4a. The output A XOR B provides the sum bit S, whereas the output A AND B provides the carry-out bit C_o.

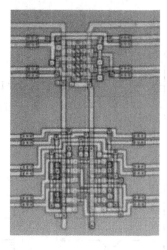

Fig. 3. Microscope photograph of the `CONTROLLED NOT` and the `CONTROLLED CONTROLLED NOT` gate

It is well known that a full adder can be constructed from two half adders and one `OR` gate. However, as one can expect, this is not the most economic implementation. Figure 4b gives a far more efficient design [4] [5] [10]. Not only we have here four blocks instead of six, but (and this is even more important) we have only four dual input lines. The circuit consists of $2 \times 8 + 2 \times 16 = 48$ transistors.

Figure 5 shows a prototype 4-bit adder chip. It contains a total of $4 \times 48 = 192$ transistors. It sums a 4-bit number (A_0, A_1, A_2, A_3) and a 4-bit number (B_0, B_1, B_2, B_3). The result is a 4-bit number (S_0, S_1, S_2, S_3). The first carry-in bit, i.e. $(C_i)_0$ is set to zero, whereas the carry-over bits ripple from one full adder to the next, the last carry-out $(C_o)_3$ yielding the overflow bit. From Figure 4b, we see that not only the sum

$$S = A + B \tag{3}$$

is calculated, but that another output recovers the value of input A:

$$S = A + B$$
$$T = A \ . \tag{4}$$

This is no surprise, as eqn (3) is not reversible (the value of S being insufficient for recovering both A and B), whereas result (4) is computationally reversible:

$$A = T$$
$$B = S - T \ .$$

See Figure 6.

Besides A, S and C_o of Figure 4b, the full adder also provides a fourth output (i.e. A `XOR` B). This result is considered as 'garbage'. The garbage outputs are the counterpart of the preset inputs.

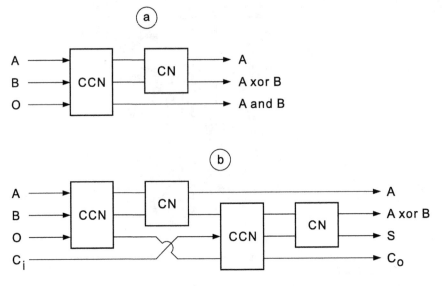

Fig. 4. Block diagram of Feynman reversible adders: (a) the half adder and (b) the full adder

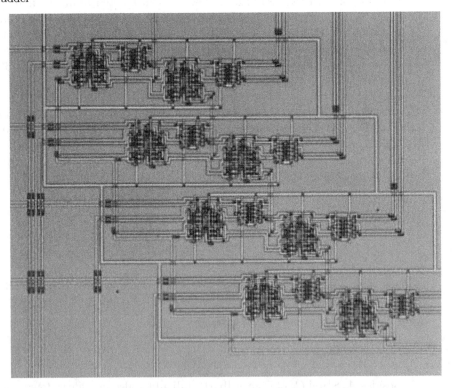

Fig. 5. Microscope photograph of a prototype c-MOS reversible Feynman four-bit adder

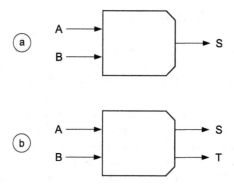

Fig. 6. Two different ways of adding two numbers: (a) irreversible way, (b) reversible way

A very important advantage of the Feynman gates, with respect to the 'hexagonal' gates, is the fact they fulfil two conditions simultaneously:

- the backward truth tables are equal to the forward truth tables (or in other words: the gates are identical to their reverse gates) and
- the electronic implementation is identical to its mirror image.

As a result, circuits,

- that are intirely built from Feynman gates and
- where these building blocks are interconnected in a symmetric way,

can compute in both directions. The inputs and the outputs of such circuits are indistinguishable. There is no need for additional hardware to implement 'electronic reversibility' [8]. The circuit can equally perform the same calculation from left to right as from right to left.

An important class of such circuits is formed by the garbageless circuits proposed by Fredkin and Toffoli [3]. Indeed, the 'undo' subcircuit is the mirror image of the 'do' part, whereas the 'spy' circuit is its own mirror image. An example is shown in Figure 7: a garbageless one-bit full adder.

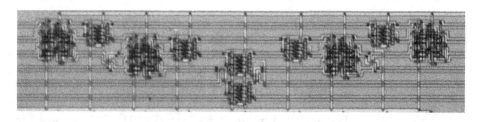

Fig. 7. Microscope photograph of a prototype garbageless Feynman one-bit full adder

4   Adiabatic Addressing

The described reversible c-MOS circuits are particularly suited for adiabatic addressing [8] [11] [12] [13] [14]. The dynamic behaviour of a 1-bit adder is examined by a change of the input (A, B, C_i). If e.g. we change (A, B, C_i) from (0,0,1) to (1,0,1), by raising V_A from $-\varphi$ to $+\varphi$ (See Figure 8a), then changes happen in a non-adiabatic way, just like in conventional c-MOS. Energy dissipation for charging an output capacitor $\mathcal{C}$ equals $\frac{1}{2}\mathcal{C}(2\varphi)^2$. The problem can easily be circumvented by introducing an intermediate state, where all three voltages V_A, V_B, and V_{C_i} equal zero (and thus the logic variables A, B, and C_i are undetermined). See Figure 8b. In the first part of the switching process all capacitive loads are discharged, sending their stored energy to the voltage sources at the input pins; in the second part of the switching process the input voltage sources recharge the output capacitors.

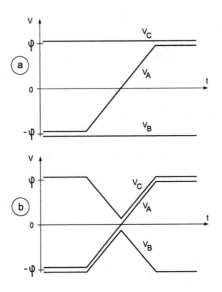

Fig. 8. Input voltages as function of time: (a) conventional addressing, (b) adiabatic addressing

Spice simulations confirm that the quasi-adiabatic charging indeed consumes less energy than the conventional one. The energy consumed for one switch and back, we call the 'energy dissipation per computational cycle' ΔW. At high speed, we find the limit value $\frac{1}{2}\mathcal{C}(2\varphi)^2$. For sufficiently slow clocks, the adiabatic switching is economic. However, ΔW decreases more slowly with addressing time τ than the τ^{-1}-law predicted by the Athas equation [11]. This is caused by the fact that a transistor is not ohmic [8]. Even more disturbing is the saturation of ΔW, for very slow switching, at a value approximately equal to $\frac{1}{2}\mathcal{C}(V_t)^2$, where V_t is the threshold voltage of the transistors. Indeed, transmission gates

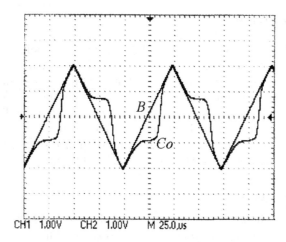

Fig. 9. Experimental transient signals on oscilloscope screen: input bit B and output bit C_o of a full adder

are turned off as long as both transistors have a gate-source voltage inferior to the threshold voltage. Figure 9 illustrates this behaviour for $\tau = 50$ μs and $\varphi = 2$ V (and $V_t = 0.9$ V). Per cycle of 100 μs, the 1-bit adder recovers 5.3 pJ, while dissipating only 0.6 pJ of non-recoverable energy.

If we aim to decrease the energy dissipation further, we have to lower either the threshold voltages V_t of the transistors or their input capacitances. This can be performed in three ways:

  - either we lower V_t by applying an appropriate bias to the bulk,
  - or we go to next generation (i.e. submicron) silicon process,
  - or we choose a drastically different technology, e.g. silicon-on-insulator.

5   Further Development

We can easily generalize Feynman's gates toward a broad class of reversible logic gates, we call control gates. Such a gate has w inputs $(A_1, A_2, ..., A_w)$ and w outputs $(P_1, P_2, ..., P_w)$, satisfying

$$P_i = \qquad\qquad A_i \qquad\qquad \text{for all } i \in \{1, 2, ..., m\}$$
$$P_i = f_i(A_1, A_2, ..., A_m) \text{ XOR } A_i \text{ for all } i \in \{m+1, m+2, ..., w\} ,$$

with $1 < m < w$ and with f_i arbitrary boolean functions. The m inputs $(A_1, A_2, ..., A_m)$ are called the controlling inputs; the $w - m$ inputs $(A_{m+1}, A_{m+2}, ..., A_w)$ are called the controlled inputs. The number w is the width of the gate.

The implementation of a control gate is straightforward: the m control lines are mere electric wires from input to output, whereas the remaining $w - m$ outputs P_i are generated from 'squares' like in Figure 1, the corresponding inputs A_i being preset to logic 0. A submicron 4-bit carry-look-ahead adder, entirely based on this principle, is in preparation.

6   Conclusion

We have demonstrated a way to build up reversible boolean functions by means of the fundamental building blocks proposed by Feynman. The electronic implementation of these logic gates is based on dual-line pass-transistor logic. Applying reversible logic introduces an overhead of circuitry, whereas the threshold behaviour of MOS prevents to approximate the adiabatic limit. Therefore our architecture is not competitive, but is particularly useful for studying the fundamentals of digital computation. We have applied our design methodology to some basic circuits. In particular a four-bit adder has been demonstrated.

Acknowledgement

The authors thank the *Invomec* division of *Imec v.z.w.* (Leuven, Belgium) and the *Europractice* organization, for processing the chips at *Alcatel Microelectronics* (Oudenaarde, Belgium). Adamski, Pietrzak, Sibiński, and Widerski acknowledge mobility under *Tempus* project JEP 11298-96 of the European Commission.

References

1. C. Bennett and R. Landauer: The fundamental physical limits of computation. Sc. American **253** (July 1985) 38–46
2. G. Stix: Riding the back of electrons. Sc. American **279** (Sept. 1998) 20–21
3. E. Fredkin and T. Toffoli: Conservative logic. Int. Journal of Theoretical Physics **21** (1982) 219–253
4. R. Feynman: Quantum mechanical computers. Optics News **11** (1985) 11–20
5. R. Feynman: Feynman lectures on computation (A. Hey and R. Allen, eds.). Addison-Wesley, Reading (1996)
6. L. Storme, A. De Vos, and G. Jacobs: Group theoretical aspects of reversible logic gates. Journal of Universal Computer Science **5** (1999) 307–321
7. A. De Vos, W. Marańda, E. Piwowarska, and A. Lejman: A chip for reversible digital computers. In: A. Napieralski and M. Turowski (eds.): Proc. 3 rd Advanced Training Course on Mixed Design of VLSI Circuits, Łódź (June 1996) 544–549
8. A. De Vos: Reversible computing. Progress in Quantum Electronics **23** (1999) 1–49
9. A. De Vos: Fundamental limits of power dissipation in digital electronics. In: A. Napieralski (ed.): Proc. 6 th Advanced Training Course on Mixed Design of VLSI Circuits, Kraków (June 1999) 27–36
10. B. Desoete, A. De Vos, M. Sibiński, and T. Widerski: Feynman's reversible logic gates, implemented in silicon. In: A. Napieralski (ed.): Proc. 6 th Advanced Training Course on Mixed Design of VLSI Circuits, Kraków (June 1999) 497–502
11. W. Athas, L. Svensson, J. Koller, N. Tzartzanis, and E. Chou: Low-power digital systems based on adiabatic-switching principles. I.E.E.E. Transactions on V.L.S.I. Systems **2** (1994) 398–407
12. J. Nossek and A. Schlaffer: Some aspects of adiabatic switching. In: A. Trullemans and J. Sparsø (eds.): Proc. 8 th Int. Workshop Patmos, Lyngby (Oct. 1998) 319–334
13. B. Desoete and A. De Vos: Optimal charging of capacitors. In: A. Trullemans and J. Sparsø (eds.): Proc. 8 th Int. Workshop Patmos, Lyngby (Oct. 1998) 335–344
14. A. De Vos and B. Desoete: Equipartition principles in finite-time thermodynamics. Journal of Non-Equilibrium Thermodynamics **25** (2000) 1–13

Modeling of Power Consumption of Adiabatic Gates versus Fan in and Comparison with Conventional Gates

M. Alioto and G. Palumbo

DEES
(Dipartimento Elettrico Elettronico e Sistemistico)
UNIVERSITA' DI CATANIA
Viale Andrea Doria 6, I-95125 CATANIA - ITALY
Phone ++39.095.7382313; Fax ++39.095.330793
malioto@dees.unict.it ; gpalumbo@dees.unict.it

Abstract. In this communication adiabatic and conventional gates with a different fan-in are modeled and analytically compared. The comparison is carried out assuming both an assigned power supply and setting it to minimize power consumption. The analysis leads to simple expressions, which allow to understand how the power advantage of adiabatic logic changes by increasing the fan-in of the implemented gate. The analytical results were validated by means of Spice simulations using a 0.8 μm CMOS technology.

1 Introduction

Power consumption reduction has become a key design aspect in ICs [1], because of the wide diffusion of portable equipment. Using the conventional CMOS design style, the most effective way to reduce power dissipation is to lower the supply voltage [11].

Recently, the Adiabatic Switching approach to reduce power dissipation in digital circuits was proposed [2]. It is to be used and verified in many digital applications [5],. [6], [7], [8]. A time-varying clocked ac power is used to slowly charge the node capacitances, and then partially recover the energy associated to that charge by slowly decreasing the supply [2], [3], [4].

Even if the interest in adiabatic logic design style and architectures is growing, comparisons between adiabatic and conventional styles are analytically carried out only in the simple case of an inverter [2], [3], [4] , [9].

In this communication, we analytically evaluate the power reduction of adiabatic logic with respect to conventional one considering gates with a different fan-in. More specifically, we analyze the inverter, NAND2, NAND3 and NAND4 gates. The resulting expressions are simple, hence they allow to understand how the advantage of adiabatic style changes by increasing the fan-in of the gate and for different load capacitances.

The comparison is carried out both assuming an assigned supply, as in the case of logic levels compatibility requirement, and setting it to minimize power consumption for a given speed requirement.

The validity of the used model is tested by Spice simulations on NAND2, NAND3 and NAND4 gates designed with a 0.8 μm technology.

D. Soudris, P. Pirsch, and E. Barke (Eds.): PATMOS 2000, LNCS 1918, pp. 265-275, 2000.

2   Adiabatic Gates Advantage over Conventional

2.1   Inverter

The adiabatic inverter is shown in Fig. 1 [2]. It is implemented by using two transmission gates and a power clock, V_ϕ, whose maximum amplitude is equal to V_{DD} and rise time equal to T. It has differential inputs and outputs, and is loaded by an external capacitance C_L.

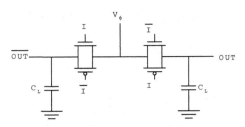

Fig. 1. Adiabatic inverter.

To evaluate the energy consumption, we approximate the transmission gate in the ON and OFF state to the linear circuits in Figs. 2a and 2b, respectively.

Fig. 2. Linear equivalent circuit of
(a) transmission gate in the ON state, (b) transmission gate in the OFF state.

Without loss of generality, we assume a linear ramp clocked power supply waveform. However, all results can simply be extended to a general power clock waveform by multiplying them by a proper shape factor, ξ, which only depends on the clock waveform [2-4]. Moreover, in the following the transmission gate parameters will be evaluated by assuming symmetrically sized transistors (i.e., $(W/L)_p = 2(W/L)_n$) and minimum sized NMOS devices. Analysis of the resulting RC circuit, considering both the energy wasted during charge and recovery, leads to the following expression of the adiabatic energy, $E_{ad,NOT}$, wasted in a cycle (i.e., for a single computation)

$$E_{ad,NOT} = 2\frac{R}{T}(C + C_L)^2 V_{DD}^2 \tag{1}$$

where the equivalent resistance of the transmission gate is $R = [\mu_n C_{OX}(W/L)_n(V_{DD} - 2V_T)]^{-1}$.

The average energy wasted by a symmetrically designed conventional inverter (i.e., $(W/L)_p = 2(W/L)_n$) with minimum sized NMOS transistor, $E_{conv,NOT}$, is roughly equal to $0.5*(1/4)*(C+C_L)V_{DD}^2$, obtained by approximating its intrinsic capacitance to that of a transmission gate[1], and assuming input values statistically independent from the others with an equal probability to be zero or one (hence, the resulting switching activity is ¼). Hence, to compare the power dissipation of adiabatic and conventional logic for an assigned supply, let us define the parameter F_{NOT} as

$$F_{NOT} = \frac{E_{ad,NOT}}{E_{conv,NOT}} = 16 \frac{1+\alpha}{T_n \left(\frac{V_{DD}}{V_T} - 2 \right)} \qquad (2)$$

where $\alpha = C_L/C$ is the load capacitance normalized to the parasitic capacitance of the transmission gate, and $T_n = TV_T \mu_n C_{OX}(W/L)_n/C$ is the normalized rise time of the adiabatic power clock, V_ϕ. To minimize power consumption for a given speed requirement, the supply of the conventional inverter is set to the minimum value, $V_{DD,op,conv}$, which satisfies its propagation delay constraint τ_{PD} [9]

$$V_{DD,op,conv} \cong V_T \left(1 + \sqrt{\frac{C + C_L}{\mu_n C_{OX} \left(\frac{W}{L} \right)_n V_T \tau_{PD}}} \right) \qquad (3)$$

where we assumed $4\mu_n C_{OX}(W/L)_n V_T \tau_{PD}/(C+C_L) \gg 1$. To minimize adiabatic energy consumption, we have to set V_{DD} simply equal to $4V_T$ [2-4]. Let us define $E_{conv,op,NOT}$ and $E_{ad,op,NOT}$ as the energy wasted by the conventional and adiabatic inverter obtained setting an optimized supply voltage given by eq. (3) and equal to $4V_T$, respectively.

Let us consider the case of supply optimized to meet a speed requirement with minimum power dissipation. To compare the energy dissipation at a defined speed, we set the transition period, T, of the adiabatic inverter equal to the propagation delay of the conventional gate (i.e., $T=\tau_{PD}$). To carry out the comparison in this optimized case, let us define the parameter $F_{op,NOT}$ as

$$F_{op,NOT} = \frac{E_{ad,op,NOT}}{E_{con,op,NOT}} = 128 \frac{1+\alpha}{T_n \left(1 + \sqrt{\frac{1+\alpha}{T_n}} \right)^2} \qquad (4)$$

2.2   NAND Gates

In Fig. 3 the topology of an n-inputs adiabatic NAND gate is shown. The analysis of the adiabatic NAND energy consumption is not as simple as that of the inverter, since its equivalent linearized circuit, obtained substituting each transmission gate with the model in Figs. 2a and 2b, is an RC ladder network if all the input values are high.

---

[1] This leads to slightly overestimate the capacitance associated to the transistors in cut-off region, but if C_L is comparable or greater the error is not significant

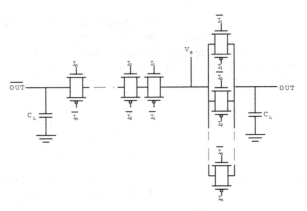

Fig. 3. Adiabatic n-inputs NAND.

Recently, in [10] a simple and general approach to analyze accurately adiabatic gates dissipation was proposed. It was demonstrated that the energy wasted by the generic resistor R_{ij} between nodes i and j in the equivalent circuit of an adiabatic gate is given by

$$E_{R_{ij}} \approx \frac{V_{DD}^2}{T} \frac{\left(T_{D_j} - T_{D_i}\right)^2}{R_{ij}} \qquad (5)$$

where T_{D_i} and T_{D_j} are the time constants associated to the nodes i and j. The time constant T_{D_i} at a given node i is given by

$$T_{D_i} = \sum_{k=1}^{no.\,of\,nodes} r_{ik} C_k, \qquad (5a)$$

where C_k and r_{ik} are the capacitance at node k and the DC transresistance gain between nodes k and i, respectively [12]. This model can be used to evaluate the energy wasted by the adiabatic NAND2 for each input value. In Figs. 4a, 4b, 4c and 4d the equivalent circuit of the adiabatic NAND2 is shown assuming an input equal to (0,0), (0,1), (1,0) and (1,1), respectively.

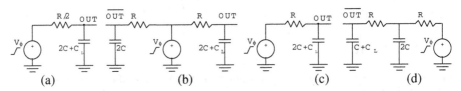

Fig. 4. Equivalent circuit of an adiabatic NAND2 for input ($I0, I1$) equal to a) (0,0), b) (0,1),c) (1,0), d) (1,1).

For each of these networks, applying eq. (5) to each resistor, summing the contributions of the resistors and multiplying the results by two to take into account the charge and recovery phases, we get

$$E_{0,0} = \frac{V_{DD}^2}{T} R(2C + C_L)^2 \tag{6a}$$

$$E_{0,1} = 2\frac{V_{DD}^2}{T} \left[R(2C + C_L)^2 + R(2C)^2 \right] \tag{6b}$$

$$E_{1,0} = 2\frac{V_{DD}^2}{T} R(2C + C_L)^2 \tag{6c}$$

$$E_{1,1} = 2\frac{V_{DD}^2}{T} \left[R(3C + C_L)^2 + R(C + C_L)^2 \right] \tag{6d}$$

The average energy wasted by the adiabatic NAND2, $E_{ad,NAND2}$, is equal to the weighed sum of each term with the probability P_{lm} of the corresponding input.

In the following, we will assume that each input of an n-inputs gate has equal probability to be zero or one, and is statistically independent from the others, hence each input value has a probability to occur equal to $1/2^n$. For the NAND2, this leads to

$$E_{ad,NAND2} = \sum_{l,m} E_{lm} P_{lm} = \frac{1}{4} \sum_{l,m} E_{lm} \tag{7}$$

which means that $E_{ad,NAND2}$ is equal to ¼ times the sum of the energy contributions associated to all of the possible input values. Hence, for the NAND2 we get

$$
\begin{aligned}
E_{ad,NAND2} &= \frac{1}{4}(E_{00} + E_{01} + E_{10} + E_{11}) \\
&= \frac{1}{2}\frac{V_{DD}^2}{T} R\left[\frac{7}{2}(2C + C_L)^2 + (C + C_L)^2 + (2C)^2 \right] \\
&\cong \frac{1}{2}\frac{V_{DD}^2}{T} R\left[\frac{9}{2}(2C + C_L)^2 \right]
\end{aligned}
\tag{8}
$$

in which we assumed C_L comparable or greater than C, and hence we neglected the terms which do not depend on C_L and in the others we increased the coefficient of C to 2 (i.e., the fan-in). The approximation leads to an error below 20% if $C_L > 0.1C$, which always holds for realistic load capacitances.

In Fig. 5 the topology of an n-inputs conventional NAND is shown.

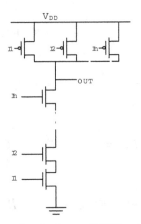

Fig. 5. Conventional n-inputs NAND gate.

The intrinsic capacitance is the sum of the capacitances of n complementary pairs of transistors. To simplify the analysis, let us approximate the capacitance of each complementary pair to that of a transmission gate, C. This only leads to a slight overestimation if C_L is comparable or greater than C. Hence, the average energy wasted by the conventional n-inputs NAND for a single computation is equal to

$$E_{conv,NAND2} = \frac{1}{2}\frac{2^n-1}{2^{2n}}\left(nC+C_L\right)V_{DD}^2 \tag{9}$$

where switching activity was evaluated assuming each input has equal probability to be zero or one and is statistically independent from the others.

To compare the power dissipation of the adiabatic and conventional NAND2 for a given load capacitance, C_L, switching frequency, assuming an assigned supply voltage, V_{DD}, let us define the parameter F_{NAND2} as

$$F_{NAND2} = \frac{E_{ad,NAND2}}{E_{conv,NAND2}} = 24\frac{2+\alpha}{T_n\left(\dfrac{V_{DD}}{V_T}-2\right)} \tag{10}$$

obtained from eqs. (8) and (9) with $n=2$.

In the same way, let us introduce the parameters $F_{NAND3}=E_{ad,NAND3}/E_{conv,NAND3}$ and $F_{NAND4}=E_{ad,NAND4}/E_{conv,NAND4}$ to compare the NAND3 and NAND4 for an assigned supply voltage. The expressions of $E_{ad,NAND3}$ and $E_{ad,NAND4}$ are easily derived following the same procedure used to obtain eq. (8):

$$E_{ad,NAND3} = \frac{1}{4}\frac{V_{DD}^2}{T}R\left[\left(5C+C_L\right)^2+\frac{35}{6}\left(3C+C_L\right)^2+\left(C+C_L\right)^2+\right.$$

$$\left.+28C^2\right] \cong 2\frac{V_{DD}^2}{T}R\left(3C+C_L\right)^2 \tag{11}$$

$$E_{ad,NAND4} = \frac{1}{8}\frac{V_{DD}^2}{T}R\left[(7C+C_L)^2+(5C+C_L)^2+\frac{103}{12}(4C+C_L)^2+\right.$$

$$\left.+(3C+C_L)^2+(C+C_L)^2+112C^2\right] \qquad (12)$$

$$\cong 1.6\frac{V_{DD}^2}{T}R(4C+C_L)^2$$

in which we introduced the same approximations as in (8), which lead to an error below 20% if $C_L > 0.3C$ and $C_L > 0.42C$, respectively. The resulting expressions of F are

$$F_{NAND3} = \frac{E_{ad,NAND3}}{E_{conv,NAND3}} = 37\frac{3+\alpha}{T_n\left(\dfrac{V_{DD}}{V_T}-2\right)} \qquad (13)$$

$$F_{NAND4} = \frac{E_{ad,NAND4}}{E_{conv,NAND4}} = 55\frac{4+\alpha}{T_n\left(\dfrac{V_{DD}}{V_T}-2\right)} \qquad (14)$$

From eqs. (10) (13) and (14), it is apparent that the power advantage of adiabatic NAND gates is proportional to $1/T$ as in the inverter case.

As done for the simple inverter, let us consider the comparison of the adiabatic and the conventional NAND2 assuming the supply optimized for minimum power consumption and a given speed constraint. The optimized adiabatic NAND2 energy consumption, $E_{ad,op,NAND2}$, is found setting $V_{DD}=4V_T$ in eq. (8). For the conventional NAND2 gate, as in the case of the inverter, we set the supply to the minimum value which allows to meet the propagation delay requirement, τ_{PD}. The delay of a conventional NANDn gate can be evaluated by substituting an equivalent NMOS device to the n series-connected minimum NMOS transistors of the pull-down network; its aspect ratio, $(W/L)_{eq}$, is given by that of a single NMOS, $(W/L)_n$, divided by the number of series transistors [11]. Hence, the optimized value of the supply for a conventional NANDn, $V_{DD,op,conv}$, is simply obtained substituting the aspect ratio of the equivalent NMOS to that of a minimum one in eq. (3) (i.e., $(W/L)_{eq}=(W/L)_n/n$):

$$V_{DD,op,conv} = V_T\left(1+\sqrt{n}\sqrt{\frac{nC+C_L}{\mu_n C_{OX}\left(\dfrac{W}{L}\right)_n V_T\tau_{PD}}}\right) \qquad (15)$$

The resulting parameter $F_{op,nand2}$ is

$$F_{op,NAND2} = \frac{E_{ad,op,NAND2}}{E_{conv,op,NAND2}} = 192\frac{2+\alpha}{T_n\left(1+\sqrt{2}\sqrt{\frac{2+\alpha}{T_n}}\right)^2} \qquad (16)$$

where eq. (15) was used setting $n=2$. Analogously, for the NAND3 and NAND4 we get

$$F_{op,NAND3} = \frac{E_{ad,op,NAND3}}{E_{conv,op,NAND3}} = 293 \frac{3+\alpha}{T_n\left(1+\sqrt{3}\sqrt{\frac{3+\alpha}{T_n}}\right)^2} \tag{17}$$

$$F_{op,NAND4} = \frac{E_{ad,op,NAND4}}{E_{conv,op,NAND4}} = 437 \frac{4+\alpha}{T_n\left(1+\sqrt{4}\sqrt{\frac{4+\alpha}{T_n}}\right)^2} \tag{18}$$

It is worth noting that the expressions comparing adiabatic and conventional logic, both for an assigned supply (eqs. 2, 10, 13 and 14) and for an optimized supply (eqs. 4, 16, 17 and 18), depend only on the normalized time, T_n, the normalized load capacitance, α, and on V_{DD} only for the first case.

3   Simulation Results

To test the validity of the proposed expressions, Spice simulations of adiabatic and conventional NAND2, NAND3 and NAND4 (for the inverter the validity was confirmed in [9]) were performed under different conditions, by using a 0.8 μm CMOS technology. In particular, values of F were evaluated by varying the transition period, T, in a range starting from a value which made F lower but close to unity. Applied inputs were statistically independent and with an equal probability to be zero or one. For each gate, we assumed a load capacitance, C_L, equal to 20 fF and 200 fF, corresponding to a fan-out of about two and twenty. The simulation runs were performed both in the case of an assigned supply equal to $V_{DD}=3.3$ V and in the case of optimized supply. For the used technology, $C=12.1$ fF, $V_t=0.74$ V, therefore the optimum supply for adiabatic gates is $V_{DD}\cong3.3$ V, and $T_n= 9.85T$.

As an example, we report in Fig. 6 and 7 the simulated and predicted values of F for a NAND4 with a load capacitance $C_L=20$ fF, assuming an assigned and optimized supply, respectively.

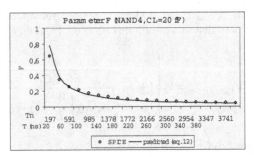

Fig. 6. Simulated and predicted (eq. 14) value of F_{NAND4} versus transition period for $C_L=20$ fF and $V_{DD}=3.3$ V.

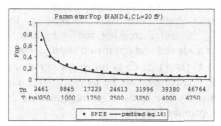

Fig. 7. Simulated and predicted (eq. 18) value of $F_{op.NAND4}$ versus transition period for $C_L=20$ fF and optimized supply.

The curves obtained in the other cases are similar. Considering all the cases, the error found is always lower than 30%, and in the case of assigned supply tends to be lower than the optimized case. Hence, the derived expressions are suitable for comparison of adiabatic and conventional gates.

4   Comparison between Adiabatic and Conventional Gates

4.1   Power Dissipation

The parameter F defined for all the considered gates in Subsec. 2 is useful to compare the performance of adiabatic and conventional gates for different fan-ins, load and supply.

Note that for an assigned supply the adiabatic advantage linearly decreases with load capacitance. This property approximately holds even for an optimized supply. In fact, $T_n \gg \sqrt{n}\sqrt{n+\alpha}$ for practical values of T_n such that $F<1$. Hence, in both cases the adiabatic advantage is inversely proportional to the transition period, T, and linearly decreases with the load capacitance.

Let us analyze how F changes varying the fan-in of the gate. Parameter F can be written as a product of two functions, one equal to $1/T_n(V_{DD}/V_T-2)$ for assigned supply and roughly equal to $1/T_n$ for the optimized case, and one depending only on parameter α, $k(\alpha)$. In Table I, function $k(\alpha)$ is shown for the considered gates both for assigned and optimised supply.

Table 1. Function $k(\alpha)$ is shown for the NOT, NAND2, NAND3 and NAND4 gates for assigned and optimized supply.

	k_{NOT}	k_{NAND2}	k_{NAND3}	k_{NAND4}
Assigned V_{DD}	$16(1+\alpha)$	$24(2+\alpha)$	$37(3+\alpha)$	$55(4+\alpha)$
Optimized V_{DD}	$128(1+\alpha)$	$192(2+\alpha)$	$293(3+\alpha)$	$437(4+\alpha)$

For low values of C_L ($\alpha \to 0$), from inspection of Table I, it is evident that advantage of adiabatic logic decreases increasing the fan-in of the gate ($F_{NOT} < F_{NAND2} < F_{NAND3} < F_{NAND4}$), and increasing fan-in by one determines more than a doubling of F, both for assigned and optimized supply. The same observations hold if C_L is comparable with C ($\alpha \approx 1$). If C_L is much greater than the parasitic capacitance ($\alpha \to \infty$), the parameter F still increases by increasing the fan-in, but at a slower rate. In fact, increasing fan-in by one leads to an increase of F by about 50%. This holds both for assigned and optimized supply.

Summarizing the results, the adiabatic performance gets worse increasing the fan-in of the gate irrespective of the load capacitance value. Moreover, the increase rate of F due to increase of the fan-in by one decreases from 100% to 50%, considering zero to high load capacitance, respectively.

4.2   Power-Delay Product

It is of interest to evaluate the power-delay product, PDP, of adiabatic and conventional gates, since it is an important figure of merit for digital circuits. It measures the efficiency of a design style in the trade-off between power and speed.

For a given supply voltage, load capacitance and switching frequency, the ratio between adiabatic and conventional power-delay product for a generic gate is

$$\frac{PDP_{ad}}{PDP_{conv}} = \frac{E_{ad}T}{E_{conv}\tau_{PD}} = F\frac{T}{\tau_{PD}} \tag{19}$$

where the conventional NAND gate delay is equal to
$\tau_{PD} = n(nC + C_L)V_{DD}/[\mu_n C_{OX}(W/L)_n(V_{DD} - V_T)^2]$.

After some simple calculations, from (19) it can be seen that, for all the considered gates, PDP_{ad}/PDP_{conv} is equal to a coefficient k_{PDP} multiplied by $[1+((V_{DD}/V_T)(V_{DD}/V_T-2))^{-1}]$, which is only slightly greater than unity for practical values of V_{DD}.

Table 2. Coefficient k_{PDP} for the NOT, NAND2, NAND3 and NAND4 gates.

k_{NOT}	k_{NAND2}	k_{NAND3}	k_{NAND4}
16	12	12.2	13.6

The coefficient k_{PDP} for the different gates is reported in Table II, which shows that the adiabatic logic PDP is always worse than conventional one for every fan-in values. Increasing fan-in, excepting for the simple inverter case, the disadvantage of adiabatic gates increases.

In the comparison with an optimized supply at a given speed ($T = \tau_{PD}$), PDP_{ad}/PDP_{conv} is simply equal to F_{op} before analyzed. Hence, in cases in which it makes sense to use adiabatic logic ($F_{op} < 1$), the adiabatic gates are always more efficient than conventional ones.

5   Conclusions

In this communication adiabatic gates were analytically compared to conventional ones for a different fan-in, load capacitance and assuming both an assigned supply and an optimized supply to minimize power consumption for a given speed constraint. Simple expressions were obtained. It was found that the power advantage of adiabatic logic linearly decreases with load capacitance and proportionally increases with the transition period of the power clock. Moreover, this advantage decreases by increasing gate fan-in for every value of load capacitance, and hence the advantage of NAND gates is always lower than that predicted for the inverter. For high values of C_L the advantage decrease due to fan-in increase worsen more slowly.

Finally, power-delay product of adiabatic and conventional logic were compared, and it was demonstrated that for a given supply adiabatic gates are always worse than conventional ones, while for an optimized supply adiabatic gates are always more efficient than conventional ones.

References

1. A. Matsuzawa, "Low-Voltage And Low-Power Circuit Design For Mixed Analog/Digital Systems In Portable Equipment," *IEEE Jour. of Solid-State Circuits*, no. 4, pp. 470-486, April 1994.
2. W. Athas, L. Svensson, J. Koller, N. Tzartzanis, E. Ying-Chin Chou, "Low-Power Digital Systems Based on Adiabatic-Switching Principles," *IEEE Trans. on VLSI*, Vol. 2, N. 4, pp. 398-407, December 1994.
3. L. Svensson, "Adiabatic Switching," in *Low Power Digital CMOS Design*, A. Chandrakasan, R. Brodersen, Kluwer Academic Publisher, 1995.
4. W. Athas, "Energy-Recovery CMOS," in *Low Power Digital Design Methodologies*, J. Rabey, M. Pedram, Eds., Kluwer Academic Publisher, 1995.
5. V. Oklobdzija, D. Maksimovic, F. Lin, "Pass-Transistor Adiabatic Logic Using Single Power-Clock Supply," *IEEE Trans on CAS part II*, Vol. 44, N. 10, pp. 842-846, Oct. 1997.
6. W. Athas, N. Tzartzanis, L. Svensson, L. Peterson, "A Low-Power Microprocessor Based on Resonant Energy," *IEEE Jour. of Solid-State Circ.*, Vol. 32, No. 11, pp. 1693-1701, Nov. 1997.
7. M. Knapp, P. Kindlmann, M. Papaefthymiou, "Design and Evaluation for Adiabatic Arithmetic Units," *Analog Integrated Circuit and Signal Processing*, Vol. 14, pp. 71-79, 1997.
8. Chun-Keung Lo, P. Chan, "Design of Low Power Differential Logic Using Adiabatic Switching Technique," *Proc. ISCAS'98*, Monterey, June 1998.
9. M. Alioto, G. Palumbo, "Adiabatic Gates: A Critical Point of View," *Proc. ECCTD'99*, Stresa, August 1999.
10. M. Alioto, G. Palumbo, "Evaluation of power consumption in adiabatic circuits," *Proc. ISCAS2000*, Genève, May 2000.
11. J. Rabaey, *Digital Integrated Circuits (A Design Perspective)*, Prentice Hall, 1996.
12. D. Standley and J. L. Wyatt, Jr. "Improved signal delay bounds for RC tree networks," *VLSI Memo*, no. 86-317, Massachusetts Institute of Technology, Cambridge, Massachusetts, May 1986.

An Adiabatic Multiplier

C. Saas, A. Schlaffer, and J.A. Nossek

Institute for Circuit Theory and Signal Processing, Munich University of Technology,
Arcisstr. 16
D-80298 München,Germany
`chsa@nws.ei.tum.de`

Abstract. Adiabatic switching might be a possibility to overcome the power losses in CMOS due to the charging of capacities. The design of adiabatic gates and registers has been examined in the past. The possibilities offered to the design of logic are evaluated in this paper.

For this purpose an array multiplier has been chosen as a representative for more complex structures. To provide the possibility of comparison, it has been realized as an adiabatic circuit as well as using a standard CMOS design. In this article special interest has been drawn to the placement of the registers in the adiabatic circuit. This was done by using a modified retiming algorithm.

Both designs were simulated using SPICE. Although the simulation results show a significant reduction of power, they have to be interpreted with caution. Based on them it is discussed whether the reduction of dissipated energy can compensate the required overhead or not.

1 Introduction

To show the possibilities of adiabatic circuits it is not sufficient to evaluate the concept on single gates. Due to the fact that the registers consume a significant part of the area and cause a major part of the power loss their number is very important. Since registers are mandatory for the function of adiabatic circuits of the proposed type the register count is strongly affected by the architecture. An array multiplier has been realized as CMOS and as an adiabatic circuit. To maintain acceptable simulation times, a word size of 3 bit has been chosen. This gives the possibility to simulate a wide range of different parameters like the ramping time T and the transistor size.

2 Structure of the Multiplier

We decided to use an array multiplier because of its regular structure [1]. This makes the placement of the registers as described in chapter 4 more effective than an unsymmetric architecture like the wallace tree multiplier. The disadvantage of having a longer delay than other possible architectures is not important in adiabatic circuits because its speed does not depend on the logic depth but on the pipeline steps needed.

D. Soudris, P. Pirsch, and E. Barke (Eds.): PATMOS 2000, LNCS 1918, pp. 276–284, 2000.

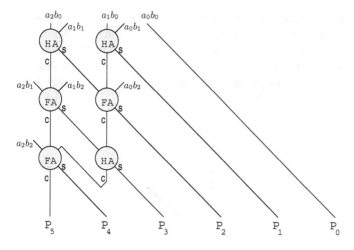

Fig. 1. 3x3 bit array multiplier

3   Gate and Register Design

Only three different gates are needed to realize the proposed multiplier: The AND gate to compute the partial products as well as full and half adders to sum them. The AND gate is used as an example to explain their design. The adders are designed accordingly.

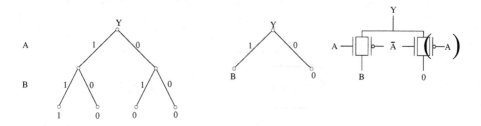

Fig. 2. Design process of an AND gate

The logic function is first represented by a binary decision diagram (BDD), which could be directly realized as a transistor schematic, but contains significantly more transistors than necessary. The BDD can be simplified by choosing an optimal order for the inputs and by including through signals in the root nodes. In addition only the n-channel transistor is required if the corresponding through signal is directly connected to ground. The whole design process is explained in detail in [4].

Note that the input signals have to be divided into two classes: Signals which are connected to the gates of the transmission gates and are therefore called *control signals*. The other class is called *through signals* and it is connected to the source contacts of the transmission gates. The control signals must have a constant value while the through signals are active. Note that all outputs of logic gates are always through signals.

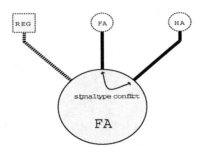

Fig. 3. Signal type conflicts at a full adder

The full adder shown in figure 3 receives two of its three input signals from other adders. As they are the outputs of a logic gate, they are through signals. The full adder itself can handle only one input signal as a through signal. So, there is a signal type conflict which can not be solved in the same clock cycle because a control signal can never be generated from a through signal. The only way to overcome this problem is to delay the control signal by one clock cycle and use a register to convert the through signal into a control signal.

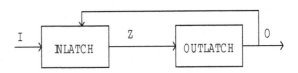

Fig. 4. Register

The registers have been designed according to [5]. In general, a register consists of two latches which both have a delay of half a clock cycle. Most of the registers have to accept a through signal at the input and have to provide a control signal at the output. The latches have been designed accordingly. The inlatch could be realized reversible, as the information about the state of the intermediate signal Z is still available in the output O. This was not possible in the outlatch because the status of the output O is not stored. In order to avoid nonadiabatic charging, diodes were included in the outlatch.

The need for registers and the need for the inverted signals lead to an overhead of transistors.

	CMOS	adiabatic
AND Gate	6	7
half adder	14	15
full adder	28	30
register	8	40
signal type conversion	0	6
3x3- multiplier	180	910

As shown in the table the increased number of transistors results mainly from the additional registers. There are none at all in the CMOS design, whereas the adiabatic circuit requires 17 of them. This leads to an increase of 5 in the area consumption.

4   Register Distribution

With a register after each logic gate it is already possible to build any logic function. The register can provide any required signal type for the following gates, but the resulting structure is fully pipelined. Of course, this is not the best solution. Rather the number of registers should be minimized by using the output signals as through signals in the next stage as often as possible. This can be achieved by using a modified retiming algorithm and a systematic distribution of signal type conflicts.

4.1   Graph of the Adder Field

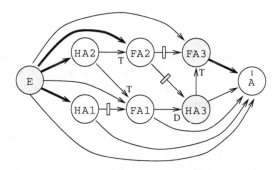

Fig. 5. Graph of the adder field with fixed registers

To perform the retiming algorithm a model of the circuit is needed. The circuit is represented by a finite, edge-weighted directed graph G [2]. The vertices

V of the graph model the functional elements of the circuit. Each edge $e \in E$ connects an output of some functional element to an input of some functional element and is weighted with a register count. The register count is the number of registers in the connection. All inputs are combined in a single vertex, as well as all outputs.

4.2   Register Placement

Due to problems in solving the arising optimizing problem there is no global optimizing algorithm available at the moment [3]. As a solution the task has to be divided into two parts. First the assignment of the signal types, which also defines the places for the fixed registers which are needed for the signal type conversion and second the optimal placement of the registers used to keep the timing correct.

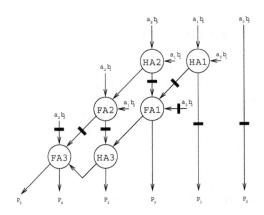

Fig. 6. Register placement

Assignment of Signal Types. The assignment of signal types to the different inputs of a logic gate is not given. During the design of a logic gate at least one input can be included in the root nodes, thus it is becoming a through signal. There are several possibilities to choose the input included. The question is which one should be chosen.

Since all delays in one path have to be included in all other pathes to a certain gate, additional latency for signal type conversion will consume a lot of registers. Therefore it is obvious that the minimization of the overall latency is a reasonable aim.

The overall latency is minimized by gradual minimizing the latency for each vertex. As they require a register, signal type conflicts are always assigned to the predecessor vertex with the lowest latency. Although the method is based on a local optimization, a global minimum for the overal latency is achieved. [3]

Optimization of the Register Distribution. Once the registers for the signal type conversion are placed, they may not be moved anymore. To place additional registers, which will ensure that the timing of the circuit is correct, a modified retiming algorithm for state minimization [2] is used. The demand for unmovable registers can easily be included in the constraints of the retiming algorithm.

The algorithm must have the freedom to increase the overall latency. Therefore the problem becomes a pipelining problem.

5   Integration in a Design Flow

The proposed methods for signal assignment and register distribution can easily be integrated in a standard design flow. As an input, a complete gate netlist is required, which will be composed by any netlist compiler, starting e. g. from a VHDL-description. The signal type assignment will then label all signals as through or through types, the following retiming algorithm will insert additional registers into the gate netlist. This modified netlist may then be handed on to a place and route tool or to a logic simulator. The place and route tool, of course, has to handle the different signal types, but this can easily be put into practice by including signal types into the names of input signals.

6   Simulation Results

6.1   Evaluation Criteria

At first it seems very easy to evaluate the measures to reduce the losses by comparing the power consumption. As soon as speed is considered, this method is no longer suggestive. In a CMOS circuit, for example, the power loss is proportional to the clock frequency. The product of the power loss of a circuit and the gate delay is called *power-delay-product*. In CMOS circuit it is proportional to the *average energy dissipation per arithmetic procedure*. Therefore it seems to be a very reasonable mesurement for the reduction of power losses. On the one hand, the dissipated energy leads to a warming up of the circuit, on the other hand, the energy reservoir might be limited. In a CMOS architecture the power-delay-product is independent of the clock frequency. In adiabatic cricuits the dissipated energy should decrease appropriate to $E_{diss} \sim \frac{1}{T}$.

6.2   Variation of Ramping Time

Figure 7 shows the average energy consumption over the ramping time T. All transistors have the minimal size of $l = 0.25 \mu m$ and $w = 0.5 \mu m$. The voltage has been set to 2.5V. The simulation results clearly show that the circuit is working adiabatically. For small values of the ramping time T the energy dissipation is indirect proportional to T.

Although the size of the adiabatic circuit is about 5 times larger than that of

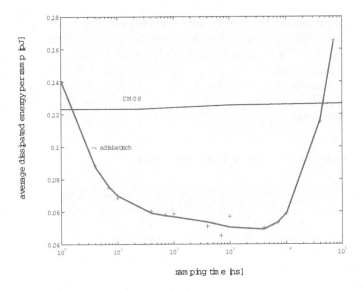

Fig. 7. Energy loss of multiplier in 0.25μ

the CMOS its energy loss is smaller for ramping times which are longer than 2ns. This corresponds to a clock frequency of 120Mhz. The minimal losses are achieved at $T \approx 1\mu s$. For this T the energy consumption is only 36% that of the CMOS circuit. If the ramping time T is longer than $10\mu s$ the static leakage currents have to be taken into account, and the dissipated energy is proportional to T.

This simulation is a rather optimistic view of the adiabatic circuit. On the one hand, ideal ramps are used to drive the clock phases. On the other hand, losses in the generation of the clock phases are not considered.

6.3   Influence of Channel Length

Another simulation was run with scaled transistors to allow a prediction for future process technologies.

The channel length was set to $l = 1\mu m$ and the width was doubled. This corresponds to an older process. Although the absolute energy dissipation is much higher because of the larger gate capacities, the adiabatic principle works even better. The minimum losses are reduced to about 18% at $T \approx 10\mu s$.

This result allows the prediction that problems will arise with further down scaling of the transistors due to the increase of short channel leakage currents. The minimal energy dissipated is depending on the ratio of channel resistance in the conducting and non conducting state as shown in formula 1.

$$E_{dissmin} \approx \sqrt{\frac{25}{18}} \hat{U}_B^2 C_L \sqrt{\frac{R_{on}}{R_{off}}} \tag{1}$$

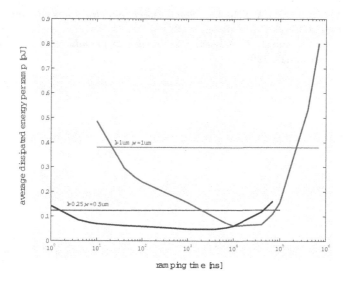

Fig. 8. Energy losses of multiplier in 1μ

This ratio has increased with the downscaling of the structures due to short channel effects. This effect can be seen in figure 8.

7   Conclusion

This work was not capable to give a final descision whether there is a chance for adiabatic charging to assert against well established techniques, but there are some hints which allow a prognosis of the possibilities offered.

The realisation of whole logic blocks using adiabatic switching is problematic because of the increased number of transistors. Therefore the adiabatic circuit has to compensate the increased energy resulting of the transistor overhead. For the example shown this was still possible for $0.25\mu m$ channel length. It has to be verified if this is still possible in the case of smaller transistors. This might be a problem because the short channel leakage currents increase if the structures are getting smaller.

To draw a conclusion it can be said that the use of adiabatic switching for the design of logic is doubtful as long as MOS transistors are used.

References

1. K. Hwang: *Computer Arithmetic.* John Wiley & Sons, 1979
2. C. E. Leiserson and J. B. Saxe: Retiming Synchronous Circuitry *Algorithmica*, pages 5-35, 1991
3. A. Schlaffer: *Entwurf von adiabatischen Schaltungen*, Ph.D. thesis, Munich University of Technology, 2000, to appear

4. J. A. Nossek and A. Schlaffer: *Some aspects on adiabatic switching* (invited paper). In PATMOS '98 Proceedings, 1998.

5. A. Schlaffer and J. A. Nossek: *Register design for adiabatic circuits*. In PATMOS '99 Proceedings, 1999.

Logarithmic Number System
for Low-Power Arithmetic

V. Paliouras and T. Stouraitis

Electrical and Computer Engineering Department
University of Patras, Greece
{paliuras,thanos}@ee.upatras.gr

Abstract. In this paper, properties of the Logarithmic Number System (LNS) are investigated which can lead to power savings in a digital system. To quantitatively establish power savings, the equivalence of an LNS to a linear fixed-point system is, initially, explored and a related theorem is introduced. It is shown that LNS leads to reduction of the average bit assertion probability by more than 50%, in certain cases, over an equivalent linear representation. Finally, the impact of LNS on hardware architecture and, by means of that, to power dissipation, is discussed.

1 Introduction

In the last years, power dissipated in an electronic system has evolved into an important design issue, mainly due to the impetus offered by the need for portable equipment, as well as the requirement for very high-speed processors [1].

Power dissipation minimization is sought at all levels of design abstraction, ranging from software/hardware partitioning down to technology-related issues. The average power dissipation in a circuit is computed via the relationship

$$P_{\text{ave}} = a f_{\text{clk}} C_L V_{\text{dd}}^2, \tag{1}$$

where f_{clk} is the clock frequency, C_L is the total switching capacitance, V_{dd} is the supply voltage, and a is the average activity in a clock period.

A wide variety of design techniques have been proposed [1], aiming to reducing the various factors of product (1). Among them, the successful selection of the number system and the proper design of arithmetic circuits has been proposed as a power dissipation minimization technique [2][3], which can affect all factors of (1) [4].

In this paper, it is shown that the adoption of the Logarithmic Number System (LNS) [5] can lead to substantial power dissipation savings, due to the reduction of average bit activity and due to the simplification of certain arithmetic operations achieved by its utilization. The concept of equivalence between LNS and linear fixed-point representation is investigated, in order to define the logarithmic word length and the base, as well as to provide a quantitative performance comparison between the two representations in the context of design

D. Soudris, P. Pirsch, and E. Barke (Eds.): PATMOS 2000, LNCS 1918, pp. 285–294, 2000.

for low power. Special attention is paid to equivalence, since in order to quantify power dissipation savings over an n-bit fixed-point system, it is necessary to derive the LNS which provides sufficient range and precision, so that the comparison results are meaningful from the application point of view.

The organization of the remainder of the paper is as follows. In section 2, the basics of the LNS encoding are briefly reviewed and its equivalence to linear representations is explored. In section 3, the activity reduction made possible via the LNS encoding is investigated. Section 4 discusses the complexity of LNS operations. Finally, conclusions are offered in section 5.

2   LNS and Equivalence to Linear Representations

The LNS representation maps a real number X to a triplet, as follows

$$X \xrightarrow{\text{LNS}} (z, s, x = \log_b |X|), \tag{2}$$

where b is the base of the logarithm, z is the zero flag, and s is the sign of X. A zero flag is required as, $\log_b X$ is not a finite number for $X = 0$. Similarly, since the logarithm of a negative number is not a real number, the sign information of X is stored in flag s. Logarithm $x = \log_b |X|$ is encoded as a binary number, and it can be written as

$$x = I.F, \tag{3}$$

where I is the integer part and F is the fractional part.

Traditionally, LNS has been considered as an alternative to floating-point representation [6][7]. However, in this paper, LNS is compared to an n-bit linear fixed-point representation and it is shown to provide substantial improvement in terms of power dissipation.

Two are the main issues in a finite word length number system, namely the *range* of the numbers which can be represented and the *precision* of the representation [6]. The representational equivalence of an n-bit linear fixed-point system and of an LNS needs to be investigated, as the two representations differ in both range and precision behavior, due to the nonlinear nature of the logarithm.

Let k and l be integers which denote the word length of the integer and fractional part of an LNS word, respectively. Let (k, l, b)-LNS denote an LNS of integer and fractional word length k and l, respectively, and of base b. The problem of equivalence between a (k, l, b)-LNS and an n-bit linear fixed-point system, is to compute k and l in such a way that the two number representations satisfy a suitably defined criterion, for a particular base b.

The relative representational error, ϵ_{rel}, of a number A encoded in a number system, is, in general, a function of value A and it is defined as

$$\epsilon_{\text{rel}} = \frac{|A - \widehat{A}|}{A}, \tag{4}$$

where A is the actual value and $\widehat{A}$ is the corresponding value representable in the system. Notice that $A \neq \widehat{A}$ due to the finite length of the words. The relative representational error $\epsilon_{rel,LNS}$, for an (k, l, b)-LNS is given by (cf. [6], for the case $b = 2$)

$$\epsilon_{rel,LNS} = b^{2^{-l}} - 1, \tag{5}$$

while for the n-bit linear fixed-point case, the corresponding $\epsilon_{rel,FXP}$ is, due to definition (4), given by

$$\epsilon_{rel,FXP} = \frac{2^{-n}}{A}, \tag{6}$$

where A denotes an n-bit fixed-point number. From (5) and (6), it can be noticed that $\epsilon_{rel,FXP}$ depends on A, while $\epsilon_{rel,LNS}$ does not. In order to overcome the particular difference and be able to compare the precision of the two representations, the following two restrictions are posed:

1. the two representations should cover equivalent data ranges and
2. the two representations should exhibit equal average representational error.

The average representational error, ϵ_{ave}, is defined as

$$\epsilon_{ave} = \frac{\sum_{A=A_{min}}^{A_{max}} \epsilon_{rel}(A)}{A_{max} - A_{min} + 1}, \tag{7}$$

where A_{min} and A_{max} define the range of representable numbers.

Due to definition (7), the average representational error for the fixed-point case, is given by

$$\epsilon_{ave,FXP} = \frac{1}{2^n - 1} \sum_{i=1}^{2^n - 1} \frac{1}{i}, \tag{8}$$

which, by computing the sum on the right-hand side, can be written as

$$\epsilon_{ave,FXP} = \frac{\psi(2^n) + \gamma}{2^n - 1}, \tag{9}$$

where γ is the Euler gamma constant and function ψ is defined through

$$\psi(x) = \frac{d}{dx} \ln \Gamma(x), \tag{10}$$

where $\Gamma(x)$ is the Euler gamma function.

In the case of the LNS, as $\epsilon_{rel,LNS}$ is constant over the range, due to (5), it occurs that

$$\epsilon_{ave,LNS} = b^{2^{-l}} - 1. \tag{11}$$

In the following, the maximum number representable in each number system is computed and utilized to compare the ranges of the representations. Notice that different figures can also be used for range comparison, such as the ratio A_{max}/A_{min}.

The maximum number representable by an n-bit linear integer is $2^n - 1$; therefore the upper bound of the fixed-point range is given by

$$A_{max}^{FXP} = 2^n - 1. \tag{12}$$

The maximum number representable by (k, l, b)-LNS encoding (2), is

$$A_{\max}^{\text{LNS}} = b^{2^k + 1 - 2^{-l}}. \tag{13}$$

Therefore, according to the equivalence criteria posed earlier, in order that an LNS is equivalent to an n-bit linear fixed-point representation, the following restrictions should be simultaneously satisfied:

$$A_{\max}^{\text{LNS}} \geq A_{\max}^{\text{FXP}} \tag{14}$$

$$\epsilon_{\text{ave,LNS}} \leq \epsilon_{\text{ave,FXP}} \tag{15}$$

Hence, from (9) and (11)–(13), it is obtained that

$$b^{2^k + 1 - 2^{-l}} \geq 2^n - 1 \tag{16}$$

$$b^{2^{-l}} - 1 \leq \frac{\psi(2^n) + \gamma}{2^n - 1}, \tag{17}$$

which, when solved for k and l, give

$$l = \left\lceil -\log_2 \log_b (1 + \frac{\psi(2^n) + \gamma}{2^n - 1}) \right\rceil \tag{18}$$

$$k = \left\lceil \log_2 \left(\log_b(2^n - 1) + 2^{-l} - 1 \right) \right\rceil. \tag{19}$$

The above analysis can be summarized by introducing the following theorem.

Theorem 1. *A (k, l, b)-LNS covers a range at least as long as an n-bit fixed-point system with an average representational error equal or smaller to that of the fixed-point system, when l and k are given by (18) and (19), respectively.*

Values of k and l that correspond to various values of n for various values of b, can be seen in Table 1.

While the word lengths k and l computed via (18) and (19) meet the posed equivalence specifications (14) and (15), LNS is capable of covering a significantly larger range than the equivalent fixed-point representation. Let n_{eq} denote the word length of a fixed-point system which can cover the range offered by an LNS defined through (18) and (19), or, equivalently, let n_{eq} be the smallest integer which satisfies

$$2^{n_{\text{eq}}} - 1 \geq b^{2^k + 1 - 2^{-l}}. \tag{20}$$

From (20) it follows that

$$n_{\text{eq}} = \left\lceil (2^k + 1 - 2^{-l}) \log_2 b \right\rceil. \tag{21}$$

It should be stressed that, when $n_{\text{eq}} \geq n$, the precision of the particular fixed-point system is better than that of the LNS derived by (18) and (19). Equation (21) reveals that the particular LNS, while meeting the precision of an n-bit linear representation, in fact, covers the range provided by an n_{eq}-bit linear system.

Table 1. Correspondence of n, k, l, and n_{eq} for various bases b.

n	$b = 1.5$			$b = 2$			$b = 2.5$		
	k	l	n_{eq}	k	l	n_{eq}	k	l	n_{eq}
5	3	2	6	3	3	9	2	3	7
6	4	3	10	3	4	9	2	4	7
7	4	4	10	3	5	9	3	5	12
8	4	5	10	3	5	9	3	6	12
9	4	5	10	4	6	17	3	7	12
10	5	6	20	4	7	17	3	7	12
11	5	7	20	4	8	17	3	8	12
12	5	8	20	4	9	17	4	9	23
13	5	9	20	4	10	17	4	10	23
14	5	10	20	4	11	17	4	11	23
15	5	11	20	4	12	17	4	12	23

3   Power Dissipation and LNS Encoding

In this section, it is shown that assuming a uniform distribution of input linear n-bit numbers, the distribution of bit assertions of the corresponding LNS words, reveals that LNS can be exploited to reduce the average activity.

Let $p_{0\to1}(i)$ be the bit assertion probabilities, i.e., the probability of the ith bit transition from 0 to 1. Assuming that data are temporaly independent, it holds that

$$p_{0\to1}(i) = p_0(i)p_1(i) = (1 - p_1(i))p_1(i), \tag{22}$$

where $p_0(i)$ and $p_1(i)$ is the probability of the ith bit being 0 or 1, respectively. Due to the assumption of uniform data distribution, it holds that

$$p_0(i) = p_1(i) = \frac{1}{2}, \tag{23}$$

which, due to (22), gives

$$p_{0\to1}(i) = \frac{1}{4}. \tag{24}$$

Therefore, all bits in the linear fixed-point representation exhibit an equal $p_{0\to1}(i)$, $i = 0, 1, \ldots, n - 1$.

Activities of the bits in an LNS-encoded word are quantified under similar assumptions. Since there is an one-to-one correspondence of linear fixed-point values to their LNS images defined by (2), the LNS values follow a probability function, identical to the fixed-point case. In fact the LNS mapping can be considered a continuous transformation of the discrete random variable X, which is a word in the linear representation, to the discrete random variable x, an LNS word. Hence the two discrete random variables follow the same probability function [8].

However, the $p_{0\to1}^{\text{LNS}}$ probabilities of bit assertions in LNS words are not constant as $p_{0\to1}(i)$ of (24); they depend on the significance of the ith bit. To evaluate the probabilities $p_{0 \to 1}^{\text{LNS}}(i)$, the following experiment is performed. For all possible values of X in a n-bit system, the corresponding $\lfloor \log_b X \rfloor$ values in a (k, l, b)-LNS format are derived and probabilities $p_1(i)$ for each bit are computed. Then, $p_{0 \to 1}^{\text{LNS}}(i)$ is computed as in (22).

The actual assertion probabilities for the bits in an LNS word, $p_{0 \to 1}^{\text{LNS}}(i)$, are depicted in Fig. 1. It can be seen that $p_{0 \to 1}(i)$ for the more significant bits is substantially lower than $p_{0 \to 1}(i)$ for the less significant bits. Also, it can be seen that $p_{0 \to 1}(i)$ depends on b. This behavior, which is due to the inherent data compression property of the logarithm function, leads to a reduction of the average activity in the entire word. Average activity savings percentage, S_{ave} is computed as

$$S_{\text{ave}} = \left(1 - \frac{\sum_{i=0}^{k+l-1} p_{0 \to 1}^{\text{LNS}}(i)}{0.25n} \right) 100\%, \tag{25}$$

where it has been used that $p_{0 \to 1}^{\text{FXP}}(i) = 1/4$ for $i = 0, 1, \ldots, n-1$, n denotes the length of the fixed-point system, and the word lengths k and l are computed via Theorem 1. Savings percentage S_{ave} is demonstrated in Fig. 2(a) for various values of n and b, and it is found to be more than 15% in certain cases.

However, as implied by the definition of n_{eq} in (21), the linear system which provides an equivalent range to that of a (k, l, b)-LNS, requires n_{eq} bits. If the reduced precision of (k, l, b)-LNS compared to n_{eq}-bit fixed-point system, is acceptable for a particular application, S'_{ave} is used to describe the relative efficiency of LNS, instead of (25), where

$$S'_{\text{ave}} = \left(1 - \frac{\sum_{i=0}^{k+l-1} p_{0 \to 1}^{\text{LNS}}(i)}{0.25 n_{\text{eq}}} \right) 100\%. \tag{26}$$

Savings percentage S'_{ave} is demonstrated in Fig. 2(b) for various values of n and b. Savings are found to exceed 50% in some cases. Notice that Fig. 2 reveals that, for a particular word length n, the proper selection of logarithm base b can significantly affect the average activity. Therefore, the choice of b is important in designing a low-power LNS-based system.

4   Power Dissipation and LNS Architecture

In the previous section, it has been shown that the LNS representation is beneficial over the fixed-point representation in terms of the average bit activity. In this section, the impact of LNS on the architecture is discussed.

LNS exploits properties of the logarithm to reduce the strength of several arithmetic operations, thus it leads to complexity savings. By reducing the area complexity of operations, the switching capacitance C_L of (1) can be reduced. Furthermore, reduction in latency allows for further reduction in supply voltage, which also reduces power dissipation [1].

Let x and y be the (k, l, b)-LNS images of the linear quantities X and Y. The transformation of operations is summarized in Table 2. Table 2 shows that n-bit

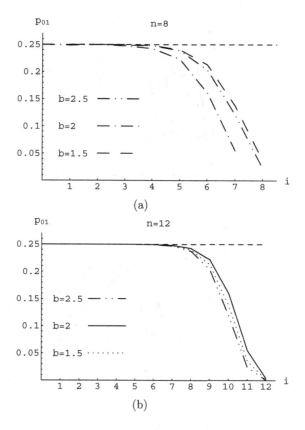

Fig. 1. Activities against bit significance i, in an LNS word, for $n = 8$ (a) and $n = 12$ (b) and various values of the base b. The horizontal dashed line is the activity of the corresponding n-bit fixed-point system.

multiplication and division are reduced to $(k + l)$-bit addition and subtraction, respectively, while the computation of roots and powers is reduced to division and multiplication by a constant, respectively. For the common cases of square root or square, the operation is reduced to left or right shift respectively. For example, assume that a n-bit carry-save array multiplier, which has a complexity of $n^2 - n$ 1-bit full adders (FAs) is replaced by an n-bit adder, which, assuming $k + l = n$, has a complexity of n FAs, for a ripple-carry implementation [6]. Therefore, multiplication complexity is reduced by a factor r_{C_L}, given as

$$r_{C_L} = \frac{n^2 - n}{n} = n - 1. \tag{27}$$

Equation (27) reveals that the reduction factor r_{C_L} grows with the word length n.

However, addition and subtraction are complicated in LNS, since they require a table look-up operation for the evaluation of $\log_b(1 \pm b^{y-x})$, although

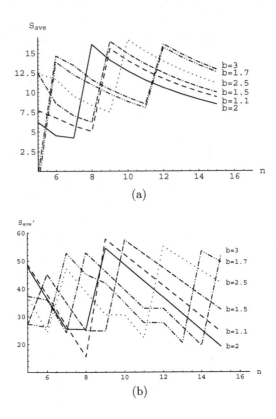

Fig. 2. Percentage of average activity reduction due to the use of LNS, compared to n-bit (a) and to n_{eq}-bit (b) linear fixed-point system, for various bases b of the logarithm. The diagram reveals that the optimal selection of b depends on n and it can lead to significant power dissipation reduction.

different approaches have been proposed in the literature [9][10]. A table look-up operation requires a ROM of $n \times 2^n$ bits, a size which can inhibit LNS utilization for large values of n. In an attempt to solve this problem, efficient table reduction techniques have been proposed [11]. As a result of the above analysis, applications with a computational load dominated by operations of simple LNS implementation, can be expected to gain power dissipation reduction due to the LNS impact on architecture complexity.

Finally, it should be noted that overhead is imposed for linear-to-logarithmic and logarithmic-to-linear conversion. Conversion overhead contributes additional area and time complexity as well as power dissipation. However, as the number of operations grows, the conversion overhead remains constant; therefore its contribution to the overall budget becomes negligible.

Table 2. Impact of LNS on arithmetic operations.

multiply	$Z = XY = b^x b^y = b^{x+y}$	$z = \log_b Z = x + y$
divide	$Z = \frac{X}{Y} = \frac{b^x}{b^y} = b^{x-y}$	$z = x - y$
root	$Z = \sqrt[m]{X} = \sqrt[m]{b^x} = b^{\frac{x}{m}}$	$z = \frac{x}{m}, \quad m, \text{integer}$
power	$Z = X^m = (b^x)^m$	$z = mx, \quad m, \text{integer}$
addition	$Z = X + Y = b^x + b^y = b^x(1 + b^{y-x})$	$z = x + \log_b(1 + b^{y-x})$
subtraction	$Z = X - Y = b^x - b^y = b^x(1 - b^{y-x})$	$z = x + \log_b(1 - b^{y-x})$

5 Conclusions

The impact of LNS onto power dissipation of a digital system, which performs arithmetic operations, has been investigated. The discussion is based on proposed conditions of equivalence between an LNS and fixed-point representations. It is shown that LNS can lead to significant average bit activity reduction. It has been found that the efficiency of the LNS representation is dominated by the choice of word lengths k and l, and—the often neglected parameter—b. Furthermore the impact of LNS onto architecture has been briefly discussed to show that architecture simplification is also possible, in certain cases.

LNS, with a combined exploitation of savings in signal activity and of savings due to architectural simplification for suitable applications, can be a successful candidate for the implementation of future low-power computationally-intensive systems.

References

1. A. P. Chandrakasan and R. W. Brodersen, *Low Power Digital CMOS Design*. Boston: Kluwer Academic Publishers, 1995.
2. B. Parhami, *Computer Arithmetic: Algorithms and Hardware Designs*. New York: Oxford University Press, 2000.
3. K. Parhi, *VLSI Digital Signal Processing Systems*. New York: Wiley - Interscience, 1999.
4. W. L. Freking and K. K. Parhi, "Low power properties of Residue Number System processors," *IEEE Transactions on Circuits and Systems – Part II*. To appear.
5. E. Swartzlander and A. Alexopoulos, "The sign/logarithm number system," *IEEE Transactions on Computers*, December 1975.
6. I. Koren, *Computer Arithmetic Algorithms*. Englewood Cliffs, NJ: Prentice-Hall, 1993.
7. T. Stouraitis, *Logarithmic Number System: Theory, Analysis and Design*. PhD thesis, University of Florida, 1986.
8. P. Z. Peebles, Jr., *Probability, Random Variables, and Random Signal Principles*. New York: McGraw-Hill, 1987.
9. I. Orginos, V. Paliouras, and T. Stouraitis, "A novel algorithm for multi-operand Logarithmic Number System addition and subtraction using polynomial approximation," in *Proceedings of the 1995 IEEE International Symposium on Circuits and Systems (ISCAS'95)*, pp. III.1992–III.1995, 1995.

10. V. Paliouras and T. Stouraitis, "A novel algorithm for accurate logarithmic number system subtraction," in *Proceedings of the 1996 IEEE Symposium on Circuits and Systems (ISCAS'96)*, vol. 4, pp. 268–271, May 1996.

11. F. Taylor, R. Gill, J. Joseph, and J. Radke, "A 20 bit Logarithmic Number System processor," *IEEE Transactions on Computers*, vol. 37, pp. 190–199, Feb. 1988.

An Application of Self-Timed Circuits to the Reduction of Switching Noise in Analog-Digital Circuits

Raúl Jiménez[1], Antonio J. Acosta[2], Eduardo J. Peralías[2], and Adoración Rueda[2]

Instituto de Microelectrónica de Sevilla / Centro Nacional de Microelectrónica
Edificio CICA, Avda. Reina Mercedes s/n, 41012-Sevilla, SPAIN
Phone: 34-95-505 66 66; FAX: 34-95-505 66 86

[1] Dep. de Ingeniería Electrónica de Sistemas Informáticos y Automática. Universidad de Huelva
[2] Dep. de Electrónica y Electromagnetismo. Universidad de Sevilla
{naharro, acojim, peralias, rueda}@imse.cnm.es

This work has been sponsored by the Esprit-IV Project No. 26354 ASTERIS

Abstract. This paper presents an application where a self-timed approach reduces the switching noise in a mixed analog-digital circuit. Switching noise is of important concern in mixed signal systems, since it limits the performances of the analog part. Specifically, the digital core of an Analog to Digital converter has been designed following both a synchronous design style and another self-timed. Comparison between both versions shows the self-timed implementation reduce up to 50% the switching noise corresponding to the synchronous implementation.

1 Introduction

CMOS integrated circuits for mixed analog-digital systems are increasing in interest and importance. There is a continuous trend toward high-frequency, high-resolution, low-power and low-voltage analog circuitry included in a common substrate with complex high-performance digital circuitry. However, due to digital switching noise, that adversely affects sensitive analog circuitry via substrate-coupling, it is difficult to realize high resolution analog circuits on the same substrate with digital circuitry [1], [2]. There exist some techniques to reduce this noise, from the "analog" point of view [3]. Only recently, this problem is being considered from the digital domain [1], [2], [4] and [5], with the aim of designing low-switching-noise digital families.

The switching noise is produced by the variation in the supply current due to transitions of digital signals. These variations can affect the analog circuitry, via substrate coupling, reducing its performances and even causing operating transient and permanent errors. A way of measuring this parameter consists in monitoring the supply current, since the variation from average level is directly proportional to this noise [1]. So, in order to measure the switching noise, we are going to use the maximum variation of the supply current as an undirected measurement.

The self-timed approach can be seen as an advantageous alternative to the synchronous circuits in this type of applications [6]. On one hand, the self-timed cells decide themselves the need of its operation without the use of a global clock, so it is easy to

D. Soudris, P. Pirsch, and E. Barke (Eds.): PATMOS 2000, LNCS 1918, pp. 295-305, 2000.
© Springer-Verlag Berlin Heidelberg 2000

avoid the operation when it is unneccesary. On the other one, the operation of the different blocks is not synchronized, meaning that current consumption is not simultaneous, being distributed throughout the time, and the switching noise will decrease. The self-timed design presented in this paper, the digital core of an A/D converter, has been realized with the structures introduced in [7]. These structures use a half-handshaking protocol and the so-called SODS-QF structure. The main advantage of this structure that it does not need any memory element to solve the early precharge problem [7], obtain a structure with less area and additional reduction of switching noise, since the memory elements are ones of the main generators of this noise.

This communication is divided as follows. In section 2 the main characteristics the application to be implemented, the digital core of an A/D converter, are shown. Section 3 deals with the synchronous design of the circuit. In section 4 we include the self timed design. Section 5 includes some results as well as a comparison between both implementations. And finally, section 6 gives the main conclusions.

2   A Pipelined A/D Converter as Example of Mixed-Signal Circuit

A general scheme of a pipeline ADC is shown in fig. 1. It is composed of k stages connected in series, each one contributing to the output code with a certain n_i number of bits. An i-th converter stage comprises a n_i-bit sub-ADC, a n_i-bit sub-DAC, and a residue interstage amplifier with a gain G_i depending on the stage resolution, the output y_{i+1} of this stage is known as residue and it is the input of the next stage in the cascade. In many practical realizations, both the sub-DAC and the residue amplifier in each stage are implemented by a unique circuit known as MDAC (multiplying digital-to-analog converter). Calibration and correction techniques are usually included to reduce effects caused by component mismatches, gain errors and nonidealities in high-speed/high-resolution converters. Calibration is mainly aimed at reducing effects caused by component mismatches and gain errors in the stage MDACs, and it is necessary in converters with more than 10 bits of effective resolution. On the other hand, the goal of digital correction is to eliminate the effects that nonidealities in the sub-ADCs have on the overall converter operation.

We have designed the digital part of a pipeline A/D converter including self-correction and self-calibration techniques. In particular, the case chosen corresponds to the prototype reported in [8], which also include Design for Test strategies.

As a mixed-signal point of view, a pipelined A/D converter has an analog part (named APB in fig. 1), performing the analog-to-digital conversion, and a digital part (named DCAD in fig. 1), performing subcodes synchronization, correction, calibration control capabilities and must be prepared for different operation modes. Basically, the synchronization block is a variable-length FIFO array, the correction logic is a set arithmetic operators, the calibration logic is a finite state machine, with arithmetic logic and RAM memories to store the error codes after calibration, and the control block clock generation and test pattern generators for the analog part are provided by finite state machines. More details of the particular architecture of DCAD can be found in [

The DCAD block in fig. 1 has been designed following a synchronous strategy (Section 3) and a self-timed approach (Section 4), in order to compare them especially

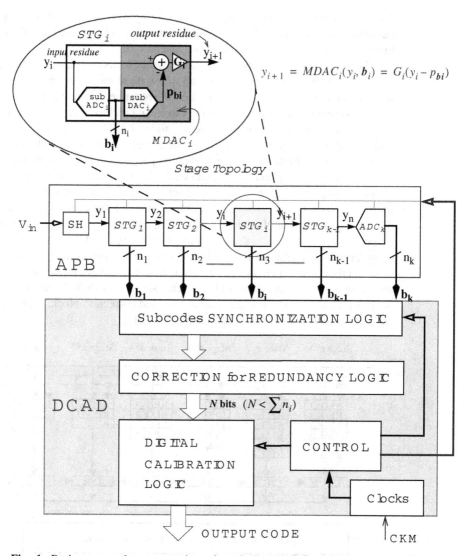

Fig. 1. Basic structural representation of a pipelined ADC with digital self-calibration/ correction capabilities

aspects regarding the generation of digital switching noise. We have selected as example a pipelined ADC with 10-bit, 10 Msamples/s, 6 stages with 2 or 3 bits of resolution (programmable) and 1 stage with 1 bit resolution, with test, self-calibration and self-correction capability.

3   A Synchronous Design of the Digital Part of the A/D Converter

The synchronous design of the DCAD block in fig. 1, has been realized following a classical top-down methodology, by using the automatic synthesis tools integrated in

Mentor Graphics. The design flow has, as starting point, the VHDL description of the circuits, performed at a RT level. This description has been satisfactorily verified following the verification methodology proposed in [9].

The scheme showed in fig. 2 only includes the blocks corresponding to the subcode synchronization and correction. We only show these blocks because they contain the main differences when comparing to the self-timed implementation, as it will be seen in the next section. The synchronization block include FIFO registers, while the correction block is a cascade of correction cells, called *CfR*. Also, we find a cell called *CfR* with a double functionality: performing the correction of the subcodes and generating an address for RAM memories for calibration purposes. Once the verification has been realized, the tool has generated automatically the netlists at a gate level.

In this implementation, the clocking scheme is of important concern. This scheme uses three clock signals: ex1, ex2 and ckb. Clocks ex1 and ex2 are used both to control the pipeline operation of analog blocks (sampling analog input and providing digital output) and to synchronize the subcodes provided by the APB. The signal ex2 can be considered as $\overline{ex1}$ and it is neccesary due to the operation of the converter, since processing takes place in both edges of signal ex1. The ckb signal is used for controlling the calibration, test and control blocks. This signal is a shifted version of ex1, and needed for processing the code coming from the correction block. More details about the synchronous implementation can be found in [9].

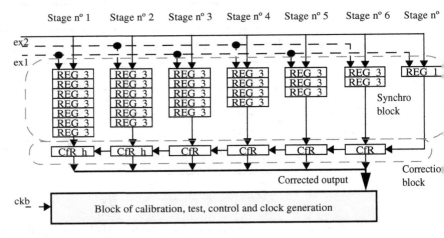

Fig. 2. Scheme of the synchronous implementation at a block level.

4   A Self-Timed Design of the Digital Part of the A/D Converter

When designing self-timed systems, one of the main problems is the overhead in hardware resources. Taking this into account, we have implemented using self-timed techniques, only those blocks that could take the main advantages of self-timed philosophy. These advantages are maximized when there is a great dependency between operation and input data [10]. So, we have selected the subcode synchronization and correction blocks to be implemented by a self-timed approach [11].

The inclusion of a self-timed circuit between two clocked systems (the APB and the digital part different of synchronization and correction) forces a serious compatibility between internal self-timed protocol signals and clocks signals for the rest of the circuit (ex1, ex2 and ckb). Signal ex1 is used to validate the data coming from the APB, while the output of the correction logic is captured by ckb. Also, we must implement an interface between both synchronous and self-timed worlds. This interface will be a set of flip-flop cells, named *bistable* in fig. 3. The control signal of these flip-flops is the ex1 signal, indicating the moment when output codes are valid.

The synchronization block, consisting of a self-timed FIFO array, has the function of ensuring the output code is generated with correct data, that is, subcodes corresponding to the same analog input value. In the self-timed implementation, we must force all input data to be valid, delaying the enabling signal (ex1) a semiperiod in the shift register on each stage. For this reason, we must add a new cell, called *init*. Because of the programmability of the converter and the need of adding the cell *init*, we can give it other functionality: determining the need of operation in a specific self-timed register column. Thus, we can reduce the switching noise and the power consumption avoiding the unnecessary operations of idle stages.

Because of the programmability in the number of output bits (2 or 3) from analog cells, we must add a new cell, called *cod_gen*. In order to minimize the hardware resources, we have substituted the last register (*Reg3*) for this new cell in every FIFO. For calibration purposes, the input data of the two most-significative correction blocks must be specifically provided by *cod_gen* cells.

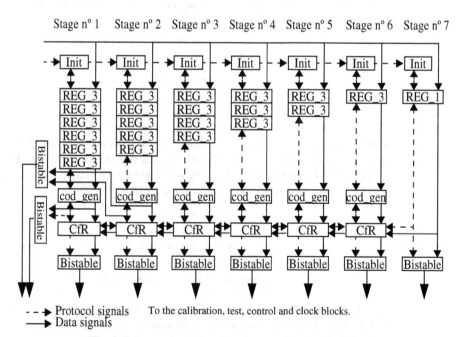

Fig. 3. Scheme of self-timed implementation at a block level.

4.1 Set of Self-Timed Cells

The cells compounding the self-timed block are *init*, *reg*, *cod_gen*, *CfR* and *bistabl*
connected to build the self-timed block as shown in fig. 3.

The cell *init* has two functions: delaying the ex1 signal from the previous stage an
determining when the current stage must operate. The operation set for *init* is (eqs. 1-4

$$A = reset(clk\_in + A) \tag{(}$$

$$Q = \bar{L}(S_i + \bar{C}) + LS_{i-1} + Q_0 \tag{(}$$

$$R_{in} = clk\_in Q \tag{(}$$

$$clk\_out = \overline{A clk\_in} \tag{(}$$

where signal *A* is an internal signal detecting the arrival of the first rising edge of ex
signal *Q* identifies the operation status of the current stage, signal R_{in} is the input reque
to this self-timed stage and signal *clk_out* is the output request for the following stage

The cell *reg* latches temporally the data coming from the analog part APB, in ord
to synchronize the subcodes. In the APB block, there are two kind of cell generatir
subcodes, *STG* and *ADCk* (fig. 1), so there would exist two kind of cell *reg* dependir
on the number of bits they have to latch. The cell *STG* has 2 or 3 three output bits, the
the cell *reg* connected to it will have to latch three bits. While the cell *ADCk* has o
output bit, then its cell *reg* will have to latch only one bit.

The cell *cod_gen* generates the input codes to the correction block. The operatic
set performed by these cells is (eqs. 5-7):

$$h_0 = b_0(b_1 + d_i) \tag{(}$$

$$h_1 = b_1 d_i + b_2 \bar{d}_i \tag{(}$$

$$h_1 = b_2 d_i \tag{(}$$

where d_i is the control signal determining the number of bits generated by cells *STG*,
is the subcode generated by the *STG* and h_i is the input to correction block.

The cell *CfR* corrects the subcode b_i generated by the analog part, and so, it is th
core of the correction block. The operation set is (eqs. 8-10):

$$o_0 = car\_in \oplus h_0 \tag{•}$$

$$o_1 = h_1 \oplus (car\_in h_0) \tag{(}$$

$$o_2 = h_2 d_i + h_1 \bar{d}_i + car\_in h_0 (\bar{d}_i + h_1) \tag{(1}$$

where h_i and d_i have the same meaning than for cell *cod_gen*, and *car_in* is the outp
o_2 of the *CfR* of the preceding stage. In the case of the first *CfR* in the chain, this inp
is the bit latched in the last *reg*.

In order to synchronize the output of self-timed blocks with the synchronous part of DCAD block, we have used the *bistable* cell. The control signal of these flip-flops is the local validation signal from the previous stage. We have ensure that there is not setup time violation in the flip-flop of *biestable* cell by including additional delay. Also, the precharge non-valid data are filtered and will not pass to the synchronous block.

4.2 Design Process of the Self-Timed Implementation

The design process has taken four phases. The first one was a verification of the self-timed circuit both at a functional level using VHDL, and an electrical level with HSPICE.

The high-level description includes behavioural modeling of cells. The verification at this level has been carried out with an extent set of input patterns. The outputs patterns have been used as input patterns to the rest synchronous blocks, while the global converter has been verified following the methodology presented in [9].

Once we have verified the circuit at a high level description, we have validated the design with HSPICE. Again, we have only verified the behaviour, since the characterization will be realized via the extraction of the layout and with the integrated prototype. The results of these simulations have shown a correct function of the global circuit.

The self-timed cells have been laid out in a full-custom style using MAGIC. The technology used was 0.6 µm CMOS with double metal layer. Our strategy was planned to draw all cells layouts with the same length or width to assembly the different blocks in rows. In the table 1, we show the size of each cell. The whole self-timed systems has been a result of assembling the cells according the schematic of fig. 3. The global circuit has been simulated with HSPICE to verify its correct behaviour, including parasitic effects.

Table 1. Size of the self-timed cells

Cell	Size (µm x µm)
init	39 x 38
reg_1	90 x 28
reg_3	71 x 50
cod_gen	75 x 50
CfR	105 x 60
bistable	85 x 18

5   Implementation, Simulation Results and Comparison

A synchronous and a self-timed version of the DCAD block have been integrated in

a 0.6 μm CMOS technology, with double metal layer. The microphotograph of both im
plementation is shown in fig. 4. The self-timed block (synchronization and correction
is highlighted, takes a 7% of the total area of the self-timed DCAD and has 3375 tran
sistors. The total core area is 2016 μm x 1689 μm for the synchronous and 2012 μm

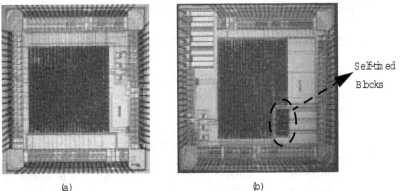

Fig. 4. The microphotographs corresponding to (a) synchronous implementation and (b) sel
timed implementation of the DCAD block.

1936 μm for the self-timed. Thus, the area overhead of the self-timed implementatio
is about 14%, when compared to the synchronous one.

A comparison in terms of speed, has not been performed since the analog part
slower than the digital part, regardless the implementation scheme used for the digit
part. Generally, analog circuits are slower than the digital ones, so speed performan
of digital part is not significative in most mixed-signal circuits, including our case
study (the APB runs at 10 Msamples/s, that is easily reached by the DCAD, able to wo
up to 100MHz).

To make a comparison in terms of power consumption, we have only considered th
synchronization and correction blocks because they are the only difference between th
synchronous and the self-timed implementations. We will suppose that the other bloc
will have a similar power consumption. The table 2 shows the power consumption
three cases: the synchronous implementation, with the self-timed implementation wi
all the stages operating and the self-timed implementation with only four stages opera
ing. The mayor average power consumption of the self-timed version is due to the stat
consumption of the SODS-QF structures during the early precharge phase, as well as
the hardware excess.

Since the switching noise is a limiting factor in mixed-signal circuits, we have d
voted great efforts to make a fair comparison. We have obtained the minimum (nega
tive) value of supply current as a direct measurement of switching noise for the se
timed block and the synchronous counterpart.

In fig. 5, we show the waveform corresponding to the supply current for the sy
chronous block. We can see that the maximum value is above from 40 mA., and acc
rately 45.8 mA. As all operations are done in the transitions of the clock signal, *ex1*
ex2, the widths of the peaks are very small. Then the peak of current, and the switchi

Table 2. Measurement related to power consumption corresponding to the self-timed (synchronization and correction blocks).

Power Consumption	Synchronous	Self-timed (all FIFOs working)	Self-timed (four FIFOs working)
Average (mW.)	18.8	37.4	16.5
Maximum (mW.)	219	118	86.8

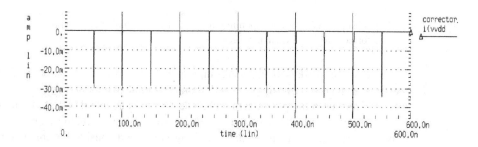

Fig. 5. Supply current corresponding to the synchronous implementation of the synchronization and correction blocks.

noise, will be high.

In the fig. 6, we can see this measurement for the self-timed block when the operation is performed in all stages, with a maximum value of 23.7 mA. When it is performed in the case of only four stages operating, the reduction in the last case is about 36%, due to the three first stages do not perform any operation and so they do not consume any supply current. Also, we can see how the current peaks are wider than in the synchronous implementation. This means that the operation is less centralized and the different blocks do not need supply current at the same time. So the maximum value of these peaks is less than in the synchronous case. We can appreciate a low-value static power consumption, due to the operation of self-timed cells in a situation of early precharge.

As a final comparison in terms of switching noise, the self-timed implementation has a better behavior than the synchronous implementation, being approximately about 50% of the synchronous measurement if we compare with the case in which all stages have to operate. In the cases in which all stages do not have to operate, this difference will be greater because the synchronous value will hold while the self-timed value will decrease.

6   Conclusions

In this paper, we have introduced the implementation of the subcode synchronization and correction logic corresponding to the digital part of a pipelined A/D converter, using two design techniques, one synchronous, and other self-timed. One of the main ob-

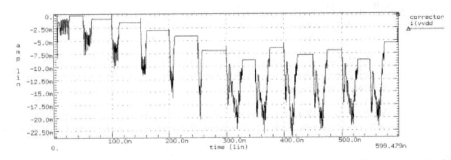

Fig. 6. Supply current corresponding to the self-timed implementation of the synchronizatic and correction block when all stages operate.

jectives of our work is to realize a comparison in the most significative parameters, suc area, speed, power consumption and, mainly, digital switching noise.

According to the parameters obtained, we can conclude that both implementatio have a quite similar characteristics. But, in mixed-signal Analog-Digital circuits, tl most restrictive parameter is the switching noise, and considering this parameter, tl best implementation is the self-timed one.

References

1. Allstot, D. J., Chee, S-H. and Shrivastawa, M.: Folded source-coupled logic v CMOS static logic for low-noise mixed-signal ICs. IEEE Transactions on Circui and Systems I, vol 40, pp 553-563, Sept. 1993.
2. Ng, H-T. and Allstot, D. J.: CMOS Current Steering Logic for Low-Voltag Mixed-Signal Integrated Circuit. IEEE Trans. on VLSI Systems, Vol. 5, pp 30 308, Sept. 1997.
3. Tsividis, Y.: Mixed Analog-Digital VLSI Design and Technology. McGraw-Hi 1995.
4. Albuquerque, E., Fernandes, J. and Silva, M.: NMOS current-balanced logi Electronics Letters, vol 32, pp 997-998, May 1996.
5. Jiménez, R., Acosta, A.J., Juan, J., Bellido, M.J. and Valencia, M.: Study ar Analysis of Low-Voltage/Low-Power CMOS Logic Families for Low Switchir Noise. Proc. of 9th Int. Workshop Power and Timing Modeling, Optimization ar Simulation (PATMOS'99), pp. 377-386, Kos Island, October 1999.
6. Gonzalez, J.L. and Rubio, A.: Low Switching Noise CMOS Circuit Design Stra egy based on Regular Self-Timed Structures. Proc. Midwest Symposium on C cuits and Systems, pp. 176-179, 1999.
7. Jiménez, R., Acosta, A.J., Barriga, A., Bellido, M.J. and Valencia, M.: Efficie Self-Timed Circuits based on weak NMOS-Trees. Proc. of 5th IEEE Int. Confe ence on Electronics, Circuits and Systems (ICECS'98), pp. 179-182, Vol. 3, L boa, September 1998.
8. Peralias, E, Rueda, A. and Huertas, J.L.: A DFT Technique for Analog-to-Digi

Converters with Digital Correction. Proc. 15th IEEE VLSI Test Symposium (VTS'97), pp. 302-307, 1997.

9. Peralias, E.J., Acosta, A.J., Rueda, A. and Huertas, J.L.: A VHDL-based Methodology for Design and Verification of Pipeline A/D Converters. Proc. Design, Automation and Test in Europe (DATE'00), pp. 534-538, March 2000.

10. Berkel, K.v., Burgess, R., Kessels, M., Schalij, F. and Peeters, A.: Asynchronous Circuits for Low Power: A DCC Error Corrector. IEEE Design and Test of Computers, Vol. 11, no. 2, pp. 22-32, Summer 1994.

11. Jiménez, R.: Una aportación al Diseño de Circuitos Integrados CMOS Autotemporizados. PhD. Thesis, Universidad de Sevilla, Julio 2000 (in Spanish).

PARCOURS – Substrate Crosstalk Analysis for Complex Mixed-Signal-Circuits

Andreas Hermann[1], Mathias Silvant[2], Jürgen Schlöffel[3], and Erich Barke[1]

[1]University of Hannover, Institute of Microelectronic Systems, Appelstr. 4, 30167
Hannover
hermann@ims.uni-hannover.de

[2]Simplex Solutions SA, ZA Le Parvis, 38500 Voiron, France

[3]Philips Semiconductors, Stresemannallee 101, 22529 Hamburg

Abstract. In integrated mixed-signal circuits signal integrity is affected by parasitic substrate coupling. Therefore, substrate crosstalk analysis has to be performed in layout verification. The PARasitic COUpling Model GeneratoR for Substrate (PARCOURS) applies a three-dimensional model for the substrate considering conductivity and permittivity if required. As a remarkable feature PARCOURS uses different levels of accuracy. The highest level integrates circuit elements with multiple substrate terminals in order to model the flow of parasitic currents in the vicinity of the die surface. The lowest level simplifies the substrate terminal as a point connection. A commercial videochip has been examined with the introduced approach.

1 Introduction

Three important factors impact the performance of today's mixed-signal integrated circuits with respect to substrate coupling. The decrease of feature size results in tighter coupling due to higher vicinity. The increase of operation speed of digital parts leads to more noise spread into the substrate and the decrease of the signal-to-noise ratio leads to a higher sensitivity against disturbances. This is why several recent publications focus on substrate coupling [1-11]. They can be divided into experimental [2,3,4], finite-element methods (FEM) [1,6,9,10] and boundary element methods (BEM) [5,7,8,11]. Most of them discuss operation frequencies below 1 GHz. Only [10] shows results for operation up to 40 GHz. The output of most of the discussed algorithms is an electrical network representing the substrate as purely resistive. For operation in the GHz range this is no longer valid. In order to handle the complexity of large circuits simplifications are used. The most important simplification is to treat the substrate as a semi-conducting semi-space with a flat surface. The devices contact the substrate through a conducting layer on top of the surface. All BEM-approaches make use of this simplification. The FEM-approaches use more complex three-dimensional models. Common to all discussed approaches is that the substrate space is modeled as a stratified medium composed of several homogeneous layers, which are characterized by their conductivity and permittivity, respectively. Our approach is able to deal with rather small (mostly analog) circuits in a complex technology (i.e. Bipolar or BiMOS) which need three-dimensional modeling and large (digital) circuits in a simpler technology (e.g. CMOS, I²L) that are complex due to the number of involved elements.

D. Soudris, P. Pirsch, and E. Barke (Eds.): PATMOS 2000, LNCS 1918, pp. 306-315, 2000.

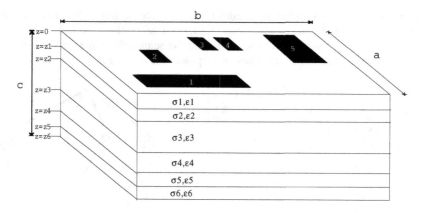

Fig. 1. Simplified Substrate Model

2   Discretization

PARCOURS uses FEM-discretization. The modelling procedure uses a stratified substrate model corresponding to the substrate doping profile. The layout data defines the topology of the mesh for each layer. Several layers form the 3D-substrate region and are stacked vertically. In order to reduce the complexity of the mesh and consequently of the derived electrical network a non-rectangular gridding algorithm is applied. The applied method is called Voronoi Tessellation and was first published in [6].

2.1   Voronoi Tessellation

Every object that is connected to the substrate like transistors (MOS and Bipolar), guard-rings, wells, or tie-downs leads to geometrical points on the surface of the top nodeplane.

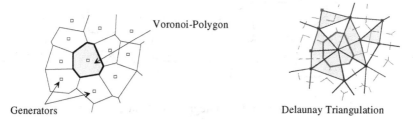

Fig. 2. Voronoi Diagram and Delaunay Triangulation

These points are generators for the Voronoi mesh. Algorithms for Voronoi Tessellation have a worst-case time complexity of O(NlogN), with N being the number of generators [12]. We chose an insertion algorithm that builds the Voronoi Diagram by inserting the generators in turn. The advantage of this algorithm is that the mesh can be extended or changed afterwards. It is quite simple to add another generator (e.g. corresponding to an additional substrate tie-down) by locally changing the former mesh. Critical, with respect to runtime is the search for the already inserted, next neighbor. In order to accelerate the search the generators are organized in a quarternary tree whose branches correspond to layout areas.

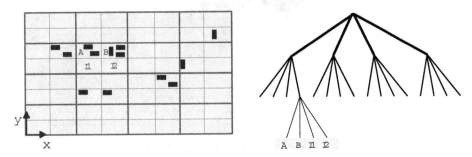

Fig. 3. Quarternary Tree Structure

Fig. 3 shows a layout with some components and the corresponding quarternary tree for some of the components. The Voronoi Tessellation leads to tiles which are used to build up the electric mesh following the box-integration method. The components are stored in the leafs of the branches (i.e. A and B in Fig. 3). Some leafs remain empty (i.e. I1 and I2).

2.2   Box-Integration Method

We assume that the electric field is homogeneous within a tile. Furthermore the conductivity σ is assumed to be constant within a tile.

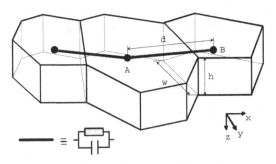

Fig. 4. Box-Integration Method

The resistance between two nodes is calculated using formula (1).

$$R_{AB} = \frac{d}{\sigma \cdot w \cdot h} \tag{1}$$

For high frequency applications as [10] it is necessary to take permittivity into account. So we assume that the permittivity ε is also constant within the tile. Then the capacitance can be calculated from (2).

$$C_{AB} = \frac{\varepsilon}{\sigma \cdot R_{AB}} \tag{2}$$

To abide by our first assumption, the homogeneous electric field inside a tile, additional generators are added to the Voronoi Diagram in zones where generator density is low.

3   Runlevel Concept

The presented system is the first approach that is capable to handle small-sized layouts with complex (analog) circuitry and rather large circuits with many components. The complex analog circuit requires a very detailed extraction whereas for large circuits a rough extraction is adequate which is accurate enough to simulate the main sources of noise.

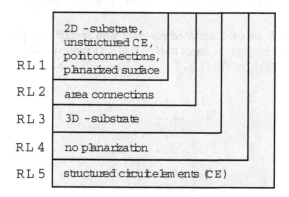

Fig. 5. Overview of Available Runlevels

PARCOURS can be used with several runlevels. The complexity of the algorithm and the necessary information rise with higher runlevels (Fig. 5). "CE" stands for "circuit element" – a transistor, a resistor, a capacitor, a well containing several MOS-transistors.

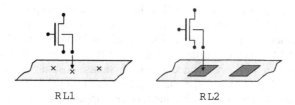

Fig. 6. Runlevel 1 and 2

Runlevel 1 (RL 1) starts with a rather rough approach, assuming a substrate with only one layer. The surface of the substrate is planarized. The circuit elements are connected by only one point connection to the substrate mesh. The resulting network is small and adequate for a fast extraction of a large number of MOS-transisors in a digital application. Runlevel 2 (RL 2) uses rectangular shapes as connecting windows to the substrate. For each well and each device on the substrate several Voronoi generators form the rectangular shape (Fig. 7). The accuracy of the electrical substrate network is higher, but the network itself is larger.

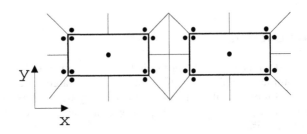

Fig. 7. Rectangular Interconnects and Corresponding Voronoi Tessellation

Runlevel 3 (RL 3) uses the stratified substrate with several nodeplanes. The surface is still planarized, Fig. 1 gives an example. The next runlevel (RL 4) uses a real three-dimensional substrate model (Fig. 8).

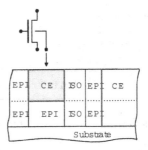

Fig. 8. Runlevel 4

"Substrate" denotes the planar substrate model of runlevel 3. "EPI" stands for an epitaxial layer, "ISO" stands for another layer in the element region, e.g. a guardring-diffusion or a trench structure. Note, that the existence of an epitaxial layer is only an option. The circuit elements (CE) normally interact with the substrate via depletion layer capacitances. Apart from wells containing several transistors, all circuit elements model this depletion layer capacitance themselves. All Voronoi generators adjacent to such a well are connected to the same substrate node of the involved circuit element. Fig. 9 shows the main feature of runlevel 5 (RL5): For a selected number of circuit devices special models are applied to model the geometrical structure of the device. Therefore, they are provided with multiple substrate terminals.

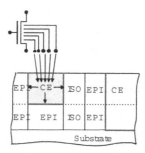

Fig. 9. Runlevel 5 Uses Directed Substrate Connections

3.1   Model Extensions

The used design environment uses BiMOS-technology and bjt503 models for the bipolar transistors and MOS9 models for the MOS transistors. The bjt503, also known as MEXTRAM model, incorporates a parasitic transistor formed by the base (P), collector (N) and the substrate (P). This parasitic PNP-transistor is modeled by a junction capacitance between collector and substrate and a current source injecting current into the substrate. A special transistor model with five substrate terminals has been programmed and is used to simulate the device in RL 5 [15].

4   Interface

PARCOURS can use the Dracula database and is now augmented to access Cadence Diva database. Fig. 10 illustrates the flow to perform a substrate coupling analysis on a design given in Cadence DIVA. PARCOURS reads the netlist extracted from Cadence Design Framework. Technology constraints and control statements are defined in a Technology File. We use the Cadence Database Access (CDBA) which is an interface that enables programs to access the internal DIVA database. Required data is taken out of the database. The content depends on the runlevel. With the collected input PARCOURS generates the equivalent electrical network for the substrate and connects it to the devices. The output is a netlist for the network simulator Spectre. For runlevel 1 to runlevel 4 it is in Spice syntax. Spectre is able to simulate regular Spice-netlists, however, additional modules using the hardware description language SpectreHDL can also be used. The extended models with additonal substrate nodes used in runlevel 5 are written in SpectreHDL. Simulating with HDL-models is more time-consuming than using the internal models written in C, but for the experimental stage they are easier to handle.

The prototype of our substrate extractor works with Cadence´s Design Framework.

5   Model Validation

The substrate model is verified by comparing simulation results obtained with the simulator Spectre to results of measurements.

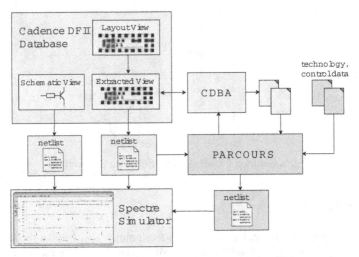

Fig. 10. Database Access to Cadence Design Framework

5.1   Distance/Size Investigations

The first set of investigations concerns substrate tie-downs. Fig. 11 shows a physical cross-section. The applied technology uses a p-doped substrate with $\sigma = 1/(2\Omega\text{cm})$ and an n-doped high-resistive epitaxial layer.

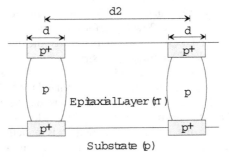

Fig. 11. Cross-Section for Distance/Size Investigations

Three rows of pairs of rectangular tie-downs have been designed varying in size d1 from row to row and in distance d2 within a row. The resistance between the pairs of tie-downs was measured with an RLC-meter. The measured results are given in Fig. 12 as well as the simulation results by PARCOURS. It shows a good correspondence between the measured and simulated results.

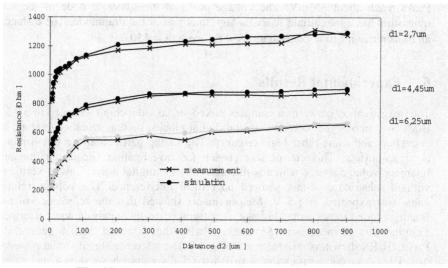

Fig. 12. Comparison: Distance/Size Investigations

5.2   Transmitter/Detector Investigations

Another set of measurements and simulations was performed in order to investigate the transmitter-receiver behavior of MOSFETs. The source of substrate noise is modeled as a chain of CMOS-inverters. The chain contains six inverters, each built with an NMOSFET (W/L = 36/1,2) and a PMOSFET (W/L = 90/1,2).

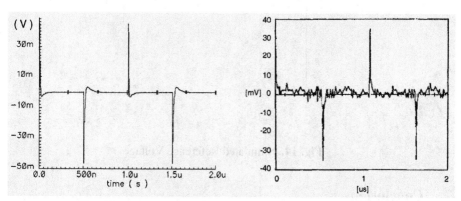

Fig. 13. Simulated Results (left) and Measured Results (right)

As output load a 3 pF capacitor is used. The chain is fed with alternating pulses produced by a signal generator. The sensor is an NMOSFET with a W/L-ratio of 36/1,2. The sensors source is grounded, the gate is biased at 2 V and its drain is connected to the supply voltage of 5 V, shunted by a resistor of 1 kΩ. Fig. 13 shows the simulated voltage at the drain node on the left side and the measured voltage on the right side. Once again, the correspondence between simulated and measured results is very good.

Peaks reach about 35 mV. The voltage peaks at the substrate node of the sensor transistors has a magnitude twice as high than that at the drainnode. The distance of all transmitter-receiver constellations was longer than 150 μ.

6   Experimental Results

The investigation concerns a complex mixed-signal videochip fabricated in a 0.5 μm BiMOS-technology. It contains several digital blocks, such as clock generators, level converters and some other logic circuitry. The analog part contains a line-driver and two gap-buffers. This circuit was chosen for investigation because it contains a reference voltage source which is threatened by the digital noise. Layout verification without substrate crosstalk showed no critical interference. The voltage reference value was expected at 1.5 V. Measurements showed that the reference voltage is floating around the expected voltage. A substrate crosstalk analysis was performed to examine this phenomenon. Simulations with the extended netlist generated by PARCOURS show this interference in Fig. 14. The reference signal contains peaks up to 0.3 V. A layout verification with PARCOURS would have shown this problem before manufacturing the device.

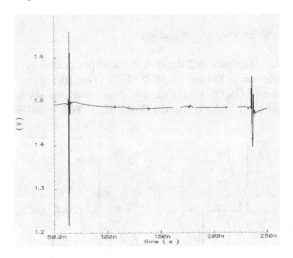

Fig. 14. Simulated Reference Voltage

7   Conclusions

We have presented a new modeling strategy for substrate crosstalk simulation. The approach uses the Voronoi Tessellation method which is known to lead to less circuit nodes than a uniform grid. Our algorithm uses a set of runlevels which extracts substrate parasites with rising accuracy. In the highest runlevel structured models with multiple substrate terminals are applied, that are actually written in SpectreHDL. Due to the run-level model the tool is both useful for the rough extraction of large digital circuits and the detailed extraction of analog or mixed analog/digital circuits. The accuracy of the linear parasitic substrate model and some simulations has been verified by measurements. The approach works with Cadence Design Framework Databases Dracula and Diva. Therefore the prototype is applicable to industrial layouts. Investigations were applied to a commercial videochip.

References

1. T.A. Johnson, R.W. Knepper, V. Marcellu, W. Wang, "Chip Substrate Resistance Modeling Technique for Integrated Circuit Design", IEEE Transactions on Computer-Aided Design of Integrated Circuits, vol. CAD-3(2), pp. 126-134, 1984
2. D.K. Su, M. Loinaz, S. Masui, B.A. Wooley: "Experimental Results and Modeling Techniques for Substrate Noise and Mixed-Signal Integrated Circuits", IEEE Journal of Solid State Circuits, vol. 28, pp. 420-430, Apr. 1993
3. B.R. Stanisic, N.K. Verghese, R.A. Rutenbar, L. R. Carley, D.J. Allstot, "Addressing Substrate Coupling in Mixed-Mode IC's: Simulation and Power Distribution Synthesis", IEEE J. Solid-State Circuits, vol. 29, pp. 226-237, Mar. 1994
4. K. Joardar, "A Simple Approach to Modeling Cross-Talk in Integrated Circuits", IEEE Journal of Solid State Circuits", vol. 29, pp. 1212-1219, Oct. 1994.
5. T. Smedes, N.P. van der Meijs, A.J. van Genderen, "Extraction of Circuit Models for Substrate Cross-Talk", Proc. of Int. Conf. Computer Aided Design 1995, pp. 199-206, Nov. 1995
6. I.L. Wemple, A.T. Yang, "Integrated Circuit Substrate Coupling Models Based on Voronoi Tessellation", IEEE Trans. Computer Aided Design, vol. 14, pp. 1459-1469, Dec. 1995
7. N.K. Verghese, D.J. Allstot, M.A. Wolfe, "Verification Techniques for Substrate Coupling and their Application to Mixed-Signal IC Design", IEEE J. Solid-State Circuits, vol. 31, pp. 254-265, Mar. 1996
8. R. Gharpurey, R.G. Meyer, "Modeling and Analysis of Substrate Coupling in Integrated Circuits", IEEE J. Solid-State Circuits, vol. 31, pp. 344-353, Mar. 1996
9. T. Blalack, J. Lau, F.J.R. Clément, B.A. Wooley, "Experimental Results and Modeling of Noise Coupling in a Lightly Doped Substrate", Proc. of IEEE 1996 IEDM, pp. 623-626, Dec. 1996
10. M. Pfost, H.-M. Rein, "Modeling and Measurement of Substrate in Si-Bipolar IC's up to 40 GHz", IEEE J. Solid-State Circuits, vol. 33, pp. 582-591, Apr. 1998
11. M. Chou, J. White, "Multilevel Integral Equation Methods for the Extraction of Substrate Coupling Parameters in Mixed-Signal IC's", Proc. ACM/IEEE Design Automation Conference, pp. 20-25, Jun. 1998
12. F.P. Preparata, M.I. Shamos, "Computational Geometry: An Introduction", Springer, New York, 1985
13. M. Klemme, E. Barke, "Accurate Junction Capacitance Modeling for Substrate Crosstalk Calculation", PATMOS '98, Oct. 1998
14. S.S. Hegedus, "Parasitic Isolation PNP Devices and Their Effect on NPN Saturation Delay", ACM/IEEE Design Automation Conference, pp. 107-111, Jun. 1980
15. M. Klemme, E. Barke, "An Extended Bipolar Transistor Model For Substrate Crosstalk Analysis", Proc. of IEEE Custom Integrated Circuit Conference 1999

Influence of Clocking Strategies on the Design of Low Switching-Noise Digital and Mixed-Signal VLSI Circuits

A.J. Acosta, R. Jiménez[1], J. Juan, M.J. Bellido, and M. Valencia

Instituto de Microelectrónica de Sevilla
Centro Nacional de Microelectrónica/ Universidad de Sevilla
Edificio CICA, Avda. Reina Mercedes s/n, 41012-Sevilla, SPAIN
Phone: 34-95-5056666; FAX: 34-95-5056686;

[1]Also with the Universidad de Huelva, Spain

{acojim, naharro, jjchico, bellido, valencia}@imse.cnm.es

Abstract. This communication shows the influence of clocking schemes on the digital switching noise generation. It will be shown how the choice of a suited clocking scheme for the digital part reduces the switching noise, thus alleviating the problematic associated to limitations of performances in mixed-signal Analog/Digital Integrated Circuits. Simulation data of a pipelined XOR chain using both a single-phase and a two-phase clocking schemes, as well as of two n-bit counters with different clocking styles lead, as conclusions, to recommend multiple clock-phase and asynchronous styles for reducing switching noise.

1   Introduction

Integration of digital and analog mixed-signal integrated circuits has taken significa advantages in the implementation of advanced electronic systems. However, the int gration of large-scale digital and high-speed analog circuits in the same monolithic implies interactions, referred to as cross talk, between both parts, and analog signal de radation problems. In these mixed-signal circuits, the switching noise created by t digital circuits passes to the analog circuits, limiting their performances -resolution A/D converters, jitter in PLLs, etc-, and making very difficult the realization of hig resolution analog circuits on the same substrate with complex digital circuitry. Su noise can be easily measured by monitoring the peak value of dynamic current provid by the supply source (Fig 1), that is proportional to the carrier injection [1].

The use of noise reduction techniques alleviates the influence of switching noise [2 to separate as much as possible the digital and the analog part; to use different supp and ground sources for analog and digital circuitry; to considerate the substrate co pling and reducing it with substrate biasing and using guard-rings, etc. All these met ods are related to layout and analog design, but do not include digital design method ogy.

Recently, some low-switching-noise digital CMOS families have been reporte CSL [3], FSCL [1] and CBL [4]. These current-mode structures work with supply cu

D. Soudris, P. Pirsch, and E. Barke (Eds.): PATMOS 2000, LNCS 1918, pp. 316-326, 2000.
© Springer-Verlag Berlin Heidelberg 2000

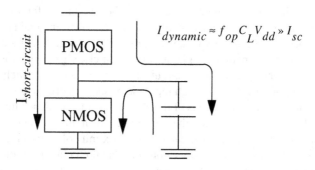

Fig. 1. Dynamic (dominant) and short-circuit current in CMOS.

rent almost constant, thus reducing variations in supply current and, hence, switching noise. However, static power consumption is the main penalty of such structures, making them unsuited for low-power applications. The use of these current-mode families is recommended only in risky-noise generation areas, while in other non-critical areas, logic should be implemented with more conventional techniques. However, the use of these current-mode logics is highly complicated, since these gates are very complex and difficult to design and test, they need current-mode to CMOS-conventional interfaces, and show static power consumption. Furthermore, additional reduction in switching noise implies higher static power consumption [5].

This communication explores additional ways of reducing switching noise from the digital domain, studying the influence of the clocking style in the digital part on the generation of switching noise, when using more conventional low-cost CMOS digital implementations.

This communication is divided as follows. Section 2 shows the theoretical influence of the clocking scheme in the switching-noise. Section 3 presents a comparison between a single-phase and a two-phase scheme as a case of study. Section 4 presents a comparison between synchronous and asynchronous counters, as example of study. Section 5 shows some simulation results. And finally, Section 6 presents the conclusions.

2   Switching Noise and Timing Schemes

The switching noise, also referred as dI/dt noise, increases when many circuits or blocks evaluate simultaneously, causing power supply fluctuations [6]. The use of an specific clock strategy when designing the digital part in a mixed-signal IC brings serious consequences relating to such noise generation. Since the timing scheme indicates the way of gates switch, and the supply current is the sum of contributions due to switching gates, as the number of synchronized gates switching increases, the peak supply current will be also increased. This is the case of Simultaneous Switching Noise (SSN) in buffer design [7] [8].

The use of a single-phase clock scheme (fig. 2a) forces that most of the transition in the system take place within a relatively small interval around (during and after) the clock active edge. By using two clock phases (fig. 2b), or a double-edge clock, switching in combinational logic, as well as in clock generator logic and flip-flops, reduce the number of gates or subcircuits that simultaneously switch, reducing the peak current of supply source. Although the logic blocks can effectively switch at any time between consecutive active edges of the clocks considered (depending on the propagation delay of combinational logic), the activity i.e., the number of nodes that switch their logic value, will be statistically greater in the proximity of active clock edges (dashed area in the activity bars in fig. 2). If we consider that both implementations (fig. 2a and fig. 2b) are identical in the sense that the same logic is used and the same nodes have the same capacitive load associated and hence, the same average current is consumed (see equation in fig. 1), the maximum current level will be given in the single-phase clock scheme (fig. 2a), since all the flip-flops and logic blocks switch (almost) simultaneously. With this reasoning, the most suited synchronous solutions for low-noise generation use more than one clock phase, although introducing clock-skew problems, decreasing the operation reliability. In such case, a trade-off between low-noise and reliability should be found.

Self-timed [9] design (fig. 2c) is an elegant cost effective means to control noise in a predictable manner. By substituting the global clock by locally-generated clock (clock1, clock2 and clock3) indicating the validity of data to be processed for the next logic block, switching of gates are unsynchronized, making that supply currents of different self-clocked blocks do not overlap, hence reducing the magnitude of the noise components. In this way, a self-timed circuit can be conceived like a k (large) clock phase system, being the operation distributed in continuous time slots rather than in discrete time instants.

3   A Case of Study: Comparison between a Single-Phase and a Two-Phase Clock Schemes

In order to verify the reasoning of Section 2, we are going to measure the switching noise in a simple system using two different clocking schemes. The system in a XOR gate array of XOR gate with pipeline at a gate level. The flip-flops used in the pipeline stage have been designed by using a TSPC approach [10]. The reason of this choice due to the more conventional master-slave flip-flops works in a equivalent two-phase configuration, so the comparison would not be fair, as we could confirm without any appreciable difference. Also TSPC are widely used in modern VLSI digital design.

In fig. 3 we show both circuits at a transistor level. In the case of single-phase clock scheme, we can distinguish two kinds of TSPC elements: TSPC NMOS, operating the rising edge of the clock, and PMOS, operating in the falling edge. While, in the case of two-phase clock scheme, we can only need TSPC NMOS flip-flops. Due to the use of the NMOS and PMOS TSPC, the output waveforms will be the same in both case, so the operation form will be identical in both case without decreasing the clock frequency for the two-phase clocking scheme.

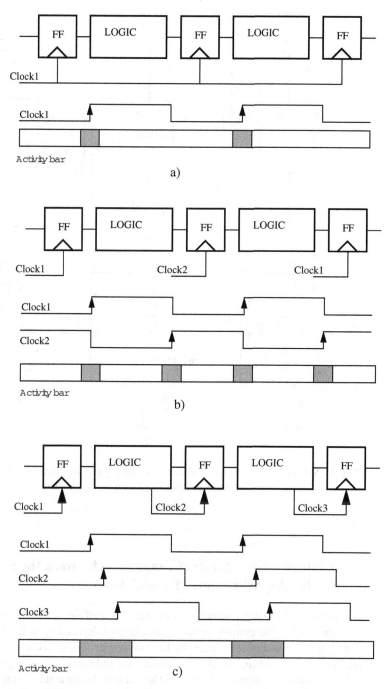

Fig. 2. Different clocking styles for a pipelined logic structure: a) Single-phase, b) Two phases, c) Self-timed. The dashed areas in the Activity bar indicate the maximum switching density.

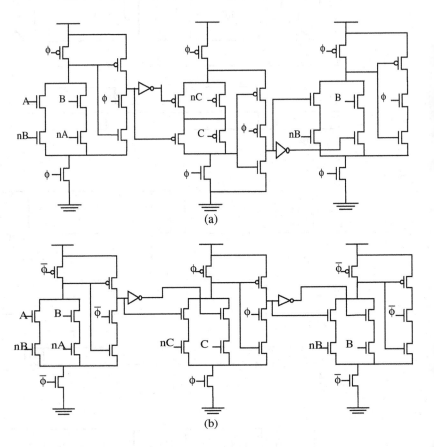

Fig. 3. Schemes at a transistor level corresponding to the array of XOR gates with a) a single phase and b) a two-phase clock schemes.

4 Another Case of Study: Comparison between the Synchronous and the Asynchronous "Ripple" Counter

Following with the demonstration of the reasoning of Section 2, let us consider a n-bit counter as a generic example to show our claim of decreasing spikes in supply current with synchronous and self-timed clocking strategies. The events counter is a sequential machine of wide use and interest in most digital and mixed-signal applications, specially for frequency division applications. The counter device counts events in the C signal increasing or decreasing the count state. Two simple implementations of a 4-bit increasing counter are shown in fig. 4. Both modular implementations use T(oggle) flip-flops as elementary memory units. The synchronous implementation (fig. 4a) uses the C signal

nal as clock of all the flip-flops, while in the ripple implementation (fig. 4b) the clock signal of each flip-flop is the output of the previous flip-flop in the counter. As it is clear, these are good examples of the different clocking strategies shown in the previous section.

In fig. 5, an HSPICE simulation of a detailed state transition (from 1111 to 0000) is shown for both counters. It can be easily seen how the transitions in Q_0, Q_1, Q_2 and Q_3 in the synchronous case are almost simultaneous, while in the asynchronous case, the transition in Q_i provokes the transition in Q_{i+1}, after the propagation delay of the flip-flop. The average supply current is approximately the same, but more "concentrated" in the synchronous case, meaning a higher maximum value and, hence, provoking greater switching noise.

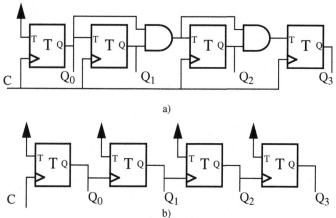

Fig. 4. 4-bit counter: a) synchronous, b) asynchronous "ripple".

5   Design and Simulation Results [11]

Simulations have been performed on a 0.7 μm standard technology. The results corresponding to the comparison between synchronous clock schemes are shown in table 1, while the results for the counters are shown in table 2.

Table 1. Simulation results of the synchronous clocking schemes for the pipelined XOR array. F=50 MHz.

	Transistors	Power (mW) Vdd=5v/3.3v	Iaverage (μA) Vdd=5v/3.3v	Ipeak (μA) Vdd=5v/3.3v
One-Phase	31	0.36 / 0.11	68.3 / 34.7	4200 / 1720
Two-Phase	31	0.36 / 0.08	68.8 / 26.4	2250 / 850

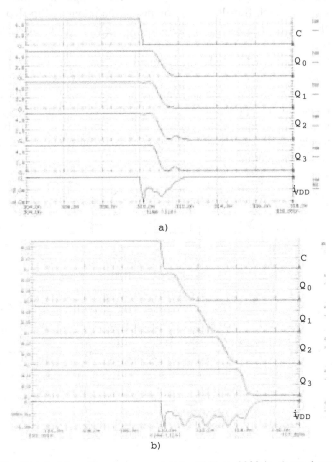

Fig. 5. Detailed transition from count state 1111 to 0000 in a) synchronous, b) asynchronous 4-bit counter.

Table 2. Simulation results for counters. F=50MHz. PDP: Power-Delay-Product.

	Transistors	PDP (pJ) Vdd=5v	Iaverage (μA) Vdd=5v/3.3v	Ipeak (μA) Vdd=5v/3.3v
4-bit synch.	116	0.17	170 / 100	4552 / 2410
4-bit asynch.	104	0.51	130 / 70	1274/ 666
8-bit synch.	244	0.24	221 / 125	9033 / 4809
8-bit asynch.	208	1.11	184 / 81.3	1421 / 708

In the case of average power consumption, we can see that there is almost any difference between both synchronous clocking schemes, being approximately the value corresponding to one-phase scheme a 105% of the corresponding to the two-phase one. In the case of counters, differences between synchronous and asynchronous are below 10%. These results can be seen in the fig. 6.

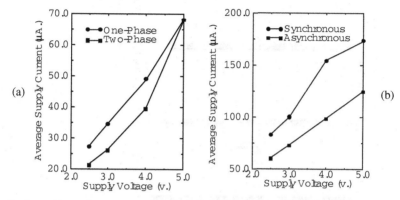

Fig. 6. Average supply current vs. supply voltage for a) the one-phase and two-phase clocking scheme and b) the 4-bit counter.

Concerning supply current peak, we can see that the peak corresponding to the single-phase is basically twice than the corresponding to the two-phase one, meaning that the two-phase scheme presents a better switching-noise behavior. In the case of counters, it is much more higher the peak value in supply current for the synchronous case (up to 4 times, depending on the Vdd value). These results can be seen in fig. 7.

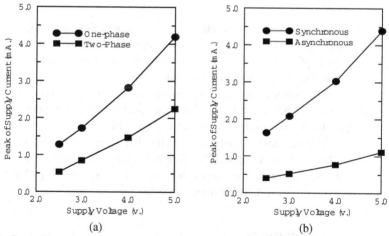

Fig. 7. Peak of supply current vs. supply voltage to (a) the one-phase and two-phase clocking scheme and (b) a 4-bit counter.

A clear measurement of the dependence of clocking schemes on peaks of supply currents is shown in fig. 8, where timing waveforms and spectra of supply current are depicted. They show how the peak values in time of the synchronous are greater, and the harmonics placed in frequencies multiple of the fundamental clock frequency (50 MHz) are considerably higher (from 4 to 11 db).

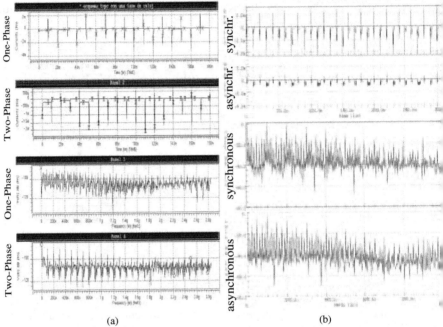

(a)   (b)

Fig. 8.   Timing waveforms and spectra of supply current for a) synchronous clocking scheme and b) the 4-bit counter, Vdd = 5v, f = 50MHz.

As counters are useful circuits, we have measured as additional parameters in the demonstrator the power-delay product. Also, we have performed a comparison with the number of stages, what is equivalent to find out the influence of the transistor-count. These results are summarized as follows:

- The power-delay product, corresponding to counters (fig. 9), is better for the synchronous approach, meaning that better performances can be found, but at the cost of extra hardware, one two-input nand gate per bit.

- The maximum supply current (fig. 10) increases linearly with the counter length for the synchronous approach, while the value for the asynchronous one is almost constant. As the number of stages increases, there are more flip-flops switching simultaneously, increasing the switching noise.

6   Conclusions

This communication has shown the influence of the clocking strategy on the switching noise generation. It will be shown how the choice of a suited clocking scheme for the digital part, alleviates the problematic associated to switching noise in mixed-signal

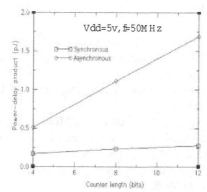

Fig. 9. Power-delay product vs counter length.

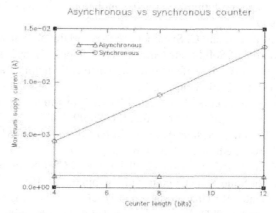

Fig. 10. Maximum supply current vs counter length.

Analog/Digital Integrated Circuits, where better timing and power performances do not necessarily imply more suitability for mixed-A/D design.

We have analyzed and simulated the switching noise generation by comparing the peak current results for two different synchronous clocking schemes (one- and two-phase clocking). Also, we have compared the results obtained for a synchronous and a asynchronous version of a common n-bit counter. Simulation data of different clocking styles have lead us to these two statements: a) Additional reduction of switching noise when using conventional digital CMOS circuits can be achieved by selecting the clock scheme suitably. b) The use of multiple clock-phase and asynchronous styles is strongly recommended. Although these solutions can introduce some problems of reliability (clock-skew), or complexity (more hardware), these are problems of minor concern in mixed-signal design, when comparing to switching noise effects.

References

1. D.J. Allstot et al., "Folded Source-Coupled Logic vs. CMOS Static Logic for Low Noise Mixed-Signal ICs", IEEE Trans. Circuits and Systems I, Vol. 40, pp. 553 563, Sept. 1993.
2. Y. Tsividis, "Mixed Analog-Digital VLSI Design and Technology". McGraw Hill, 1995.
3. H-T. Ng and D.J. Allstot, "CMOS Current Steering Logic for Low-Voltage Mixed-Signal Integrated Circuit", IEEE Trans. VLSI Systems, Vol. 5, pp. 301 308, Sept. 1997.
4. E. Albuquerque et al., "NMOS Current-Balanced Logic", Electronics Letters, Vol 32, pp. 997-998, May 1996.
5. R. Jiménez et al., "Study and Analysis of Low-Voltage/Low-Power CMOS Logic Families for Low Switching Noise", 9th Internat. Workshop on Power and Timing Modeling, Optimization and Simulation PATMOS'99, Kos Island, Greece, Oc 1999.
6. P. Larsson and C. Svensson, "Noise in Digital Dynamic CMOS Circuits", IEEE Journal of Solid-State Circuits, Vol. 29, pp. 655-662, June 1994.
7. S.R. Vemuru, "Effects of Simultaneous Switching Noise on the Tapered Buffer Design", IEEE Trans. VLSI Systems, Vol. 5, pp. 290-300, Sept. 1997.
8. S.W. Song et al., "Accurate Modeling of Simultaneous Switching Noise in Low Voltage Digital VLSI", Proc. ISCAS'99, Vol VI, pp. 210-213. 1999.
9. C.L. Seitz, "System Timing", in *Introduction to VLSI Systems*, Cap. 7, Mead an Conway, editors. Addison-Wesley, 1980.
10. Yuan, J. and Svensson, C., "High-Speed CMOS Circuits Technique", IEEE Journal of Solid State Circuits, vol. 24, pp. 62-70, 1989.
11. Jiménez, R.: Una aportación al Diseño de Circuitos Integrados CMOS Autotem porizados. PhD. Thesis, Universidad de Sevilla, Julio 2000 (in Spanish).

Computer Aided Generation of Analytic Models for Nonlinear Function Blocks

Tim Wichmann[1] and Manfred Thole[1,2]

[1] ITWM - Institute for Industrial Mathematics, Erwin-Schrödinger-Str.,
67663 Kaiserslautern, Germany, wichmann@itwm.uni-kl.de
[2] now at: Infineon Technologies AG, DAT DF AMS, P.O.Box 800949,
81609 Munich, Germany, manfred.thole@infineon.com

Abstract. In this paper we present an application of nonlinear symbolic simplification techniques to analog circuits using Analog Insydes. The goal is to get insights into the circuits' behavior and to generate efficient behavioral models. After describing the different simplification techniques and the ranking methods we explain how to generate a pin-compatible macro model. In an example, the algorithm is applied to a nonlinear square root function block.

1 Introduction

The behavior of a nonlinear analog circuit can be described by a set of nonlinear differential and algebraic equations (DAE system) in symbolic form. This system is usually far too complex to be human-interpretable and understandable. To get an interpretable symbolic expression describing the circuits' behavior and parameter dependencies it is thus necessary to apply symbolic simplification methods to the DAE system. Additionally, the simplification routines can be used to generate a macro model which can be simulated more efficiently than the original system. In contrast to simplifications by hand the proposed algorithm provides error control, i.e., the deviation of the observed input/output behavior is assured not to exceed a user given error bound.

The first version of the algorithm was presented in [3]. Several extensions of this algorithm have been developed since then, for example towards multi-input/multi-output systems, new simplification methods, or new analysis methods. We refer to [9,8,10,13,12] for a description of the enhancements.

At ITWM the algorithm is being implemented as part of Analog Insydes [7], a Mathematica [14] add-on toolbox for symbolic analysis and approximation of analog circuits.

2 Simplification Techniques

To obtain a simplified system of equations, several simplifications are applied to the system. A *simplification* can either be an algebraic manipulation or a modification of the equations which results in a new, approximative system. The latter

D. Soudris, P. Pirsch, and E. Barke (Eds.): PATMOS 2000, LNCS 1918, pp. 327–335, 2000.

requires a numeric simulation to determine the error caused by the modification. Algebraic manipulations are exact operations, thus no error tracking is needed here.

The first group of simplification techniques resides from the observation, that some variables of the DAE system do not influence the input-output behavior. That includes the elimination of variables and equations and the deletion of variables' time derivatives. In the notions of above, the first one is an algebraic manipulation.

For the second group of simplification techniques, all equations of the DAE system are expanded to sum-of-products form, where each part of this expanded sum is called a *term*. The observation, that some terms of a summation contribute a very small part to the whole sum and thus can be simplified or even neglected, motivates the following modifications on terms: Deletion of terms, substitution of terms by constant numeric values, and linearization of terms.

Each modification step is followed by a numerical error calculation to measure the real influence on the input-output behavior due to the modification. To calculate the error, a numerical simulation of the system is performed. It depends on the given problem which simulation method has to be adopted (DC, AC, transient, etc.) – it is even possible to combine different simulations. This numerical calculation yields a set of numerical values for all output variables (for example, DC points) which have to be combined through an appropriate norm to a single error value. Which norm to use (relative norm, maximum norm, etc.) again depends on the current problem. In Sect. 5, for example, we apply a multiple DC analysis combined with a maximum norm as error calculation.

Since we are working on multi-output systems, the deviation of each output variable has to be taken into account. If the error on one of the output variables exceeds the given error bound for this variable, the modification is undone.

3   Ranking Methods

The application order of the methods described in Sect. 2 influences the number of possible simplifications until the error bound is reached and an optimal order depends on the given problem. The implementation of the algorithm in Analog Insydes allows to change this order.

Within one simplification method the number of possible simplifications is also influenced by the order in which the simplifications are applied, for example, the order of terms in which they are deleted. An optimized order is desirable to maximize the number of simplifications and to minimize the number of error calculations. An algorithm that predicts the influence of a simplification on the output is called *ranking method*. Ranking algorithms are described in [9] for cancellation of terms and in [13] for substitution of terms by constant numeric values.

To handle multi-output systems the ranking algorithm must be able to predict the error for each output variable separately. Afterwards these values have to be combined with the user given error bound to an overall error prediction.

For this assume, that $\varepsilon_1, \ldots, \varepsilon_n$ are the given error bounds for the output variables $v_1, \ldots, v_n$ and $\lambda_1, \ldots, \lambda_n$ are the predicted influences of the modification on the output variables. Then one way to compute the overall error prediction is given by

$$\lambda = \frac{1}{n} \sum_{i=1}^{n} \frac{\lambda_i}{\varepsilon_i} \quad . \tag{1}$$

This is done for each part of the DAE system giving a list of error predictions. Then the parts of the DAE system are processed in the order given by increasing error prediction. What is meant by *part* in this context depends on the simplification method: For cancellation of terms, for example, part denotes a single summand of the equations, for removing of derivative terms, part denotes a summand involving derivatives. Note, that for multi-output systems the ranking order depends on the given error bounds (see Eq. (1)).

4   Model Generation

Generating a nonlinear model is quite different from the classical 2-port analysis technique as described for example in [5]: The parameters of a linear 2-port are determined by stimulating one port with an independent source while setting the current or voltage of the other port to zero by using a short or open circuit. Afterwards the complete 2-port description is set-up by superimposing the results of four of these measurements.

For numerical simulations this technique is suitable, but it fails for nonlinear model generation, for which superposition does not hold. Therefore we have to determine the complete 2-port description at once, which can be done for linear and nonlinear n-ports using symbolic analysis:

For each port choose a voltage or a current as input - the other one as output. For linear n-ports the input and output values are determined by the kind of n-port description, e. g., hybrid-parameters; for nonlinear n-ports the output and input values are given by the circuit functional behavior. Afterwards stimulate all inputs with corresponding independent sources, then set-up the circuit equations and eliminate all variables which are not needed to describe the output values. The elimination of variables is always possible for a linear circuit. Therefore the n-port parameters can be extracted directly of the resulting system of equations. For nonlinear circuits it is impossible to eliminate needless variables or to solve for the output values explicitly in most cases. Therefore, in general a nonlinear model will be an implicit system of equations.

To use the model in a numerical simulator, this system of nonlinear equations has to be converted into a simulator specific model description. In addition, an electrical interface has to be provided to gain access to the input and output values without disturbing the system of equations.

5   Example

As an example the algorithm is applied to a bipolar square root function circuit shown in Fig. 1 [6]. In this example we consider the DC input/output behavior of

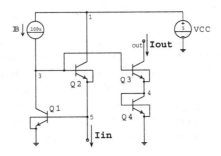

Fig. 1. Schematics of square root function block.

the circuit, i.e., we treat it as a static system. Thus the underlying DAE system here degenerates to a nonlinear equation system without any differential equations. The output current Iout is proportional to the square root of the input current Iin. The task is to generate a simplified symbolic formula describing this functional dependency and afterwards to create a parametric behavioral model as a two-port description of the circuit.

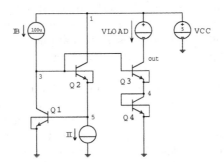

Fig. 2. Square root function block with stimulus.

This problem will be solved according to the symbolic analysis work flow described in [11]. As stated in Sect. 4 we choose Iin as an input value and Iout as an output value. Therefore, we apply a current source II and a voltage source VLOAD as shown in Fig. 2. The value of II is sweeped from $20\,\mu$A to 1 mA, the value of VLOAD is varied from 0 V to 3.5 V. We measure the node voltage V\$5 at node 5 and the current I\$VLOAD through the voltage source VLOAD. Figure 3 shows the result of the simulation of the circuit within Saber [2]. The arrow denotes the sweeping of VLOAD. But as it can be seen, the plots for different values of VLOAD are identical: Obviously the value of VLOAD has no influence on the observed output values.

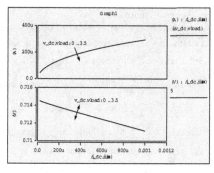

Fig. 3. Saber simulation result of `I$VLOAD` and `V$5` (Saber notation `i(v_dc.vload)` and `5`).

After finishing the numerical reference simulation within Saber, all succeeding steps including numerical simulations are now performed using Analog Insydes. For this, the Saber netlist is automatically imported into Analog Insydes. Additionally, the Saber simulation data is read in as reference for further comparisons.

Analog Insydes has the ability to switch between different transistor models. Applying both the Gummel-Poon and the Ebers-Moll bipolar transistor model gives no visible difference to the Saber reference simulation (Figure 4 shows the simulation using the Ebers-Moll model). Thus we choose the Ebers-Moll model which is much simpler than the Gummel-Poon model – the resulting DAE system as shown in Fig. 5 consists of 19 equations with 69 terms instead of 43 equations with 143 terms.

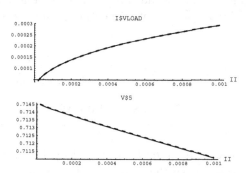

Fig. 4. Saber reference (dashed) and Analog Insydes simulation (solid) of `I$VLOAD` and `V$5`.

Once the DAE system is set-up, the nonlinear simplification algorithm can be applied. The error of the simplified DAE system will be computed for `I$VLOAD` and `V$5` on a discrete grid for `II` and `VLOAD`, where the above given sweep

$$\{IB - I\$BC\$Q2 + I\$VCC + I\$VLOAD == 0,$$
$$-IB - I\$BC\$Q1 + I\$BC\$Q2 + I\$BC\$Q3 + I\$BE\$Q2 + I\$BE\$Q3 + I\$BS\$Q2 + I\$BS\$Q3 == 0,$$
$$-I\$BE\$Q3 + I\$BE\$Q4 + I\$BS\$Q4 == 0,$$
$$II + I\$BC\$Q1 - I\$BE\$Q1 - I\$BE\$Q2 + I\$BS\$Q1 == 0, \ -I\$BC\$Q3 - I\$VLOAD == 0,$$

$$\frac{IS\$Q3}{BR\$Q3} + e^{\frac{VS3-VS4}{VT}} IS\$Q3 - e^{\frac{VS3-VSOUT}{VT}} IS\$Q3 - \frac{e^{\frac{VS3-VSOUT}{VT}} IS\$Q3}{BR\$Q3} + I\$BC\$Q3 == 0,$$

$$\frac{IS\$Q3}{BF\$Q3} - e^{\frac{VS3-VS4}{VT}} IS\$Q3 - \frac{e^{\frac{VS3-VS4}{VT}} IS\$Q3}{BF\$Q3} + e^{\frac{VS3-VSOUT}{VT}} IS\$Q3 + I\$BE\$Q3 == 0,$$

$$I\$BS\$Q3 == 0, \ \frac{IS\$Q1}{BR\$Q1} + e^{\frac{VS5}{VT}} IS\$Q1 - e^{\frac{-VS3-VS5}{VT}} IS\$Q1 - \frac{e^{\frac{-VS3-VS5}{VT}} IS\$Q1}{BR\$Q1} + I\$BC\$Q1 == 0,$$

$$\frac{IS\$Q1}{BF\$Q1} - e^{\frac{VS5}{VT}} IS\$Q1 - \frac{e^{\frac{VS5}{VT}} IS\$Q1}{BF\$Q1} + e^{\frac{-VS3-VS5}{VT}} IS\$Q1 + I\$BE\$Q1 == 0,$$

$$I\$BS\$Q1 == 0, \ -IS\$Q4 + e^{\frac{VS4}{VT}} IS\$Q4 + I\$BC\$Q4 == 0,$$

$$IS\$Q4 + \frac{IS\$Q4}{BF\$Q4} - e^{\frac{VS4}{VT}} IS\$Q4 - \frac{e^{\frac{VS4}{VT}} IS\$Q4}{BF\$Q4} + I\$BE\$Q4 == 0,$$

$$I\$BS\$Q4 == 0, \ -VCC + V\$1 == 0,$$

$$\frac{IS\$Q2}{BR\$Q2} - e^{\frac{-VS1-VS3}{VT}} IS\$Q2 - \frac{e^{\frac{-VS1-VS3}{VT}} IS\$Q2}{BR\$Q2} + e^{\frac{VS3-VS5}{VT}} IS\$Q2 + I\$BC\$Q2 == 0,$$

$$\frac{IS\$Q2}{BF\$Q2} + e^{\frac{-VS1-VS3}{VT}} IS\$Q2 - e^{\frac{VS3-VS5}{VT}} IS\$Q2 - \frac{e^{\frac{VS3-VS5}{VT}} IS\$Q2}{BF\$Q2} + I\$BE\$Q2 == 0,$$

$$I\$BS\$Q2 == 0, \ -VLOAD + V\$1 - V\$OUT == 0\}$$

Fig. 5. DAE system of the square root function block.

intervals are uniformly divided into 6 steps. The maximum error is set to an absolute deviation of $50 \, \mu A$ for I\$VLOAD and $10 \, mV$ for V\$5.

At first the DAE system is simplified algebraically by eliminating variables. This reduces the number of equations to 4 with a total number of 40 terms. Note, that this is a mathematical exact reduction, no error calculation has to be done here. In the next step cancellation of terms is applied as described in Sect. 2 up to the error bound given above. Of course, this does not change the number of equations, but reduces the total number of terms down to 11. Further algebraic elimination finally ends up in a DAE system consisting of 4 equations with 8 terms (Fig. 6). Note, that as mentioned above, the output of the original

$$\{-IB + e^{\frac{VS5}{VT}} IS\$Q1 == 0, \ -e^{\frac{VS3}{VT} - \frac{VS4}{VT}} IS\$Q3 + e^{\frac{VS4}{VT}} IS\$Q4 == 0,$$
$$II - e^{\frac{VS3}{VT} - \frac{VS5}{VT}} IS\$Q2 == 0, \ e^{\frac{VS3}{VT} - \frac{VS4}{VT}} IS\$Q3 - I\$VLOAD == 0\}$$

Fig. 6. Simplified DAE system.

system does not depend on VLOAD, so the algorithm automatically removes any occurrences of VLOAD from the original DAE system.

The equation system shown in Fig. 6 is an implicit equation system in the output variables V\$5 and I\$VLOAD and the internal variables V\$3 and V\$4. Fortunately, in this example it is possible to eliminate the internal variables and to solve the remaining equations explicitly for the output variables. This can be achieved using standard Mathematica functions. As result (Fig. 7), two explicit symbolic equations are obtained which depend on the input value II, the parameters IB and VT, and the saturation current parameters IS\$Q1, ..., IS\$Q4

$$\left\{ V\$5 == VT \, Log\left[\frac{IB}{IS\$Q1} \right], \ I\$VLOAD == \sqrt{II} \, \sqrt{\frac{IB \, IS\$Q3 \, IS\$Q4}{IS\$Q1 \, IS\$Q2}} \ \right\}$$

Fig. 7. Explicit solution of the output variables.

for each transistor. This is exactly the formula stated in [6]. But note, that this result was obtained automatically under full error control. Since it is already simple enough, no further simplification steps will be applied. The two symbolic equations shown in Fig. 7 describe the desired input/output behavior of the circuit. Figure 8 shows the comparison of the output of the Saber simulation and the simplified system. It can be seen that the error bound is fulfilled.

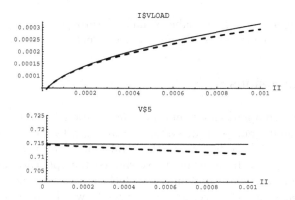

Fig. 8. Saber reference (dashed) and simulation of simplified DAE system (solid) of I\$VLOAD and V\$5.

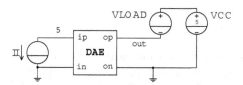

Fig. 9. Behavioral model, replacing the square root function block, with stimulus.

The last step is to generate a macro model using the simplified set of equations (Fig. 7). We choose the branch between node 5 and ground as the input port and the branch between node *out* and ground as the output port (see Fig. 9).

The Analog Insydes command `WriteModel` is used to translate the system into a Saber MAST [1] template. Afterwards this template is used as a replacement for the square root function block. The numerical simulation result computed by Saber can be seen in Fig. 10.

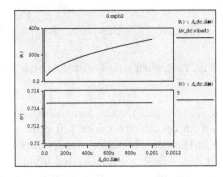

Fig. 10. Saber simulation result of I$VLOAD and V$5 (Saber notation i(v_dc.vload) and 5) using the behavioral model.

Although we used the Saber simulator throughout this example, the application of the algorithm is of course independent of a specific circuit simulator.

6   Conclusions

The presented approach extents the simplification techniques of Analog Insydes to multi-input/multi-output systems. Starting with a netlist on transistor level it is now possible to generate behavioral models automatically in a simulator independent way. In an example we showed the application of Analog Insydes to a nonlinear square root function block. It was possible to derive a human-interpretable parameterized symbolic formula, which – in contrast to calculations by hand – assures a user given error bound. Furthermore, we automatically generated a Saber MAST template of the simplified formula which can be used as a pin-compatible behavioral model replacement for the square root function block.

Acknowledgments

This work has been carried out within the MEDEA project A409 "Systematic Analog Design Environment" (SADE), supported by the German *Ministerium für Bildung, Wissenschaft, Forschung und Technologie* under contract no. 01M3037F and by Infineon Technologies.

References

1. Analogy, Inc. *MAST Reference Manual*, 1999.
2. Analogy, Inc. *Saber 5.0 Documentation*, 1999.
3. C. Borchers. The symbolic behavioral model generation of nonlinear analog circuits. In *IEEE Transactions on Circuits and Systems*, volume 45, pages 1362–1371, Oct. 1998.

4. K. E. Brenan, S. L. Campbell, and L. R. Petzold. *The Numerical Solution of Initial Value Problems in Ordinary Differential-Algebraic Equations.* North Holland Publishing Co., 1989.

5. Wai-Kai Chen, editor. *The Circuits and Filters Handbook.* CSC Press, Inc., 1995.

6. P. R. Gray and R. G. Meyer. *Analysis and Design of Analog Integrated Circuits.* John Wiley & Sons, Inc., 3rd edition, 1993.

7. ITWM – Analog Insydes home page. `www.itwm.uni-kl.de/as/products/ai/`.

8. L. Näthke, R. Popp, L. Hedrich, and E. Barke. Using term ordering to improve symbolic behavioral model generation of nonlinear dynamic analog circuits. In *Proc. European Conference on Circuit Theory and Design (ECCTD '99)*, Stresa, Italy, 1999.

9. R. Popp, W. Hartong, L. Hedrich, and E. Barke. Error estimation on symbolic behavioral models of nonlinear analog circuits. In *Proc. Fifth International Workshop on Symbolic Methods and Applications in Circuit Design (SMACD '98)*, Kaiserslautern, 1998.

10. R. Popp, L. Näthke, and C. Borchers. Automatische Erzeugung symbolischer Verhaltensmodelle für nichtlineare Analogschaltungen im transienten Großsignalbetrieb. In *Proc. 5. ITG/GMM Diskussionssitzung (Analog '99)*, München, Germany, 1999.

11. R. Sommer, E. Hennig, M. Thole, T. Halfmann, and T. Wichmann. Symbolic modeling and analysis of analog integrated circuits. In *Proceedings of the European Conference on Circuit an Circuit Theory and Design ECCTD '99*, volume I, pages 66–69, 1999.

12. T. Wichmann. Computer aided generation of approximate DAE systems for symbolic analog circuit design. In *ZAMM, Proc. Annual Meeting GAMM 2000, Göttingen (to appear).*

13. T. Wichmann, R. Popp, W. Hartong, and L. Hedrich. On the simplification of nonlinear dae systems in analog circuit design. In *Computer Algebra in Scientific Computing, CASC'99*, pages 485–498, Munich, June 1999.

14. S. Wolfram. *The Mathematica Book.* Wolfram Media/Cambridge University Press, 4th edition, 1999.

Author Index